Advances in Carbohydrate Chemistry and Biochemistry

Volume 66

Advances in Carbohydrate Chemistry and Biochemistry

Editor
DEREK HORTON

Board of Advisors

DAVID C. BAKER
DAVID R. BUNDLE
STEPHEN HANESSIAN
YURIY A. KNIREL
TODD L. LOWARY

SERGE PÉREZ
PETER H. SEEBERGER
J.F.G. VLIEGENTHART
ARNOLD A. STÜTZ

Volume 66

Amsterdam • Boston • Heidelberg • London • New York • Oxford
Paris • San Diego • San Francisco • Singapore • Sydney • Tokyo
Academic Press is an imprint of Elsevier

Academic Press is an imprint of Elsevier
Radarweg 29, PO BOX 211, 1000 AE Amsterdam, The Netherlands
Linacre House, Jordan Hill, Oxford OX2 8DP, UK
32 Jamestown Road, London NW1 7BY, UK
225 Wyman Street, Waltham, MA 02451, USA
525 B Street, Suite 1900, San Diego, CA 92101-4495, USA

First edition 2011

Copyright © 2011 Elsevier Inc. All rights reserved

No part of this publication may be reproduced, stored in a retrieval system or transmitted in any form or by any means electronic, mechanical, photocopying, recording or otherwise without the prior written permission of the publisher

Permissions may be sought directly from Elsevier's Science & Technology Rights Department in Oxford, UK: phone (+44) (0) 1865 843830; fax (+44) (0) 1865 853333; email: permissions @elsevier.com. Alternatively you can submit your request online by visiting the Elsevier web site at http://elsevier.com/locate/permissions, and selecting *Obtaining permission to use Elsevier material*

Notice
No responsibility is assumed by the publisher for any injury and/or damage to persons or property as a matter of products liability, negligence or otherwise, or from any use or operation of any methods, products, instructions or ideas contained in the material herein. Because of rapid advances in the medical sciences, in particular, independent verification of diagnoses and drug dosages should be made

ISBN: 978-0-12-385518-3
ISSN: 0065-2318

British Library Cataloguing in Publication Data
A catalogue record for this book is available from the British Library

Library of Congress Cataloging-in-Publication Data
A catalog record for this book is available from the Library of Congress

For information on all Academic Press publications
visit our website at books.elsevierdirect.com

Printed and bound in USA

11 12 13 14   10 9 8 7 6 5 4 3 2 1

Working together to grow
libraries in developing countries

www.elsevier.com | www.bookaid.org | www.sabre.org

ELSEVIER   BOOK AID International   Sabre Foundation

CONTENTS

CONTRIBUTORS vii

PREFACE ix

Anthony Charles Richardson (1935–2011)
KARL J. HALE

Detection and Structural Characterization of Oxo-Chromium(V)–Sugar Complexes by Electron Paramagnetic Resonance
LUIS F. SALA, JUAN C. GONZÁLEZ, SILVIA I. GARCÍA, MARÍA I. FRASCAROLI, AND SABINE VAN DOORSLAER

I. Introduction...	70
1. Introduction to Chromium (Bio)chemistry	70
2. State-of-the-Art EPR Techniques	72
II. General Aspects of EPR of Oxo-Cr(V) Complexes	73
III. Systematic Structural Characterization of Oxo-Cr(V)–Sugar Complexes Using EPR	77
1. Complexes with Alditols...............................	77
2. Complexes with Aldohexoses and Aldopentoses	86
3. Complexes with Methyl Glycosides.....................	94
4. Oxo-Cr(V) Complexes with Other Sugars................	98
5. Complexes with Hydroxy Acids: Alduronic and Glycaric Acids	100
6. Complexes with Oligosaccharides and Polysaccharides.....	108
IV. Conclusions...	114
Acknowledgments..	115
References..	115

Synthesis and Properties of Septanose Carbohydrates
JAIDEEP SAHA AND MARK W. PECZUH

I. Introduction...	122
II. Synthesis of Septanose-Containing Mono-, Di-, and Oligo-Saccharides..............	127
1. Cyclization via C—O Bond Formation	127
2. Preparation and Reactions of Carbohydrate-Based Oxepines	138
3. Miscellaneous Septanose Syntheses.....................	149
4. Transannular Reactions of Septanose Sugars.............	152
III. Conformational Analysis of Septanose Monosaccharides	154

	1. Observed Low-Energy Conformations	156
	2. Determination of Septanose Conformations	161
IV.	Biochemical and Biological Investigations Using Septanose Carbohydrates	163
	1. Olgonucleotides Containing Septanosyl Nucleotides	164
	2. Protein–Septanose Interactions	172
	3. Additional Reports	176
V.	Outlook	177
	Acknowledgments	178
	References	178

Imino Sugars and Glycosyl Hydrolases: Historical Context, Current Aspects, Emerging Trends

Arnold E. Stütz and Tanja M. Wrodnigg

I.	Introduction	188
	1. Discovery of Imino Sugars as Natural Products	189
	2. Early Synthetic Efforts	192
II.	Glycoside Hydrolases	193
	1. General Features and Means of Classification	193
	2. Historical Background and Development	194
III.	Glycon Structure, Functional Groups, and Catalysis	197
	1. β-Glucosidases and Nonnatural Substrates	197
	2. β-Galactosidases and Nonnatural Substrates	199
	3. α-Glucosidases and Nonnatural Substrates	200
	4. α-Galactosidases and Nonnatural Substrates	202
	5. α-Mannosidases and Nonnatural Substrates	202
IV.	Glycosidase Inhibitors	203
	1. High-Molecular-Weight Inhibitors	203
	2. Small Molecules: Covalent ("Irreversible"), Competitive (Reversible), and Noncompetitive (Allosteric) Inhibitors	203
	3. Nonsugar Inhibitors	204
	4. Carbohydrate-Related Inhibitors	206
V.	Imino Sugars	211
	1. pK_a Value and State of Ionization Approaching the Active Site	212
	2. Stereochemical Considerations Concerning the Catalytic Protonation Process	214
	3. Significance of Individual Functional Groups to Recognition and Binding	216
	4. Deoxyfluoro Derivatives as Active-Site Probes	217
	5. Imino Sugars as Active-Site Ligands for Structure and Catalysis Visualization	224
	6. "Non-Glycon" Binding Sites and Inhibitor Activity	246
VI.	Conclusions	256
VII.	Paradigm Changes and Emerging Topics	258
VIII.	Table of PDB-Entries: Enzyme–Inhibitor Complexes of Imino Sugars and Selected Other Ligands	259
	Acknowledgment	276
	References	276
	AUTHOR INDEX	299
	SUBJECT INDEX	323

CONTRIBUTORS

María I. Frascaroli, Departamento de Químico Física—Área Química General, Facultad de Ciencias Bioquímicas y Farmacéuticas, Universidad Nacional de Rosario, Rosario, Santa Fe, Argentina

Silvia I. García, Departamento de Químico Física—Área Química General, Facultad de Ciencias Bioquímicas y Farmacéuticas, Universidad Nacional de Rosario, Rosario, Santa Fe, Argentina

Juan C. González, Departamento de Químico Física—Área Química General, Facultad de Ciencias Bioquímicas y Farmacéuticas, Universidad Nacional de Rosario, Rosario, Santa Fe, Argentina

Karl J. Hale, The School of Chemistry & Chemical Engineering, and the Centre for Cancer Research & Cell Biology (CCRCB), Queen's University Belfast, Belfast BT9 5AG, Northern Ireland, UK

Mark W. Peczuh, Department of Chemistry, University of Connecticut, Storrs, Connecticut, USA

Jaideep Saha, Department of Chemistry, University of Connecticut, Storrs, Connecticut, USA

Luis F. Sala, Departamento de Químico Física—Área Química General, Facultad de Ciencias Bioquímicas y Farmacéuticas, Universidad Nacional de Rosario, Rosario, Santa Fe, Argentina

Arnold E. Stütz, Glycogroup, Institut für Organische Chemie, Technische Universität Graz, Graz, Austria

Sabine Van Doorslaer, Physics Department, University of Antwerp, Antwerp-Wilrijk, Belgium

Tanja M. Wrodnigg, Glycogroup, Institut für Organische Chemie, Technische Universität Graz, Graz, Austria

PREFACE

In this 66th volume of *Advances*, an extended tribute to the life and work of Anthony C. Richardson (Dick) is provided by Hale (Belfast). In the present day, when the successful young academic usually expects to abandon the laboratory bench and progress to become the leader of a large research group, Richardson was an anomaly in remaining an outstanding synthetic chemist who spent his entire career doing the laboratory work he loved. He had a brilliant and incisive mind coupled with a warm and nurturing personality, but he never sought the limelight or public recognition beyond the satisfaction from working with a small group and achieving a remarkable range of accomplishments in carbohydrate synthesis. These justly merit the extended documentation presented here in Hale's account. From elegantly conceived syntheses of (−)-swainsonine to Richardson's rules for predicting the outcome of sulfonate displacement by nucleophiles, to the noncaloric sucrose-derived sweetener Splenda® (trichloro-galacto-sucrose) with the Hough laboratory, and the provision of many valuable synthetic intermediates, he has enriched the carbohydrate field with a very substantial legacy.

The coordination behavior of sugars and their derivatives with inorganic cations has been largely "under the radar" of mainstream carbohydrate science in recent years, given the strong focus of many of today's researchers on glycobiology targets. However, the complexation of carbohydrate derivatives with the element chromium, in particular, has important implications in both human and animal health, and in problems of environmental damage from industrial pollutants. The toxicity and carcinogenicity of chromium is well recognized, and the use of microorganisms or plants for bioremediation of contaminated soils requires careful evaluation. The unpaired d-subshell electrons in the multiple valence states exhibited by chromium lend themselves ideally to studies of the complexes by electron paramagnetic resonance. This chapter by Sala and colleagues (Rosario, Argentina) details current knowledge gleaned from use of traditional continuous-wave EPR spectrometers and addresses the potential of newer pulsed and high-field instruments for significant advancement of our understanding.

While the pyranose and furanose ring forms of the sugars dominate the carbohydrate literature, the uncommon septanose ring forms have long intrigued sugar chemists, with particularly notable contributions by Stevens in Australia. The chapter in this volume from Saha and Peczuh (Storrs, Connecticut) provides a comprehensive overview of the subject from both the synthetic and structural viewpoints and presents a detailed analysis of the conformational behavior of these ring forms. It may be noted that the structures of the seven-membered rings and their acyclic precursors are most conveniently depicted, respectively, by Mills-type formulas and the supposedly old-fashioned Fischer-type formulas, rather than the Haworth conformational formulas commonly favored for five- and six-membered rings. This chapter offers intriguing prospects for involvement of these ring forms in biological applications, especially with regard to antisense oligonucleotides.

A most important variant of the monosaccharide structures is the class of sugar derivatives wherein nitrogen replaces the ring-oxygen atom, namely, the imino sugars. In the comprehensive overview presented here by Stütz and Wrodnigg (Graz, Austria), they integrate the wide range of imino sugar analogues now known to occur in Nature with their remarkable functions as potent inhibitors of the glycosidase enzymes. Complementary work on numerous synthetic analogues has added important new understanding of the mode of action of glycosidases in general. The authors include a detailed structural tabulation of all such inhibitors currently known, along with the Protein Data Bank links to the enzymes that they inhibit, and offer exciting prospects for the therapeutic potential of these inhibitors in modulating essential metabolic processes.

The deaths are noted with regret of two leading carbohydrate biochemists, Nathan Sharon (June 17, 2011) and Saul Roseman (July 2, 2011). Dr. Sharon served with distinction for many years as a member of the Board of Advisors of this *Advances* series, and his advice and input will be sadly missed. His work will be recognized in an upcoming volume of the series.

With this volume Professor Arnold E. Stütz is welcomed as a new member of the Board of Advisors.

DEREK HORTON

Washington, DC
October 2011

ANTHONY CHARLES RICHARDSON

(1935–2011)

Anthony Charles Richardson won universal respect, admiration, and distinction as a chemist for the many pioneering research contributions that he made to synthetic carbohydrate chemistry and natural product total synthesis over an illustrious 40-year career. His coauthorship, with Leslie Hough, of the now legendary treatise *Rodd's Chemistry of Carbon Compounds, 2nd Edition Vol 1F* also earned him a place in chemical history because of the highly authoritative coverage it provided of the entire field of monosaccharide chemistry, all the way back to Emil Fischer's time. Even today, more than 44 years after it was first written, the Hough–Richardson 595-page chemistry monograph remains one of the primary reference sources for all literature on monosaccharide chemistry published before 1967. It continues to be relied upon by modern generations of organic chemists for the accurate and detailed information it provides. In many ways, *Rodd's 2nd Edition Vol 1F* has become the monosaccharide chemist's Bible, because of the remarkable insights it expounds and the confidence and trust that users can place in the accuracy of its content. The integrity of judgment displayed by both authors in deciding what to omit and what to include helped to create an enduring and significant scholarly work.

The numerous excellent synthetic procedures published by A.C. Richardson also continue to be relied upon by contemporary organic chemists because of their great practicality and reproducibility. Indeed, a recent highly cited enantiospecific total synthesis of the antitumor agent, (−)-agelastatin A, justly testifies to this fact; it exploits the remarkably useful Hough–Richardson aziridine as a key synthetic intermediate in the eventual published route to this complex natural product.

Society has also greatly benefitted from A.C. Richardson's research, most especially from his discovery (with Les Hough, Anthony Fairclough, and Shashi Phadnis) of the noncaloric, high-intensity sweetener known as $4,1',6'$-trichloro-$4,1',6'$-trideoxy-galacto-sucrose (TGS, sucralose), which is now sold globally under the commercial name of "Splenda." In a time of increasing obesity and high incidence of dental disease, this outstanding research contribution is now helping to counteract the development, progression, and spread of several lifestyle diseases that currently contribute significantly to morbidity in the modern world.

Richardson also devised many novel reagents and chromogenic substances that are now used in various diagnostic kits of value for the rapid and accurate visual detection of human disease, as well as for the identification of pathogenic organisms associated with disease. For chemical posterity, it is therefore considered important to present a detailed account of Richardson's numerous research contributions.

A.C. Richardson (more commonly known as "Dick" Richardson in carbohydrate chemistry circles, or just plain "Tony" to his family and friends) was born in Catford, southeast London, on April 21, 1935. His father, Charles Richardson, was a mechanical engineer who worked at a nearby aeronautical factory, having diverged from the long-standing family tradition of working as London chimneysweeps to adopt a profession that was far less hazardous and which offered greater opportunities for originality of thought and personal creativity. His mother, Doris Richardson (née Doris White), worked as a milliner. Although Richardson's parents were working class and did not themselves benefit from a University education, they did still appreciate the career opportunities that such an education could offer. They therefore went to great lengths to encourage their young son, Anthony, and his elder sister, June, to work industriously and excel academically to achieve eminence in whatever field/discipline they ultimately chose to pursue. In Richardson's case, he entered the world of synthetic carbohydrate chemistry, and what a major turning point that proved to be for the development of this branch of the subject!

A.C. Richardson's early childhood was like that of many other young boys who grew up in southeast London as World War II began. The sustained aerial "Blitz" against the city led to many of the nation's children being evacuated to more-rural parts of the country having lesser commercial and military significance. In young Richardson's case, it led to separation of him at the tender age of 4, along with his sister, June, from their parents, and they were evacuated to Exeter, in the southwest of England. Despite the upheaval, trauma, and worry that this undoubtedly must have caused, young Richardson still managed to excel academically at school. By the time he left Exeter in 1945, his teachers lamented to his rather stunned parents over the great loss to the school of its "star pupil"!

Not long after he returned to London, Richardson passed the all-important 11-plus examination with flying colors to win a much sought-after place at Brockley County Grammar School, where he remained until he was 18. He left Brockley Grammar in 1953 to take up an undergraduate place at the University of Bristol, where he read Chemistry. It was there that he met his future mentor and lifelong colleague, Professor

Les Hough, who was then a dynamic young Reader in Organic Chemistry, and who was carving out an excellent reputation for himself in the up-and-coming field of synthetic carbohydrate chemistry.

Two years after he arrived in Bristol, and while attending a local ballroom dance at St. Peter's Church Hall, Dick met his future wife, Jane Dwyer, and in the final year of his undergraduate studies, they became engaged. In the summer of 1956, not long after Richardson had graduated with a First-Class Honors degree in Chemistry, he opted to stay on at Bristol University to study for the Ph.D. degree with Les Hough.

In June 1957, within a year of embarking on his Ph.D. studies, Dick married Jane. Jane was herself from the Bristol area and was 3 years younger than Dick, having been born on January 30, 1938. Although Jane's mother was a Bristolian, her father, quite by chance, had originated from Catford (like Dick). Mr. Dwyer had worked in his early years as a theatrical performer.

According to Les Hough, Richardson first caught his attention at Bristol University when he recorded an exceptional undergraduate exam performance in mathematics and physics. In Hough's opinion, prowess in these subjects was always a sure indicator of future academic attainment in chemical research. This clearly proved to be the case, as it ultimately led to the creation of a powerful research partnership in carbohydrate chemistry that would endure for more than 40 years. This alliance between two future leaders of the subject would move the field forward to almost the same extent as the area of physical organic chemistry had been advanced by Hughes and Ingold at University College, London, during the period 1930–1961.

Because of his superlative academic record at Bristol University, Dick was awarded the highly prestigious Fulbright Scholarship to work in the United States at the University of California in Berkeley, under the guidance of Professor Hermann O.L. Fischer, son of the legendary Emil Fischer. H.O.L. Fischer was himself a celebrated carbohydrate chemist in the mold of his father, and he made many significant contributions to the subject, as well as to the fledgling discipline of biochemistry. Richardson worked productively with Hermann Fischer for almost a year before his great mentor rather suddenly took ill; Fischer died soon afterward on March 9, 1960, at the age of 71.

Not long after finishing up at Berkeley, Dick and Jane returned to Britain in September 1960 and, 3 weeks later, their first child, Caroline, was born. With the support of an ICI Fellowship, Dick then took up a 2-year research appointment at the University of Bristol. The temporary character of this appointment did not seem to perturb him any, for during this time he and Jane bought a good-sized property in Henbury, just outside Bristol, where they went on to have two more children: Miles in 1962 and Marcus in 1963. Apparently, when Richardson's Head of School at the University of Bristol heard of this (at the time) rather extravagant purchase of a family home in one of the better parts of Bristol, he immediately summoned Richardson to his office to remind him of the fact that "this appointment to the Bristol staff is only temporary!" Of course, when Dick subsequently managed to sell this property in 1963 at quite a substantial profit, he made sure that the "wonderful news" of this sound investment reached the Head of Department's ears. He, despite the positive outcome, continued to believe that the young Richardson was still rather prodigal in his outlook!

In September 1963, Richardson accepted the recently advertised Lectureship in Organic Chemistry at the University of Reading, and shortly afterward the Richardson family moved to the nearby town of Woodley. He remained at Reading University for 4 years before being reunited in 1967 with Les Hough, then newly appointed as Professor at Queen Elizabeth College (QEC). QEC was a constituent college of the world-famous University of London, and where Les occupied not only the Chair of Organic Chemistry, but also the Headship of the Chemistry Department, which was located in the upmarket Campden Hill Road part of Kensington. At QEC, Richardson and Hough jointly established an internationally recognized school of synthetic carbohydrate chemistry that was to produce more than 130 original research papers under their joint authorship, as well as many independent papers.

In early 1972, with Jane expecting their fourth child, the Richardson's moved to Badgers Walk in the village of Shiplake, near Henley-on-Thames, Oxfordshire. Three months later, their anticipated "fourth child" actually turned out to be twins, Annabelle and Adam. In August 1974, Dick and Jane temporarily moved with their five children to Palo Alto, California, where Dick spent a year working with John Moffatt and Julian Verheyden in the laboratories of Syntex SA. After a highly productive research spell there, the Richardson family returned to Britain in September 1975, and they moved in March 1977 from Badger's Walk to a house in Norman Avenue, Henley, which remains the family home to this day.

Shortly before Richardson went to Syntex, he was promoted to a Readership in Organic Chemistry at QEC, having obtained the D.Sc. degree of the University of London on March 14, 1972; the latter degree was then a mandatory College requirement for any academic to be promoted to a Readership or Chair. In 1985, the three University of London institutions of King's College London (KCL), Chelsea College, and QEC all merged, and A.C. Richardson was officially transferred to the staff of KCL as a Reader in Organic Chemistry. After a further 8 years of research and teaching he eventually retired from university life in 1993 at the age of 58, whereupon the title of Emeritus Reader in Organic Chemistry at KCL was conferred upon him.

It must be emphasized that Dick did not actually retire from chemistry per se, for he and two KCL/QEC colleagues (Bob Price and Percy Praill) had already set up a joint business, PPR Diagnostics, which conducted research that would lead to the development of numerous diagnostic kits and products that are now helping to transform modern diagnostic medicine. Dick therefore decided to devote his full-time effort toward doing chemistry research for PPR, having formally retired from the academic world. This period was one of the happiest in his life, as this author recalls, and it still led to academic papers.

Having been taught organic chemistry by A.C. Richardson, both as an undergraduate and as a postgraduate at QEC and KCL, this author can comment accurately on his teaching skills. His lectures were always stimulating and never anything short of excellent; his ability to teach organic chemistry was outstanding, and he always loved to have students stop him in his lectures and ask thoughtful questions. He was very approachable and could think rapidly on his feet. A particular feature that made Richardson stand out as a teacher was the fact that he could always modulate his approach according to the type of student he was dealing with; he challenged the more able, but also took a step backward with some students if he thought that he might be overreaching. His lectures always seemed like beautifully rehearsed sermons; they were models of clarity and perfection, and well reflected his authoritative understanding of the subject. His lectures were delivered in a highly mechanistic, arrow-pushing style that was very pictorial and illustrative, and he loved nothing more than to reminisce about his own personal experiences of doing practical organic chemistry. Many of these personal recollections could be interwoven with the odd corny joke, which frequently made them more chemically meaningful as well as enjoyable! It is accurate to say that most undergraduates felt that they were in the presence of a true chemistry oracle when they were around A.C. Richardson. His overall chemical knowledge and his inherent intuition always made him stand out from most of his contemporaries. Notwithstanding his mastery of chemistry and his remarkable intellect, Richardson never got carried away with his own importance, and he always exhibited the noble qualities of kindness, devotion, and a fatherly "duty of care" with all of his postgraduate students. Put simply, he was a great teacher and mentor, and a genuine and humane friend to all of his students.

Even though Dick did a considerable amount of undergraduate teaching and examination grading at QEC and KCL, he always found the time to spend long hours in his office/laboratory doing synthetic chemistry. His lab office had all of the trappings and accoutrements of both types of facility, which often led to his being seen within its confines with a cup of his favorite Jamaican Blue Mountain coffee in one hand and a chemical reaction flask in the other, all with donned lab coat and safety spectacles, of course! Richardson displayed exceptional laboratory skills and always recorded his results with great accuracy and diligence. He maintained an impeccable laboratory notebook in which all his experiments were beautifully written up in a most endearing style of handwriting that was a joy to behold. The clarity and level of detail of his observations were marvelous to witness, and they undoubtedly set an example to all of his students. His high productivity in the laboratory, fitted in during the limited time slots that he could take away from undergraduate teaching, clearly reflected his superb organizational skills. He used every moment to maximum effect, and while chemical reactions were taking place, he would be busy doing other work, either setting up more experiments, writing up his lab notebook, or simply doing paperwork or some teaching-oriented task. He was a superb academic who was constantly at work.

His warmth of approach toward new Ph.D. students would put even the most nervous of them at their ease. He loved to engage in day-to-day banter with his research students, as well as discussions of chemistry; this arose from his inherent interest in people and life as a whole, and his excellent sense of humor. As a consequence, new students and research assistants quickly settled into their new lab environment, notwithstanding the challenges of the research projects in carbohydrate chemistry they had been set. His excellence of leadership and his ability to rescue students stranded on floundering research projects were all traits that made him a Ph.D. supervisor par excellence and a leader who inspired his students.

Another way in which Richardson would help the novice Ph.D. student settle into his lab was through the regular luncheon invitations that he and Les Hough issued for the group to come for a get-together at the Churchill Arms in Kensington Church Street on a Friday afternoon, usually once or twice a term. These pub trips were often finished off with a visit to the Monsieur Pechon cake shop on the same street, where Dick would buy a huge box of gourmet cakes for consumption by the group in his lab office. They were generally washed down with a few cups of that now famous Jamaican Blue Mountain coffee, or maybe something stronger on the odd occasion! These afternoons left a great impression on the new students for they could now see how Richardson really ticked; they recognized that he was a warm and caring human being of great personal integrity. Other regular haunts where the group would hang out with Dick and Les would be the Standard Curry House in Bayswater, or Khan's Restaurant immediately next door to the Standard, and sometimes the Mayflower Chinese Restaurant in Shaftesbury Avenue (in London's Chinatown). These group outings with Dick and Les would build up a remarkable team atmosphere within the lab, and undoubtedly they helped seal the enduring friendships that Dick and Les enjoyed with the majority of their Ph.D. students long after they had moved on from their group.

Dick and Les especially liked to take their entire research group to the annual Carbohydrate Chemistry Conferences of the Royal Society of Chemistry, which were held at one of the UK's leading research universities. These were always grand occasions, not only for the beautiful expositions of carbohydrate chemistry that could be heard from the likes of Les Hough, Gordon Chittenden, Neil Baggett, Hans Paulsen, Fraser Stoddart, and Stephen Hanessian but also for the frivolity of many of the evening social events. These ranged from wild whisky tastings up in Scotland to beer-fueled sing-songs around the piano in Sheffield, with musical guests of honor such as Bert Fraser-Reid, and these were combined with enjoyable dinners and socials at the conference bar. When one of these events was staged at QEC in the 1980s, Dick appointed himself as Head Barman, which led to even greater merriment among the troops because of all of the free beer and wine that was profusely served. All of these events helped secure a great lab spirit and an enjoyable chemistry environment for the new student.

Dick Richardson was like a father to many in his laboratory. His students always felt that he was their "Dad in the lab," a mentor who would continually look after them and advise whenever help was needed, while endeavoring to convey that independence of thought that would turn out a self-propelling chemist equipped for survival in a future career. He would always encourage the student to come up with his own independent and elegant solutions to the problems that were set, and while trying to get from A to B on the assigned problem he would continually watch over from a distance. He wanted the student to learn how to solve complex research problems independently, and there would be beautiful smiles coming over his face whenever he saw a young postgraduate enter his office with his latest new result. He would greatly relish the times that he spent talking with students about their projects.

From a personal viewpoint, it was a truly wonderful experience working around someone so decisive in thought and one who had such excellent knowledge. With Dick as your supervisor, you always felt on top of the world whenever you left his office. You felt that you were continually making progress, albeit slow. He was a model supervisor, and very few people in academia could hope to match him in that respect.

Naturally, given all of A.C. Richardson's important achievements in carbohydrate chemistry, it might be expected that he would be the recipient of numerous academic awards and research prizes. However, those of us who knew Dick Richardson were well aware that he never liked to be nominated for awards, nor receive public recognition for his work; he held the opinion that great science should be done for its own sake, and not to win prizes. Indeed, the only time that Richardson ever allowed himself to be nominated for, or to receive, such recognition was when it was an award that would permit him to engage in further high-caliber scientific endeavor. Thus, he considered the Fulbright Scholarship to be a prize worth seeking, as success in that competition would enable him to do high-quality science with a truly great scientist. Likewise, when he applied for the ICI Fellowship at Bristol University, he again considered this to be a highly worthwhile endeavor, as the award would permit him to work independently. It came as no surprise to his students when they learned that he had accepted in the mid-1980s a prestigious Invitation Research Fellowship of the Japan Society for the Promotion of Science, as again this gave him the funding and professional opportunity to do high-level science with a top-drawer senior Japanese scientist, Professor Tetsuo Suami, who was at that time working at Keio University.

For Richardson, the greatest reward that any scientist could ever obtain (or expect to obtain) was to have his research paper accepted and published in a major scientific journal. To his way of thinking, great scientists should focus primarily on contributing to knowledge and publishing the discoveries of their work, rather than gaining plaudits or accolades from their colleagues. As time went on, however, and universities started to change their direction, becoming much more corporate and strategically focused, he became more pragmatic on this point, recognizing that all of his former chemistry students in academia would have to win major awards in order to further advance their careers.

Much as E.D. Hughes was always comfortable to stand in the shadow of Ingold at UCL, so too did A.C. Richardson prefer to work as a protégé and lieutenant of Les Hough, supporting the work and achievements of his senior mentor. Indeed Richardson considered Les to be his closest personal friend and colleague, and the feeling was mutual. While Leslie Hough did receive high honors for all of the carbohydrate work they jointly did over a 30-year period (RSC Haworth Medalist in 1985, Claude S. Hudson Award of the American Chemical Society in 1986, to name but two major recognitions), the pleasure that this gave Richardson came primarily from the chemistry community's recognition and commendation of a leader he had long revered and admired, and had worked with both as a colleague and as a student. While all of these awards could equally well have been bestowed on A.C. Richardson, it is doubtful that he would have accepted any of them, as he never wanted recognition or public adulation for himself. He always preferred to stand in the background, enjoying Leslie's public success. Such was the great humility and meekness of A.C. Richardson as a man and chemist.

As regards chemical success that has had true societal impact and lasting benefit, Richardson probably received his greatest accolade when the $4,1'6'$-trichloro-$4,1'6'$-trideoxy-galacto-sucrose (TGS, sucralose) that he and Les first synthesized back in 1975 finally made it to market in the early 1990s as a noncaloric, high-intensity sweetener. Indeed, Richardson was one of the primary discoverers and developers of the TGS sweetener, in conjunction with Les Hough, Anthony Fairclough, and Shashi

Phadnis. This major contribution to the world gave both of these senior chemists (LH and ACR) enormous satisfaction. They knew that they had contributed significantly to the global fight against human obesity and dental disease, while also improving the UK's financial position as a nation.

It also gave Richardson profound satisfaction to learn that many of the synthetic intermediates that he had prepared in his younger days continued to prove useful for later researchers developing synthetic routes toward target molecules of potential relevance to the treatment of human disease. Dick Richardson and Les Hough were extremely happy to see other workers use and apply their research.

In 2003, when Dick was diagnosed with an aggressive form of Parkinson's disease, he remained stoic and in good spirits throughout the next 4 years while he was still able to travel independently. He wanted all of his friends to know that despite the diagnosis it was very much "business as usual," even though he personally knew that much physical and mental hardship would soon have to be endured. Without complaint, he carried on as normal, remaining ever buoyant, never acknowledging that he was deteriorating bodily, even though he realized that he was slowing down. In 2005, as his condition became worse, he reluctantly took the decision to cease doing practical chemistry at PPR, and he concurrently resigned from his post. During the next 2 years, he still made his regular 2-weekly trip into London unaccompanied, to have a pint of beer and lunch with Les Hough, Robert Poller, Bob Price, and the writer at the Mulberry Bush Pub near London's Festival Hall, and he also attended the annual QEC chemists reunion. However, in late 2007, much to the distress of his loving family and friends, his condition took a dramatic turn for the worse, and he was obliged to stop attending these lunches and retire completely from public life. His decline continued relentlessly until his death on the morning of Monday, March 28, 2011.

In the end, it was not the Parkinson's disease that he had fought for so long that eventually wrested him away from us. Rather, it was the fourth successive bout of pneumonia to which he had fallen victim since December 2010 that caused him to slip away that spring morning.

A.C. Richardson was very much admired by colleagues and students alike throughout his long academic career, not only for his warm and engaging personality but also for his joviality. Notwithstanding his shining career and the wide admiration and respect in which he was held, he remained first and foremost a family man, putting his wife and children, and later on his grandchildren, above everything else. Those who knew Dick will long remember his attributes and will continue to be touched by his greatness both as a chemist and as a man of honor. He was a bright star among stars, who helped in his own inimitable way to make the world a better place for all. He will be sorely missed by all who knew him.

Today's younger carbohydrate chemists will not have had the privilege of knowing this great carbohydrate chemist in person but can have the opportunity to learn more about the significant contributions of A.C. Richardson to chemistry from the remainder of this biographical memoir, which presents an overview of his extensive chemical legacy. In giving an account of Richardson's scientific work, a decision had to be taken as to whether to deal with it chronologically or simply by subject matter, focusing primarily on his main contributions to carbohydrate chemistry. This memoir traverses both paths, as it shows how Richardson's main research themes evolved over time and how one area of carbohydrate chemistry naturally led into another.

Richardson's first spell of research work with Les Hough was done at Bristol University during the period 1956–1959. His Ph.D. thesis was completed during that time, and it dealt with the oxidative degradation of sugar diethylsulfonyl dithioacetals under basic conditions. In the case of the pyranosyldiethylsulfonylmethane shown in Scheme 1, Richardson observed that the C-glycosyl compound was readily cleaved by dilute aqueous ammonia to give D-lyxose as the final product. Although it might initially be tempting to think that this reaction might proceed mechanistically as shown, this would be incorrect, for when Richardson studied the corresponding base-induced cleavage of the 2,3-isopropylidene acetal, he saw that no cleavage reaction occurred (Scheme 2).

In order to satisfactorily account for the results, Richardson then put forward the counterintuitive proposal that the degradation had to be proceeding by a dissociative ion-pair S_N1-type mechanism, wherein the polar reaction medium was inducing formation of an O(2)-stabilized solvated ion pair, followed by eventual glycosylcation capture by hydroxide or water (see Scheme 3). In the case shown in Scheme 2, because of conformational and electronic constraints imposed by the *cis*-fused *O*-isopropylidene acetal, the requisite glycosyl cation could not readily form. The end result, of course, was the proposal that this C-glycoside hydrolysis proceeded by a truly remarkable S_N1 mechanism that many of us would not have initially thought about! I think that this remarkable mechanistic construct, developed in the late 1950s, gives a measure of the astuteness of chemical thought of A.C. Richardson.

When Richardson arrived at Berkeley in 1959, Professor Fischer quickly put him to work on the now celebrated nitromethane cyclization reaction of sugar dialdehydes. This was a powerful new reaction that had just been developed by Fischer for the construction of rare 3-amino sugars. Such sugars adorn a wide range of important bioactive natural products, and the discovery of this valuable new process for their synthesis greatly facilitated the potential for assembling many of them synthetically. Fischer asked Richardson to apply the procedure to the dialdehyde derived from 1,6-anhydro-β-D-glucopyranose (levoglucosan) and then attempt the isolation of any nitro

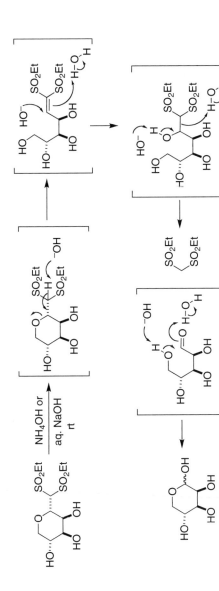

SCHEME 1. A plausible, but incorrect, mechanism for Richardson's oxidative degradation of sugar diethylsulfonyl dithioacetals under basic conditions.

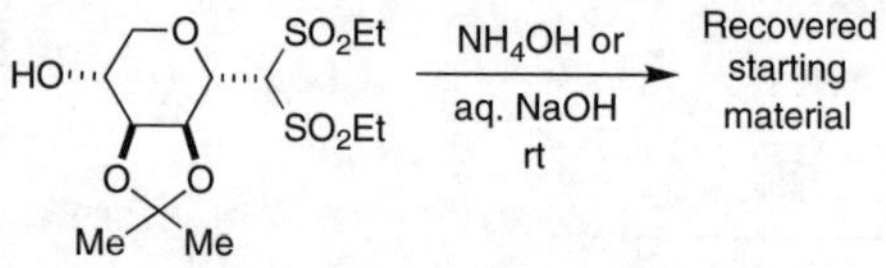

SCHEME 2. Richardson's unsuccessful attempted oxidative degradation of a D-lyxose diethylsulfonyl dithioacetal possessing a fused acetal (1959).

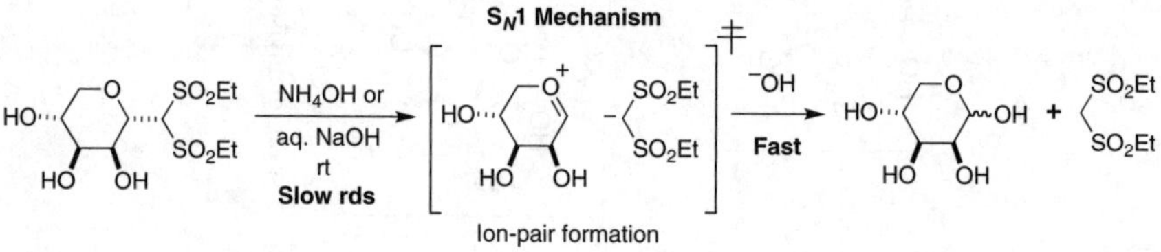

SCHEME 3. Richardson and Hough's mechanistic proposal for this hydrolysis reaction (1959).

sugars that crystallized readily in pure form, and thereafter proceed toward converting these nitro sugars into the corresponding 3-amino sugars. He also asked Richardson to take any noncrystallizable mother-liquor mixtures of nitro sugars through to the 3-amino sugar stage and then see what could be crystallized. The end results of Richardson's labors are to be found in Scheme 4.

Through this protocol, and the specific use of levoglucosan as the starting material, Richardson was able to secure access to some uncommon 3-amino-3-deoxy sugars, including 3-amino-3-deoxy-D-gulose, -D-altrose, and -D-idose derivatives, taking advantage of the well-known tendency of 3-nitropyranosides to adopt structures having the nitro group in the equatorial orientation. The individual sugars having a trans relationship of the 3-acetamido group with the adjacent groups at C-2 and C-4 were then differentiated by means of an inversion reaction of adjacent mesyloxy groups, brought about by heating with sodium acetate in 95% aqueous 2-methoxyethanol at reflux, which results in neighboring-group participation by the 3-acetamido group and attack by water on the intermediary oxazolinium intermediate that is formed. Two papers appeared on this topic in 1961 in *Proceedings of the Chemical Society of London* and in the *Journal of the American Chemical Society*. Undoubtedly even more would have been forthcoming had it not been for the premature death of Hermann Fischer in early 1960.

Upon returning to Bristol on an ICI Fellowship in late 1960, Richardson decided to work independently from Leslie Hough, so as to build up his own personal reputation

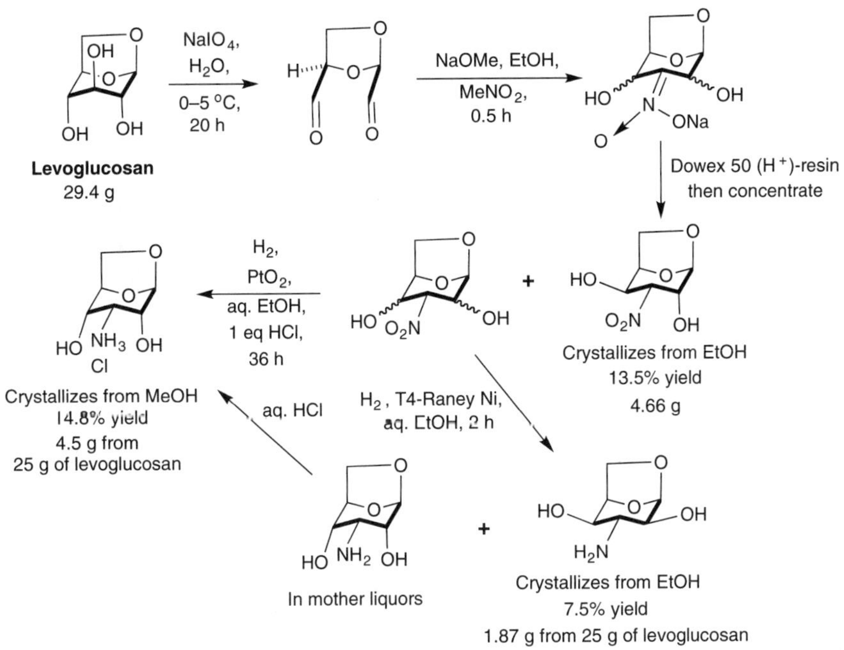

SCHEME 4. Richardson's application of the Fischer nitromethane cyclization to the sugar dialdehyde derived from the periodate cleavage of levoglucosan (1960).

for high-level chemistry and to help improve his prospects for securing a permanent lectureship position in organic chemistry. While at Bristol, he continued to develop the Fischer nitromethane cyclization work which he had begun at Berkeley, applying this reaction to the dialdehyde formed by periodate cleavage of methyl α-L-rhamnopyranoside, thereby providing a highly effective avenue to derivatives of 3-amino-3,6-dideoxy-L-glucose as well as 3-amino-3,6-dideoxy-L-talose (Scheme 5).

Richardson next became interested in applying the Fischer nitromethane cyclization to construction of the mycaminose component of the antibiotic magnamycin. On the basis of differences in the pKa values of various mycaminose acetate derivatives, Woodward had, in 1957, assigned the D-*altro* stereochemistry to mycaminose. However, when Richardson examined the reported molecular rotations of the two anomeric mycaminose triacetates (+6020° and +30,100°) and found that they deviated significantly from the values reported for the β- and α-D-altrose pentaacetates (−17,200° and +24,500°), he immediately recognized that the assigned D-*altro* configuration for the 2-, 3-, and 4-substituents must have been incorrect. Given that

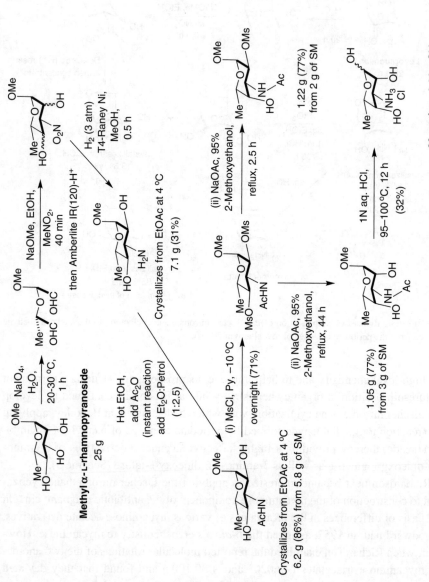

SCHEME 5. Richardson's application of the Fischer nitromethane cyclization to the sugar dialdehyde derived from periodate cleavage methyl α-L-rhamnopyranoside (1961).

he had just prepared methyl 3-amino-3,6-dideoxy-α-L-glucopyranoside from methyl α-L-rhamnoside, he next set about the synthesis of 3,6-dideoxy-3-dimethylamino-D-glucose in the manner shown in Scheme 6, using an Eschweiler–Clarke reductive amination with paraformaldehyde and formic acid to obtain the dimethylamino-pyranoside, which was then hydrolyzed with concentrated aqueous HCl at reflux to afford 3,6-dideoxy-3-dimethylamino-α-D-glucose hydrochloride. Comparison of its IR spectrum, mixed melting point, and optical rotation with values for mycaminose

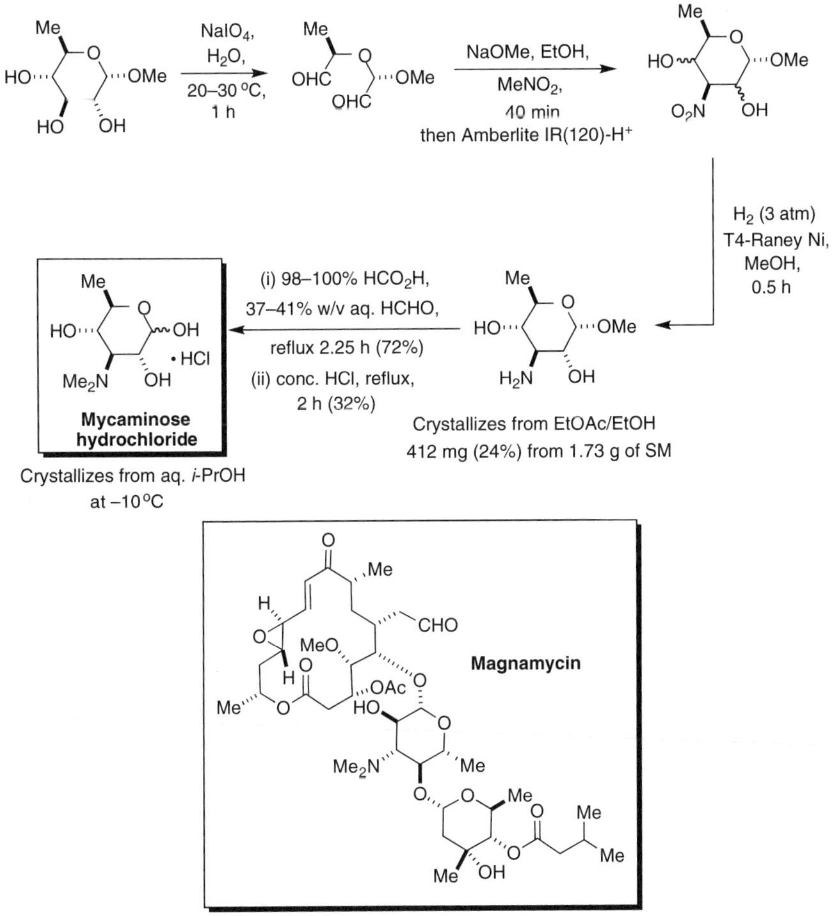

SCHEME 6. Richardson's proof of the absolute stereostructure of mycaminose in 1961 and his contribution to R.B. Woodward's final structure elucidation of magnamycin.

hydrochloride conclusively demonstrated that the latter had the D-*gluco* configuration as shown in Scheme 6. The L enantiomer, which had been prepared similarly (according to Schemes 5 and 6) gave an IR spectrum that was identical with that of the D isomer, but critically, its optical rotation value was of identical magnitude but of opposite sign. Thus, Richardson had resolved a stereochemical assignment issue that had troubled the great R.B. Woodward, a fact that Woodward gladly acknowledged in a short but most gracious personal letter that he wrote to ACR not long after Richardson's communication in *Proceedings of the Chemical Society* appeared in 1961.

Richardson kept this letter from R.B. Woodward throughout his career, and the writer was fortunate enough to see it shortly before Richardson retired. In this letter, Woodward freely admitted that he had not considered using molecular rotations to help secure the stereochemical assignment of mycaminose, and he openly stated that he had always been uncomfortable with the idea of using pKa values to make such an assignment. He also commended young Richardson for his elegant solution to this challenging problem; a solution that he himself had failed to see. He stated that this was an excellent way of solving this stereochemical problem and was most pleased that Richardson had finally brought this matter to a successful conclusion with a superb piece of work. Such was R.B. Woodward's personal greatness as a scientist and as a human being that he took no personal slight at being shown by a younger colleague to have made an incorrect assignment. The beautiful letter that he wrote to ACR not only said a lot about the qualities of A.C. Richardson as a chemist but it also reflected the natural good spirit and friendliness of R.B. Woodward. Not many scientists of such eminence would take a correction of their work in such a comfortable way. However, Woodward always wanted science to be the winner, and as he had so many extraordinary achievements to his name, he had the strength of character to thank Richardson for the significant contribution that he had made. Some years later, this work gave considerable aid to Woodward in assigning the full and correct structure of magnamycin.

Following the publication of this outstanding work, the University of Reading elected to appoint A.C. Richardson onto their Chemistry faculty. During his time there, the quality of his research did not diminish in any way, but it started to move into totally new and exciting directions. Although the synthesis of 3-amino sugars remained a consistent theme of his group during the next decade, as evidenced by Richardson's synthesis of desosamine, he opened up several new lines of research during this period, including a study of the synthesis and chemical reactions of several very useful monosaccharide aziridines, and the selective O-acylation of various readily available methyl glycosides of monosaccharides.

The synthesis of desosamine hydrochloride served as the touch paper for much subsequent work in the area of nucleophilic displacement of carbohydrate sulfonic esters. Such a reaction featured in a critical step of the Richardson synthesis of this target molecule. Richardson's route to desosamine (Scheme 7) commenced with the oxidation of methyl 4,6-*O*-benzylidene-α-D-glucopyranoside by aqueous sodium periodate and subsequent Guthrie cyclization of the resultant sugar dialdehyde with phenylhydrazine. This led to a 3-phenylazo sugar of the D-*gluco* configuration, which was then subjected to hydrogenolytic reduction and hydrazine cleavage with Raney nickel to give, after exhaustive O-acetylation, methyl 3-acetamido-2-*O*-acetyl-4,6-*O*-benzylidene-3-deoxy-α-glucopyranoside. Debenzylidenation of this product with aqueous acetic acid and subsequent O-mesylation gave a 4,6-dimesylate, which was subjected to nucleophilic displacement with an excess of sodium iodide in boiling 2-butanone. Significantly, the S_N2 displacement process at C-4 proceeded with neighboring-group participation from the 3-acetamido group, and led to a net

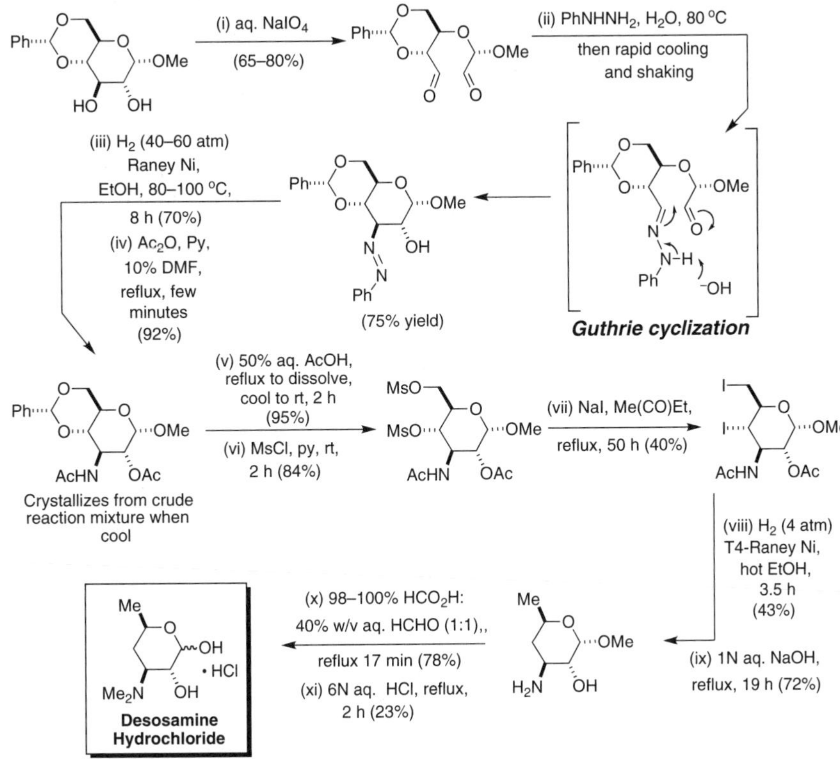

SCHEME 7. Richardson's application of the Guthrie phenylhydrazine cyclization to the synthesis of desosamine hydrochloride (1963).

double inversion (and overall retention) process being observed, and the D-gluco-4,6-diiodide product was obtained. Dehalogenation was then effected with Raney nickel followed by base hydrolysis, Eschweiler–Clarke reductive amination, and glycoside hydrolysis to furnish desosamine hydrochloride by a most elegant route.

Another research program that continued at Reading, but which had been initiated in Bristol, concerned the development of effective synthetic strategies to various monosaccharide aziridines, most notably, methyl 2,3-*N*-acetylepimino-4,6-*O*-benzylidene-2,3-dideoxy-α-D-allopyranoside and the corresponding 2,3-*N*-acetylepimino-α-D-mannopyranoside derivative, as well as methyl 3,4,6-trideoxy-3,4-epimino-α-L-galactopyranoside. This work was allied with a study of the reactions of these intermediates for the preparation of various rare amino sugars. This elegant and important work led to seven papers being published over a 5-year period at Bristol, Reading, and QEC, involving the coworkers D.H. Buss, C.F. Gibbs, A.D. Barford, J. Tjebbes, Y. Ali, and of course, Les Hough. The 1965 synthesis of the D-allo-epimine, now widely known as the Hough–Richardson aziridine (Scheme 8), is illustrative of Richardson's work in this area; it was published in *Carbohydrate Research* with C.F. Gibbs and Les Hough. The synthesis is shown in full in Scheme 8, starting off with D-glucosamine hydrochloride. The latter was first N-benzoylated with benzoic anhydride in methanol, followed by Fischer glycosidation and subsequent O-benzylidenation. Although the original Hough–Richardson synthetic protocol actually used zinc chloride and benzaldehyde for this step, Hale

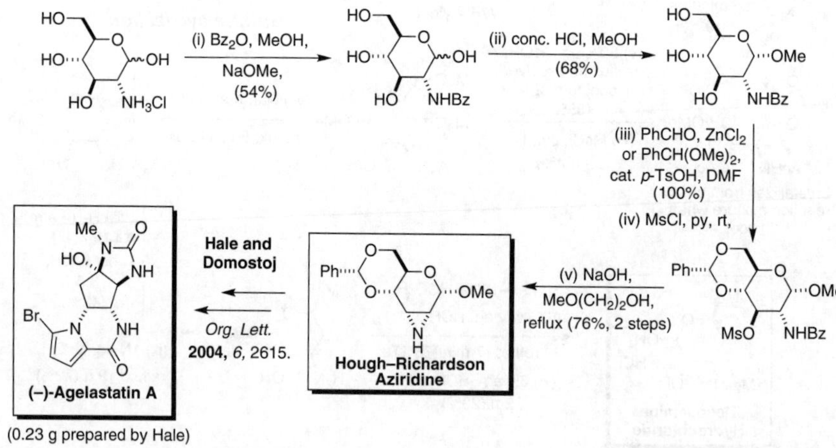

SCHEME 8. Hale and Domostoj's use of the Hough–Richardson aziridine in their 2004 enantiospecific total synthesis of the antitumor alkaloid (−)-agelastatin A.

and Domostoj later found that this transformation could be performed in higher yield by acetal exchange. O-Mesylation and treatment with base then completed the synthesis of the aforementioned allo-aziridine. Significantly, Hough and Richardson observed that once the N-acyl aziridine had been prepared, its N-acyl moiety became very susceptible to cleavage with base. This has since turned out to be synthetically advantageous, for it allows a more-stable urethane or other protecting group to be introduced on the epimine for subsequent ring-opening reactions. Using this intermediate, which is readily prepared on the kilogram scale, Hale and Domostoj in 2004 successfully synthesized almost a quarter of a gram of the powerful, broad-spectrum, antitumor agent (−)-agelastatin A. When Richardson learned of this later total synthesis during a luncheon date in the Mulberry Bush pub in London, both he and Les Hough were delighted to hear that modern-day chemists were still relying on the procedures they had developed more than 40 years earlier!

During his time at Reading, Richardson began his first successful incursions into the selective esterification of many common sugars, but it was at QEC that this work really took off, and the procedure was applied to many disaccharides, often with remarkable results. While space constraints prevent a complete discussion of all Richardson's work on the selective esterification of mono- and disaccharides, some highlights are recorded here.

Of particular note was Richardson's high-yielding (70%) selective O-acetylation of methyl β-maltoside with acetyl chloride in pyridine and toluene, which furnished the hexa-O-acetylated disaccharide having the 3′-hydroxyl group free. This beautifully set up the resulting molecule for a regiospecific O-mesylation at that position and the synthesis of a disaccharide with a 2′,3′-allo-epoxide group, which subsequently permitted 3′-deoxy-3′-halo-maltosides to be prepared by "anomalous" non-Furst–Plattner epoxide ring-opening (Scheme 9).

Another representative highlight of the regioselective O-acylation work undertaken at QEC can be found in Richardson's selective O-pivaloylation of sucrose, which led to two papers with Les Hough and Manjit Chowdhary. One of these was on the regioselective O-pivaloylation itself, the other on the applications of the pivalate products in the preparation of various ring-modified sucrose derivatives.

One of the early breakthroughs in these pivaloylation studies was the Richardson–Hough discovery that, when a solution of sucrose in dry pyridine is treated with 20 equivalents of pivaloyl chloride (a sterically demanding reagent) at −70 °C for 7 h and the reactants are later warmed to room temperature, 1,3,4,6-tetra-O-pivaloyl-β-D-fructofuranosyl 2,3,6-tri-O-pivaloyl-α-D-glucopyranoside is readily obtained in 52% yield along with a 40% yield of the per-O-pivaloylated sucrose (Scheme 10). This important finding permitted the synthesis of a host of different 4-substituted sucrose

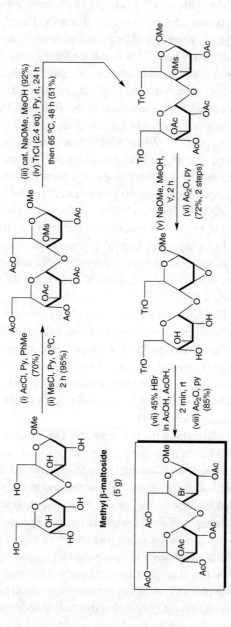

SCHEME 9. Durette, Hough, and Richardson's synthesis of methyl 3-bromo-3-deoxy-β-maltoside via opening of the 2,3-allo-epoxide derived from the 1,2,2′,3′,4′,6,6′-heptaacetate of methyl β-maltopyranoside (1974).

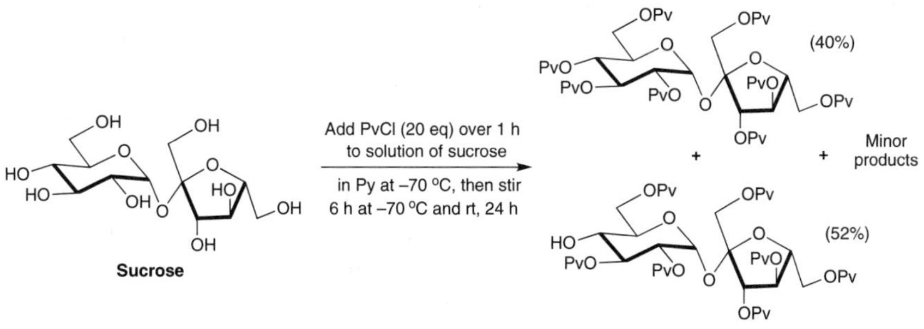

SCHEME 10. Hough, Richardson, and Chowdhary's high-yielding selective heptapivaloylation of sucrose (1984).

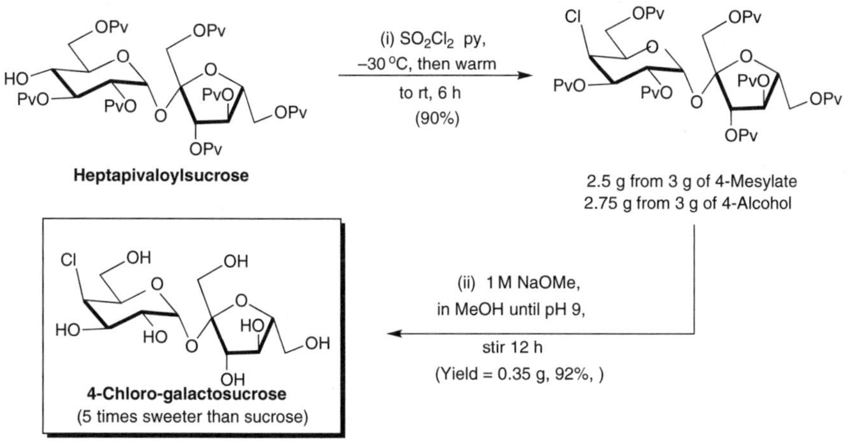

SCHEME 11. The Chowdhary, Hough, and Richardson synthesis of 4-chloro-galactosucrose from 2,3,6,1'3',4',6'-heptapivaloylsucrose (1984).

analogues, including the 4-C-methyl- and 4-C-allyl-derivatives of sucrose (work done with Albert Chiu, Stanley Dziedzic, and Gordon Birch). It also led to the first preparation by Chowdhary of 4-chloro-galacto-sucrose (Scheme 11), which subsequently proved to be five times sweeter than sucrose itself. Of course, this molecule and its per-ester derivatives are useful synthetic intermediates, in their own right, of utility for the synthetic preparation of many unusual 4-C-modified sucrose derivatives; however, this point is not discussed further here.

Another important result was the finding that, when the aforementioned esterification was conducted at $-40\ °C$ during 8 h with 7 equivalents of PvCl, a different reaction course was followed. Thus, the 3,6,1′,3′,4′,6′-hexa-O-pivaloyl sucrose was now formed in 45% yield alongside a 35% yield of the 2,4,6,1′,4′,6′-hexa-O-pivaloyl sucrose (Scheme 12).

The latter reaction set up the possibility of making such exotic derivatives as 2,3-anhydro-4-chloro-talo-sucrose (Scheme 13) by a four-step sequence involving O-mesylation of 3,6,1′,3′,4′,6′-hexa-O-pivaloylsucrose, high-temperature LiCl displacement, O-depivaloylation with concomitant epoxide formation, and finally O-acetylation.

Access to 2,3,6,1′,4′,6′-hexa-O-pivaloylsucrose as a minor reaction product (4% yield) from the selective O-pivaloylation of sucrose with 20 equivalents of PvCl at $-40\ °C$ over a 2 h period opened up the added possibility of accessing a novel galacto-sucrose derivative having an epoxide group in the fructofuranosyl ring (Scheme 14). The latter was available via O-mesylation, selective displacement at C-4, and base treatment followed by O-acetylation. Notwithstanding the (low) 4% yield of this particular hexapivalate, it was still a fairly simple matter to obtain 1 g of this hexapivalate starting material from a 10-g run of the aforementioned O-pivaloylation reaction, which delivered the 2,3,6,1′,3′4′,6′-heptapivalate (25%) and the 3,6,1′,3′4′,6′-hexapivalate (33%) as the primary reaction products (Scheme 15).

Further decreasing the number of equivalents of pivaloyl chloride to five (Scheme 16), and operating at $-70\ °C$ in a chloroform–pyridine mix for 1.5 h, led to a partially pivaloylated sucrose with the 3-, 4-, and 3′-hydroxyl groups free (25%) in addition to a derivative in 35% yield having the 2-, 3-, 4-, and 3′-hydroxyl groups free.

Similarly, when the selective O-pivaloylation of trehalose was investigated by Richardson, Hough, and Cortes-Garcia, a number of useful derivatives were also obtained in quantities adequate to be of preparative value. Thus when 12 equivalents of pivaloyl chloride were employed for this esterification at $-20\ °C$, and the reactants were thereafter stored at room temperature for three days, a 61% yield of the 2,2′3,3′,4′,6,6′-heptapivalate was obtained, and following invertive chlorination with sulfuryl chloride, a displacement with different nucleophiles gave access to C-4-modified trehaloses, one example of which is shown in Scheme 17.

The Richardson–Hough research partnership also observed that, if the quantity of pivaloyl chloride was raised to 7 equivalents and similar reaction conditions were applied, access to a partially pivaloylated trehalose having the 3- and 4′-hydroxyl groups free was gained in 38% yield, together with a symmetrical 2,3,6,2′3′,6′-hexapivalate, which was formed in 31% yield (Scheme 18).

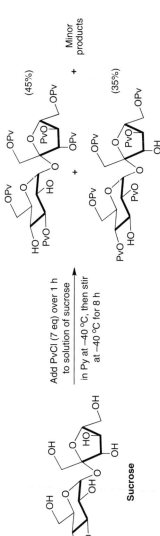

SCHEME 12. Hough, Richardson, and Chowdhary's selective hepta- and hexa-pivaloylations of sucrose (1984).

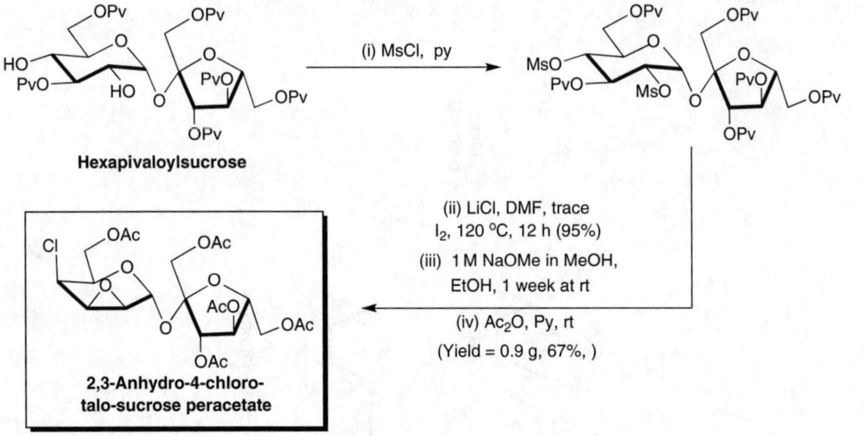

SCHEME 13. Chowdhary, Hough, and Richardson's conversion of 3,6,1′,3′,4′,6′-hexa-*O*-pivaloylsucrose into the 2,3-anhydro-4-chloro-talo-sucrose peracetate (1986).

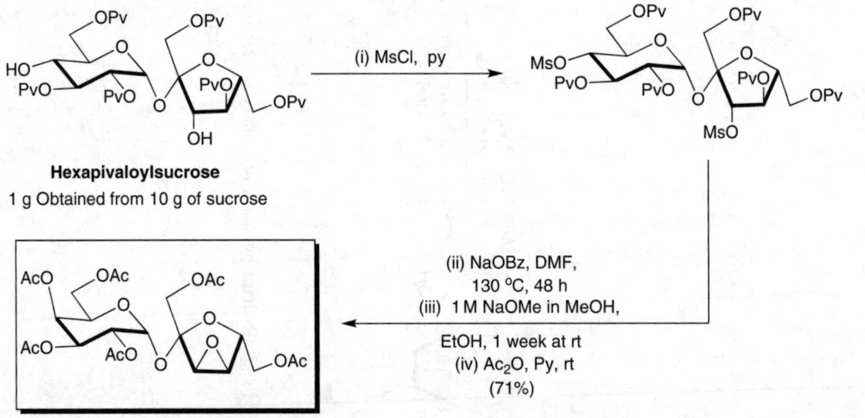

SCHEME 14. Chowdhary, Hough, and Richardson's conversion of 2,3,6,1′,4′,6′-hexa-*O*-pivaloylsucrose into the ribo-3′,4′-anhydro-galactosucrose peracetate (1986).

In entirely analogous fashion, the selective esterification of various other common disaccharides by other acylating agents was investigated. One notable example is the selective benzoylation of methyl β-lactoside with benzoyl chloride, which again produced many interesting results that are recorded in a series of papers with Ram S. Bhatt and Les Hough. There are several highlights in the *JCS Perkin Transaction 1*

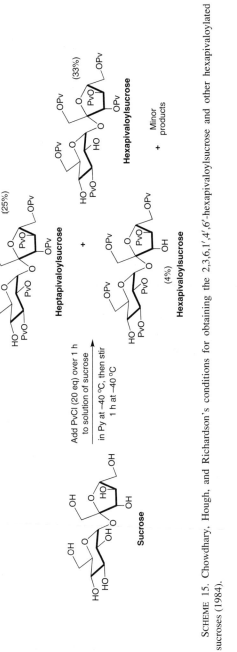

SCHEME 15. Chowdhary, Hough, and Richardson's conditions for obtaining the 2,3,6,1',4',6'-hexapivaloylsucrose and other hexapivaloylated sucroses (1984).

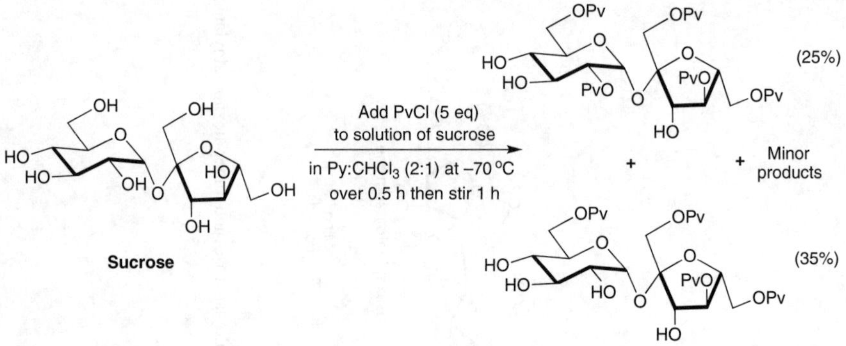

SCHEME 16. Other useful pivalate esters of sucrose obtained by Hough, Richardson, and Chowdhary (1984).

paper that appeared with these two coauthors in 1977, including the finding that the galactopyranosyl ring is the more reactive, furnishing the 3,6-dibenzoate in 31% yield when using 2.5 equivalents of benzoyl chloride and warming from −20 °C to room temperature (Scheme 19).

By way of contrast, when 6.8 equivalents of BzCl was used under identical circumstances, in a manner similar to that observed with methyl β-maltoside, a hexabenzoate with the 3'-OH free was isolated in 33% yield, and this could be chlorinated in high yield with S_N2 inversion by using sulfuryl chloride and pyridine (Scheme 20).

With Ram Bhatt and Leslie Hough, Richardson also O-mesylated the 3'-OH group of this particular 2,3,4,6,2',6'-hexabenzoate and subsequently the 3'-OMs group was displaced by heating with NaOBz in hexamethylphosphoric triamide at 105 °C for three days. The resulting heptabenzoate (obtained in 62% yield) was then subjected to acetolysis and O-deacetylation to give the 3'-allo-epimer of lactose, namely, 4-O-β-D-galactopyranosyl-D-allose.

Limiting the amount of BzCl added to 5.2 equivalents, while maintaining similar experimental conditions, led in contrast to a mixture containing four main components, the two major ones being the 3,6,2'6'-tetrabenzoate and the 3,6,6'-tribenzoate, isolated in 21% and 29% yields, respectively (Scheme 21).

Another recurrent theme of the Richardson–Hough team was the selective sulfonylation of disaccharide derivatives, which led to many interesting papers, some involving trehalose and others sucrose. For example, in 1986, they reported (with Mukund K. Gurjar and Lee V. Sincharoenkul) that selective tosylation of 6,1'6'-tri-O-tritylsucrose at room temperature for 2 days afforded the 2-tosyl ester in 52% yield

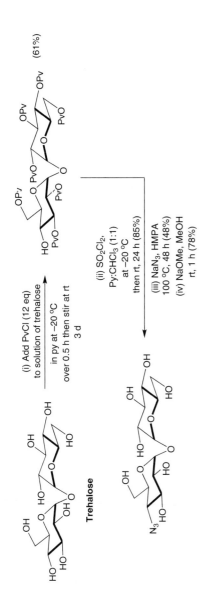

SCHEME 17. The Cortes-Garcia, Hough, and Richardson selective pivaloylation of trehalose and a subsequent application in the synthesis of 4-azido-trehalose (1990).

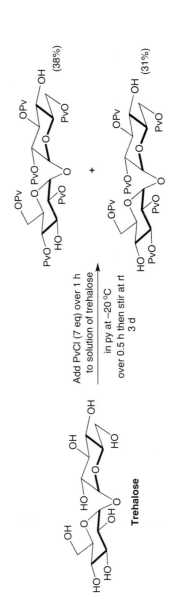

SCHEME 18. Other symmetric and nonsymmetric hexapivalates formed via the selective pivaloylation of trehalose by Cortes-Garcia, Hough, and Richardson (1990).

SCHEME 19. One representative example of Hough, Richardson, and Bhatt's selective benzoylation of methyl β-lactoside (1975).

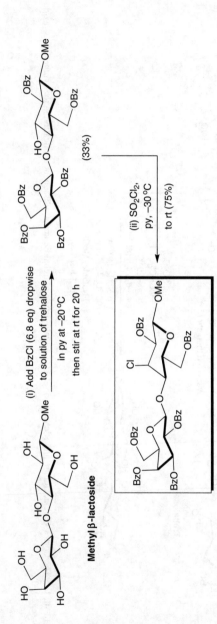

SCHEME 20. Bhatt, Hough, and Richardson's regioselective chlorination of the 3′-OH of methyl β-lactoside following regioselective hexabenzoylation.

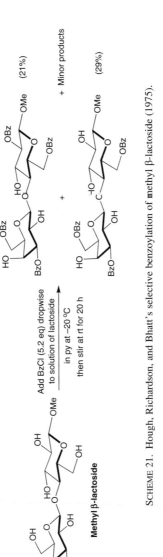

SCHEME 21. Hough, Richardson, and Bhatt's selective benzoylation of methyl β-lactoside (1975).

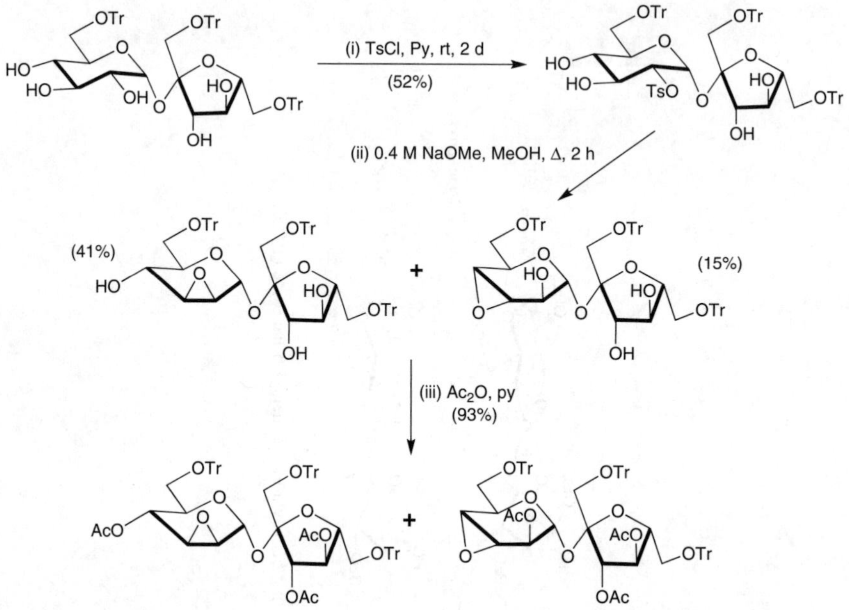

SCHEME 22. A new pathway to the 2,3-manno-epoxide derivative of sucrose (1986).

(Scheme 22) and not the 3-tosyl ester that had originally been claimed in 1971 by Jezo as the major product of this reaction. Indeed, when Richardson repeated Jezo's exact conditions, which involved heating 6,1'6'-tri-O-tritylsucrose with TsCl and pyridine at 55 °C, the product obtained had identical spectroscopic properties to those reported for the Jezo product, and this material was, in turn, shown to be identical to the 2-tosyl ester prepared in improved yield under Richardson's new conditions (Scheme 22).

Unambiguous confirmation that tosylation had occurred at O-2 rather than at O-3 was obtained when the 2-tosylate was treated with NaOMe in MeOH at reflux for 2 h, when it was converted primarily into the 2,3-manno-epoxide in 41% yield, alongside a small proportion (15%) of the 3,4-altro-epoxide (Scheme 22); the latter product had arisen from a Payne-type epoxide rearrangement. The ^{1}H NMR spectrum of the O-acetylated manno-epoxide in CDCl$_3$ showed the signal for H-1 as a singlet at δ 5.51, that of H-4 as a doublet at δ 5.10 ($J_{4,5} = 10$ Hz), with the H-2 resonance appearing at δ 3.00 and that of H-3 at δ 3.14. The $J_{2,3}$ value of 2.5 Hz, the $J_{1,2}$ value of zero Hz, and the large 10 Hz coupling between H-3 and H-4 all clearly indicated that it had the manno-configuration. Likewise, analysis of the J values for the O-acetylated altro-epoxide again confirmed its configuration. The NMR data thus proved

Richardson and Hough's assertion that it was the 2-tosyl ester that had been prepared, and not the 3-tosylate claimed by Jezo.

Richardson then went on to examine the epoxide ring-opening reactions of the 2,3-manno-sucrose epoxide with a range of nucleophiles (Scheme 23). As might now be expected, *trans*-diaxial ring-opening of the epoxide occurred on treatment with sodium azide in boiling aqueous ethanol in the presence of ammonium chloride and afforded the crystalline 3-azido-altro-sucrose tetraacetate in 92% yield after acetylation of the initial product. With an excess of lithium iodide in boiling ether for 3.5 days, the epoxide opening took a most unexpected course. It appeared that initial epoxide ring-opening occurred in *trans*-diaxial manner to give the 3-iodo-α-D-altro product, and this then underwent further nucleophilic displacement to afford the 3-iodo-manno-sucrose derivative in 37% yield. Presumably this particular equilibration occurred in order to relieve the significant 1,3-diaxial interaction with the fructofuranoside group at O-1 that would be generated as soon as the altro-iodide was formed. Another interesting product obtained from the opening of this manno-sucrose-2,3-epoxide with ammonium thiocyanate was the 2,3-allo-episulfide, which was formed via the Hough reaction.

Working on modifications of trehalose, Richardson (with P.A. Monroe and Les Hough) demonstrated the selective tosylation at O-2 of 4,6,4′,6′-di-*O*-benzylidene-α,α-trehalose to afford a symmetrical 2,2′-ditosylate, which could be converted into the bis-manno-epoxide by the action of base (Scheme 24). The latter underwent *trans*-diaxial ring-opening with nucleophiles at C-3 and C-3′, and many synthetic applications of this intermediate have now been recorded. One of these is shown in Scheme 24, from the Richardson–Hough laboratory, performed by postdoctoral fellow, Edward Tarelli, who contributed voluminously with Richardson to the exploration and development of trehalose chemistry.

In similar vein, Richardson, Hough, and Monroe devised conditions that permitted the selective monotosylation of 4,6,4′,6′-di-*O*-benzylidene-α,α-trehalose at the 2-hydroxyl group in low but, nevertheless preparatively useful, 23% yield (Scheme 25). Given the ready availability of the starting disaccharide on a large scale, this provided a viable intermediate for use in synthesis. With this 2-tosylate, access to a trehalose derivative having a 2,3-manno-epoxide functionality in just one of the pyranoside rings was made possible for the first time.

Richardson, Hough, and Monroe also devised a synthesis of the corresponding 4,6,4′,6′-di-*O*-benzylidene-α,α-trehalose derivative having an allo-configured epoxide in both sugar rings (Scheme 26). It did not involve selective sulfonylation, but nevertheless, it furnished another highly useful epoxide intermediate of the trehalose family. The synthesis of the corresponding manno-allo bisepoxide again relied upon the dimesylation of a manno-configured trehalose epoxide, followed by base treatment.

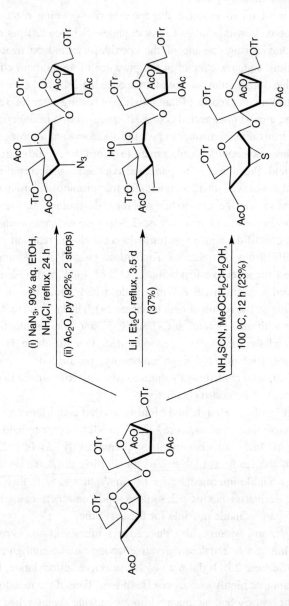

SCHEME 23. Some of the novel derivatives of sucrose derived from the 2,3-manno-sucrose epoxide by Hough, Richardson, Gurjar, and Sincharoenkul (1986).

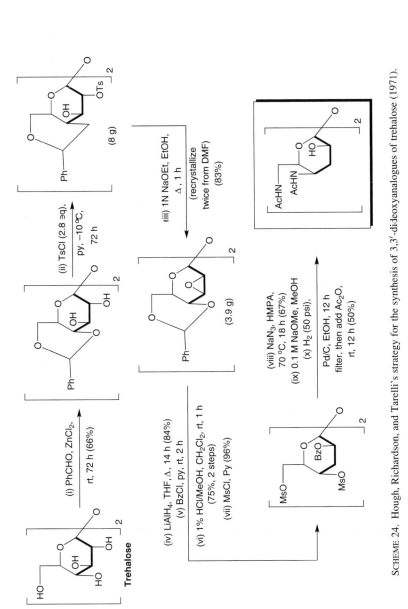

SCHEME 24. Hough, Richardson, and Tarelli's strategy for the synthesis of 3,3'-dideoxyanalogues of trehalose (1971).

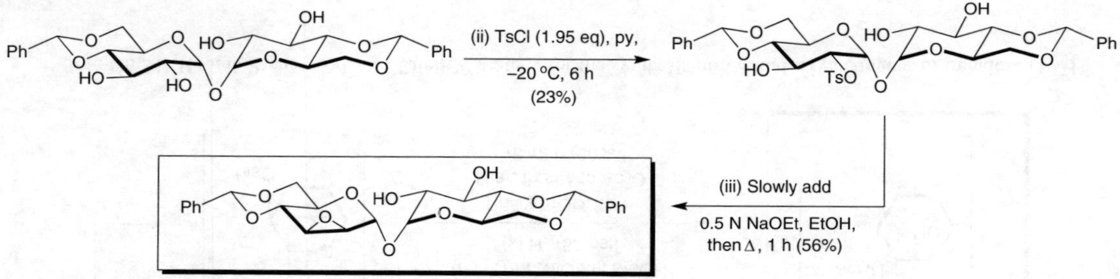

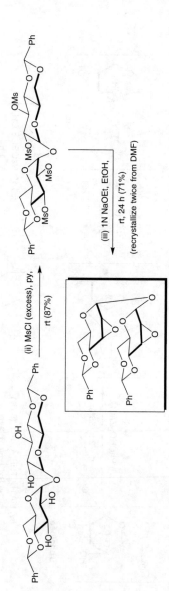

SCHEME 25. Hough, Monroe, and Richardson's synthesis of the 2,3-anhydro-4,6-O-benzylidene-α-D-mannopyranosyl 4,6-O-benzylidene-α-D-glucopyranoside from trehalose (1971).

SCHEME 26. Hough, Monroe, and Richardson's synthesis of the corresponding 2,3-allo-bisepoxide (1971).

Richardson's interest in these trehalose epoxides stemmed from his desire to prepare a wide range of modified trehalose analogues for evaluation as inhibitors of the trehalase of insects, as trehalose is known to be the storage carbohydrate of many insects and good inhibitors of this processing enzyme could potentially function as "green" insecticides.

As already noted, A.C. Richardson and Les Hough had a long-standing fascination with the selective halogenation of mono- and oligosaccharides, and together they made many significant contributions to this area. Their primary interests were in chlorination and fluorination, but bromination and iodination also caught their attention.

The first efforts toward the regioselective chlorination of sucrose came with John M. Ballard and Peter H. Fairclough in 1973; their results are presented in Scheme 27. Respectable yields of the 6,6'-dichloride and the 6'-chloride were obtained.

This was then followed, 2 years later, with the landmark synthesis of 4,1',6'-trideoxy-component (TGS) by Ph.D. student Peter H. Fairclough, which capitalized upon an acetate-group migration from O-4 to O-6 during the detritylation of 6,1',6'-tri-O-tritylsucrose pentaacetate with boiling acetic acid (Scheme 28). Low-temperature chlorination of the resulting 4,1',6'-triol with sulfuryl chloride in pyridine and chloroform at -75 °C thereupon provided 4,1',6'-trichloro-galacto-sucrose as its pentaacetate. Deacetylation with sodium methoxide in methanol gave the free sugar in 69% yield. Although this compound was initially prepared for nothing more than pure chemical exploration, what transpired in 1976 is now the stuff of legend. A telephone call to the Hough–Richardson lab from Riaz Khan at Tate and Lyle was verbally misunderstood by the young graduate student, Shashi P. Phadnis. Riaz's request for "chlorinated sucroses for testing" (in various biological assays) was mistakenly heard by Phadnis as "chlorinated sucroses for tasting"! He then dutifully went into the laboratory and tasted one of these derivatives himself, and soon discovered (to his immense surprise) that it was intensely sweet, which stood in stark contrast to the intense bitterness of 4,6,4'6'-tetrachloro-galacto-trehalose. This discovery then led to a major effort within the Tate & Lyle Research Centre in Reading, aimed at synthesizing and subsequently tasting a wide range of chlorinated sucroses.

After much effort, Tate & Lyle eventually selected 4,1',6'-trichloro-4,1',6'-trideoxy-galacto-sucrose (TGS) as the product that they would develop as a high-intensity, noncaloric sweetener. The Tate & Lyle selection was based upon the ease with which this chlorinated sucrose could be produced industrially in high yield, the high-intensity of sweetness of the molecule (650 times sweeter than sucrose), its excellent shelf-life, and the fact that it was essentially nontoxic and unmetabolized by animals (and man). It was also devoid of the bitter or peculiar after-tastes of many

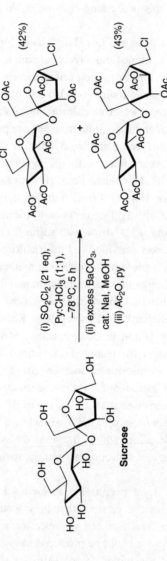

SCHEME 27. Ballard, Fairclough, Hough, and Richardson's study of the selective low-temperature chlorination of sucrose with sulfuryl chloride (1973).

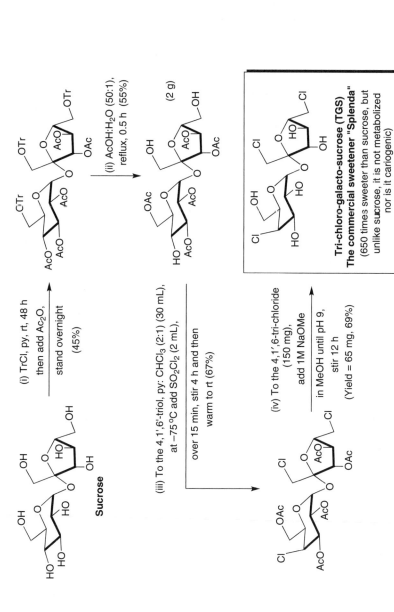

SCHEME 28. Fairclough, Hough, and Richardson's elegant synthetic pathway to the commercial nonnutritive sweetener TGS (1975).

other commercial noncaloric sweeteners, and if given at low concentrations, its sweetness quickly dissipated. Although serendipity did play a substantial role in this discovery, as freely acknowledged by both Les Hough and Anthony Richardson in various reviews and lectures, this brilliant discovery would never have been made without their curiosity-driven synthetic efforts of the early 1970s.

Given the enormous commercial significance of this discovery, it was quickly recognized by Tate & Lyle that back-up patents would almost certainly be needed to cement their market position. The fact that raffinose, an abundant trisaccharide, is also sweet, albeit not as sweet as sucrose, prompted Richardson and Hough to investigate its regioselective chlorination to see if anything equal or superior to TGS could be identified as a possible alternative or back-up compound. Raffinose has the sucrose structure with an added α-D-galactopyranosyl group at C-6. It is a very tricky compound to work with, but nevertheless, Richardson and Hough put their student, Mohamed A. Salam, on the problem. To everyone's great surprise, Salam's first efforts on this project did not go quite according to plan. He followed a series of earlier experimental procedures from the Hough–Richardson lab for the regioselective chlorination of other disaccharides. In so doing, he managed to produce two chlorinated fructofuranosides in 27% and 24% yields, respectively, following O-acetylation (Scheme 29).

This unanticipated event was useful for accessing these two novel fructofuranosides, but it did not solve the task at hand, and Salam therefore made another attempt to chlorinate raffinose regioselectively, this time quenching the reaction at low temperature. The results are shown in Scheme 30; the $6'''$-hydroxyl group of the fructofuranoside ring proved to be the most reactive, and chlorination went selectively to give, after acetylation, the $6'''$-chloro-$6'''$-deoxyraffinose peracetate in 43% yield. Minor quantities of the peracetylated $6,6'''$-dichloride and the $4,6,6'''$-trichlorides were also formed, and these were subsequently converted into various other derivatives of raffinose. Evidently nothing superior to the sweetener TGS came out of this raffinose chlorination program.

Richardson and Hough's efforts on the chlorination of sucrose ran in parallel with their studies on the selective chlorination of other readily available disaccharides. One notable piece of work done with Philippe L. Durette involved a study of the regioselective chlorination of methyl β-maltoside (Scheme 31), which was undertaken with the long-term objective of possibly developing a future synthesis of the natural antitumor agent, amicetin.

In Durette's study, he treated methyl β-maltoside with an excess of sulfuryl chloride at low temperature in chloroform–pyridine and thereafter warmed the mixture to room temperature. After dechlorosulfation with a catalytic amount of sodium iodide in methanol and O-acetylation, the stereochemically inverted $4,6,3'6'$-

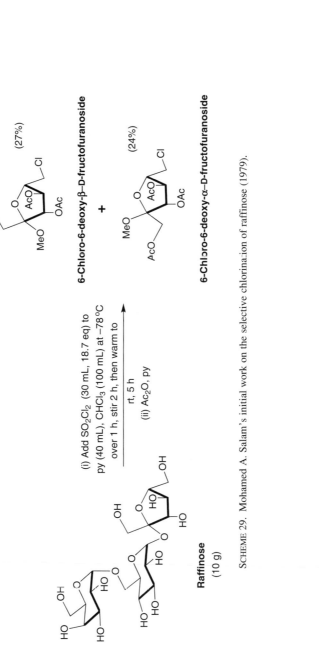

SCHEME 29. Mohamed A. Salam's initial work on the selective chlorination of raffinose (1979).

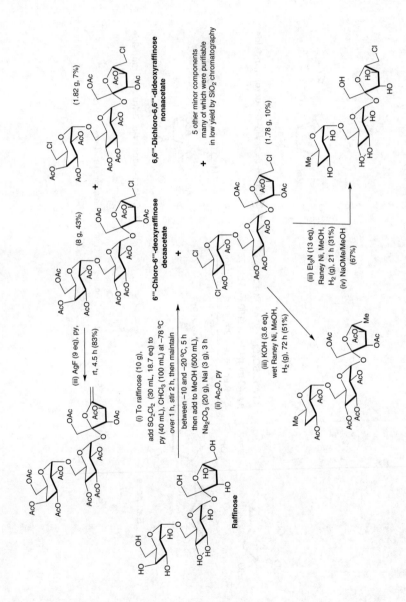

SCHEME 30. Hough, Richardson, and Salam's work on the selective chlorination of raffinose (1979).

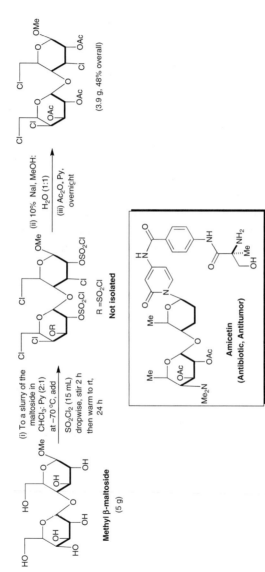

SCHEME 31. Durette, Hough, and Richardson's study of the selective chlorination of methyl β-maltoside with sulfuryl chloride (1973).

tetrachloride was formed. This, conceivably, could be converted through to the amicetin disaccharide, although no further work on this was reported from the Hough– Richardson team.

The great commercial interest in chloro-sucroses led the research leaders at Tate & Lyle to ask Hough and Richardson to explore the synthesis of various fluorinated sucrose derivatives, in a quest for high-intensity sweeteners that would be even more effective than TGS. One published example of their joint work, performed with A.K.M.S. Kabir, is shown in Scheme 32. It involved the preparation of 5-fluoro-galacto-sucrose from 2,3,1',3'4',6'-hexa-O-benzylsucrose by selective tritylation of the 6-hydroxyl group and mesylation at O-4, all accomplished in a one-pot process. The 4-mesylate was then subjected to S_N2 displacement by fluoride, using "anhydrous" solid tetra-n-butylammonium fluoride (TBAF) in boiling acetonitrile. However, none of the fluoro-sucroses prepared turned out to be significantly sweeter than TGS.

The work on fluoro-sucroses relied heavily on technology originally introduced by the Hough–Richardson laboratory for the synthesis of fluoro-trehaloses during the period 1972–1973. Their student A.K. Palmer successfully prepared 6,6'-dideoxy-6,6'-difluoro-α,α-trehalose and its galacto analogue via sulfonic ester displacement. Using similar tactics, he also prepared several 4,4'-difluoro- and 4,4',6,6'-tetrafluoro-analogues. With student Anthony F. Hadfield, 6-deoxy-6-fluoro-α,α-trehalose and related analogues were later published in 1978. Critical to this particular endeavor was the selective acid-catalyzed methanolysis of 2,3,2',3'-tetra-O-benzyl-4,6:4',6'-di-O-benzylidene-α,α-trehalose to produce the monobenzylidene derivative, which was converted into the corresponding dimesylate and thereafter subjected to selective displacement with TBAF in boiling acetonitrile. The mesyl ester group was then cleaved from O-4 with methanolic sodium methoxide, and the remaining protecting groups were detached by catalytic hydrogenolysis in 1% methanolic HCl (Scheme 33). In addition, 4,6-difluoro-galacto-trehalose was prepared by double nucleophilic displacement and catalytic hydrogenolysis. Likewise, 4-fluoro-galacto-trehalose was prepared by tritylation of the 4,6-diol, O-mesylation, S_N2-displacement by fluoride, and catalytic hydrogenolysis (Scheme 33).

In later work with Anna A.E. Penglis, Richardson and Hough prepared a derivative of 3-amino-2,3-dideoxy-2-fluoro-D-altrose by ring-opening of the N-benzoylated Hough–Richardson allo-aziridine, and using ideas similar to those advanced by Hadfield, the syntheses of various fluoro derivatives of 2-amino-2-deoxy-D-galactose and -D-glucose were completed (Scheme 34).

Aside from his fundamental and pioneering efforts in the chemistry of monosaccharides, A.C. Richardson was very active in the total synthesis of complex molecules from monosaccharide starting materials. His work in this area was particularly

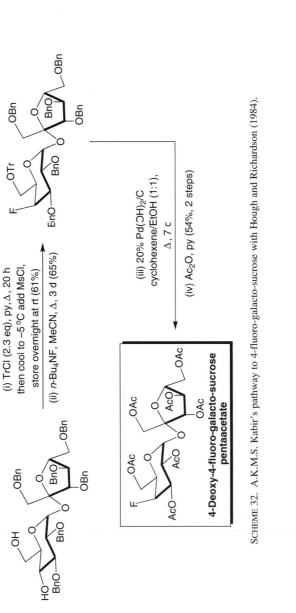

SCHEME 32. A.K.M.S. Kabir's pathway to 4-fluoro-galacto-sucrose with Hough and Richardson (1984).

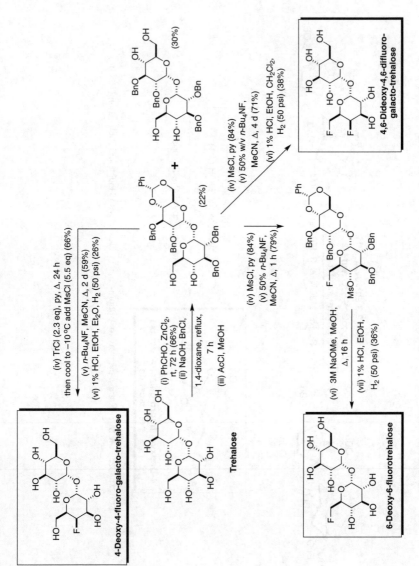

SCHEME 33. Fluoro-trehalose derivatives prepared by Hough, Richardson, and Hadfield (1978).

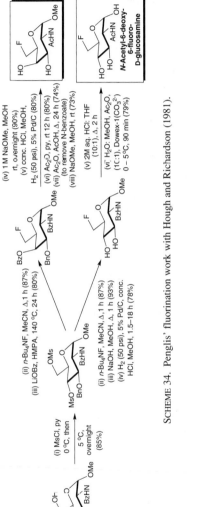

SCHEME 34. Penglis' fluorination work with Hough and Richardson (1981).

original and creative, and it always stood out for its conceptual novelty and the unusual disconnections that his great mind uncovered. Although he always had an interest in multistep target synthesis since his ICI Fellowship days at Bristol, this effort took on a new dimension after he had spent his sabbatical year at Syntex. There, he masterminded elegant total syntheses of the natural, optically active, forms of the antibiotics (−)-anisomycin and (−)-pentenomycin I, using simple chiral starting materials derived from D-glucose. This work was eventually published in *Pure and Applied Chemistry* in 1978.

The (−)-anisomycin work is presented in Scheme 35. Its key step centered around the formation of a pyrrolidine ring that possessed all three of the asymmetric centers present in the target; this was done by nucleophilic displacement of a 3-tosyloxy function in an appropriately functionalized 6-amino-6-deoxy-β-L-talose derivative, whose 1,2-diol was later released and oxidatively cleaved with sodium periodate. Grignard coupling, O-acetylation, and catalytic hydrogenation then furnished the desired natural-product target.

Richardson's synthesis of (−)-pentenomycin I displayed similar levels of conceptual elegance and was also very well received. In the mid-1970s, synthetic organic chemistry was far less refined that it is currently, and the synthesis of such chiral molecules as (−)-pentenomycin looked extremely formidable. However, Richardson was never put off by the degree of difficulty of any problem he faced; he was always determined to overcome that problem. His account of the (−)-pentenomycin I synthesis reveals the many traps and snares in which he was successively caught during that synthetic endeavor, before the successful synthesis shown in Scheme 36 was eventually accomplished.

In this venture (Scheme 36), Richardson oxidized 1,2:5,6-di-*O*-isopropylidene-D-glucofuranose to its 3-keto derivative, added nitromethane, and then elaborated the resulting nitro sugar to a 3-benzyloxymethyl derivative, which was further processed toward the 3-substituted derivative of methyl 5-deoxy-2,3-*O*-isopropylidene-β-D-*erythro*-pent-4-enofuranoside. This compound, at step xiv, was converted through to an aldos-4-ulose derivative that underwent intramolecular aldol condensation when heated on solid alumina (step xv). Selective acetolysis of the *O*-benzyl group and lipase-mediated deesterification thereafter furnished an isopropylidene acetal that readily underwent hydrolysis to give the target compound. At the time (1975) when this work was done, it was a major achievement.

One day in early 1983, when this author was a Ph.D. student in the Hough–Richardson group, Richardson came into the laboratory and asked whether Mezher Ali, then a second-year doctoral student, would be willing to undertake an enantiospecific

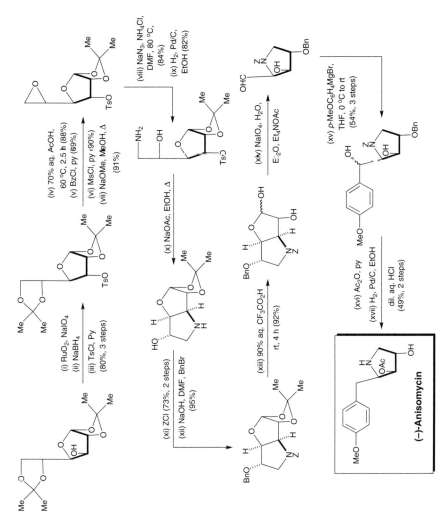

SCHEME 35. Richardson's total synthesis of (−)-anisomycin with Ram Bhatt, John Moffatt, and Julian Verheyden (and others) at Syntex in 1975.

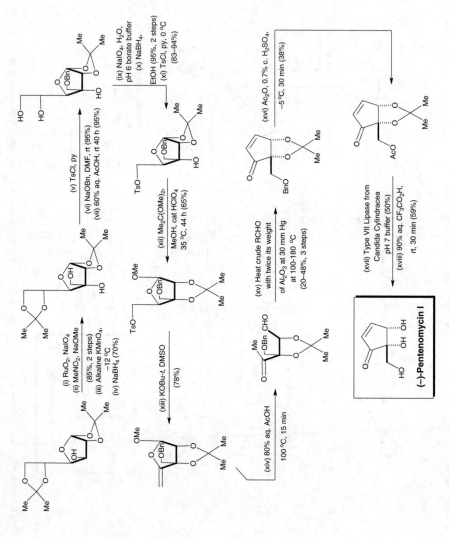

SCHEME 36. Richardson's total synthesis of (−)-pentenomycin I with Ram Bhatt, John Moffatt, and Julian Verheyden (and others) at Syntex in 1975.

synthesis of the newly discovered indolizidine alkaloid, (−)-swainsonine, as his main Ph. D. project.

How this came to pass stemmed from a chance discussion with Dr. Brian Winchester, Richardson's long-standing friend and colleague in the QEC Biochemistry Department, who one morning over coffee showed Richardson the structure of this novel alkaloid, which was present in locoweed. He related how cattle that had inadvertently ingested locoweed developed mannosidosis, an ultimately fatal disease. He indicated that swainsonine would be a much sort-after biochemical tool for the study of lysosomal storage diseases in man, if it could be obtained by total synthesis. Winchester stated that there was already a great race going on within the chemistry community to complete the first synthesis. When Richardson heard this, he immediately became interested, and after he saw its enticing structure, his lightning-fast mind saw a potential route to it. Within a few hours, he had already hatched a bold initial synthetic plan.

Mezher Ali was specifically picked for this flagship project by Richardson because of his great experimental skill in the laboratory. He had already made many notable synthetic contributions to the chemistry of methyl α-lactoside with Richardson and Hough, and both supervisors had great confidence and faith in Mezher Ali's experimental technique. It transpired that Mezher had been personally trained in practical organic chemistry in Iraq by one of Richardson's former Ph.D. students, Professor Yousif Ali (President of the Iraqi Chemists Union), a person whom Dick rated very highly indeed. The die was thus cast and, in late 1983, Richardson, Hough, and Ali completed the first reported total synthesis of (−)-swainsonine via the route shown in Scheme 37. *Chemical Communications* was their choice of vehicle for publication, and when their article appeared in early 1984, it was the first synthesis in print. However, it was very quickly joined by the two equally elegant total syntheses, one by George W.J. Fleet at Oxford University and the other by Tetsuo Suami at Keio University. Their groups reported conceptually different carbohydrate-based approaches virtually simultaneously.

The Richardson–Hough synthesis of (−)-swainsonine (Scheme 37) set off from methyl 3-amino-3-deoxy-α-D-mannopyranoside, which was readily prepared on large scale by Richardson's published 1962 procedure involving Fischer nitromethane cyclization and reduction. Richardson, Hough, and Ali then converted this starting compound into a 3,6-bicyclic pyrrolidino-mannoside, which itself was transformed into 2,4,5-tri-*O*-acetyl-3,6-benzyloxycarbonylimino-3,6-dideoxy-*aldehydo*-D-mannose via the dithioacetal (according to steps vii to xi). The aldehyde was condensed with ethoxycarbonylmethylenetriphenylphosphorane to give the enoate in good yield, and this underwent catalytic hydrogenation and tandem lactamization to afford the desired product, together with an O- to N-acetyl migration product,

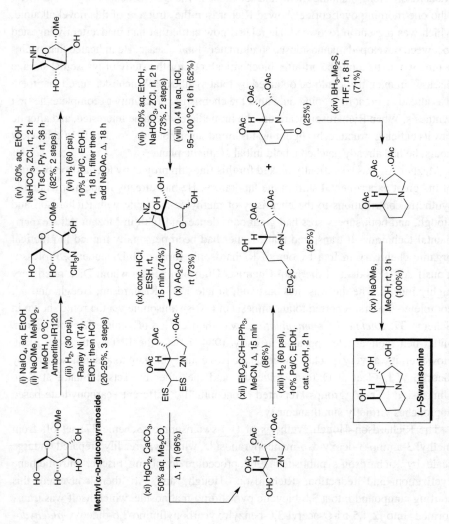

SCHEME 37. The Hough, Richardson, and Mezher Ali synthesis of (−)-swainsonine (1984).

which could be separated by chromatography. Reduction of the lactam with the borane—dimethyl sulfide complex thereafter afforded swainsonine triacetate, from which the final target alkaloid could be obtained. This was another major achievement at the time, and it made a substantial impact the world over.

This was only the beginning of alkaloid research for Richardson and Hough. Together with such talented students as David Hendry and Kai Aahmlid, they accomplished significant synthetic work on related structures. Most of this effort centered around castanospermine and its derivatives. With Hendry, an enantiospecific synthesis of 1-deoxycastanospermine was completed in 1987, together with various other castanospermine analogues. Syntheses of 1-deoxy-6-epicastanospermine and 1-deoxy-6,8a-diepicastanospermine (with Kai Aahmlid) followed in a 1990 report.

Alongside this effort in the chiral synthesis of bioactive natural-product alkaloids, Richardson and Hough, with student, Jonathan Y.C. Chan, also enunciated the idea of potentially gaining access to novel naturally occurring spiroacetal derivatives from a new glycoside of D-fructose, namely, the highly crystalline 2-chloroethyl β-D-fructopyranoside, a compound available in one step in 90% yield without recourse to chromatography. Some of the resulting work is shown in Scheme 38.

In the case of the spirobi-1,4-dioxane derivatives, Richardson also demonstrated that it was possible to prepare the opposite enantiomer by the modification shown in Scheme 39.

The spiroacetal morpholine work of Scheme 38 subsequently inspired research on the preparation of a novel family of morpholino-glycosides from sucrose via lead tetraacetate cleavage and reductive amination (Scheme 40). Significantly, the latter work, which was done with the present author, unveiled a completely new structural class of sweeteners, more intense than sucrose but having a similar taste profile.

Concurrent with the reductive amination work, Richardson, Hough, and Hale studied the nitromethane cyclization of the di- and tetraaldehydes derived from sucrose, and this again provided synthetic access to a novel series of trehalose derivatives, on this occasion, with the rare α,β glycosidic linkage (Schemes 40 and 41). Clearly, both at the beginning of Richardson's career, as well as toward its end, Fischer's beautiful nitromethane cyclization reaction still found its way into his work. Of course, as the decades passed, our ability to do synthetic chemistry was greatly facilitated by the advances in high-field NMR and silica gel chromatography, and the detailed study of these more complex examples of the Fischer cyclization eventually became possible.

Finally, this obituary and retrospective survey could not close without a significant mention of Richardson's celebrated rules for nucleophilic replacement of tosylates, mesylates, and halo groups by charged nucleophiles in carbohydrate pyranoid derivatives. They emanated out of his early studies at Reading on the preparation of

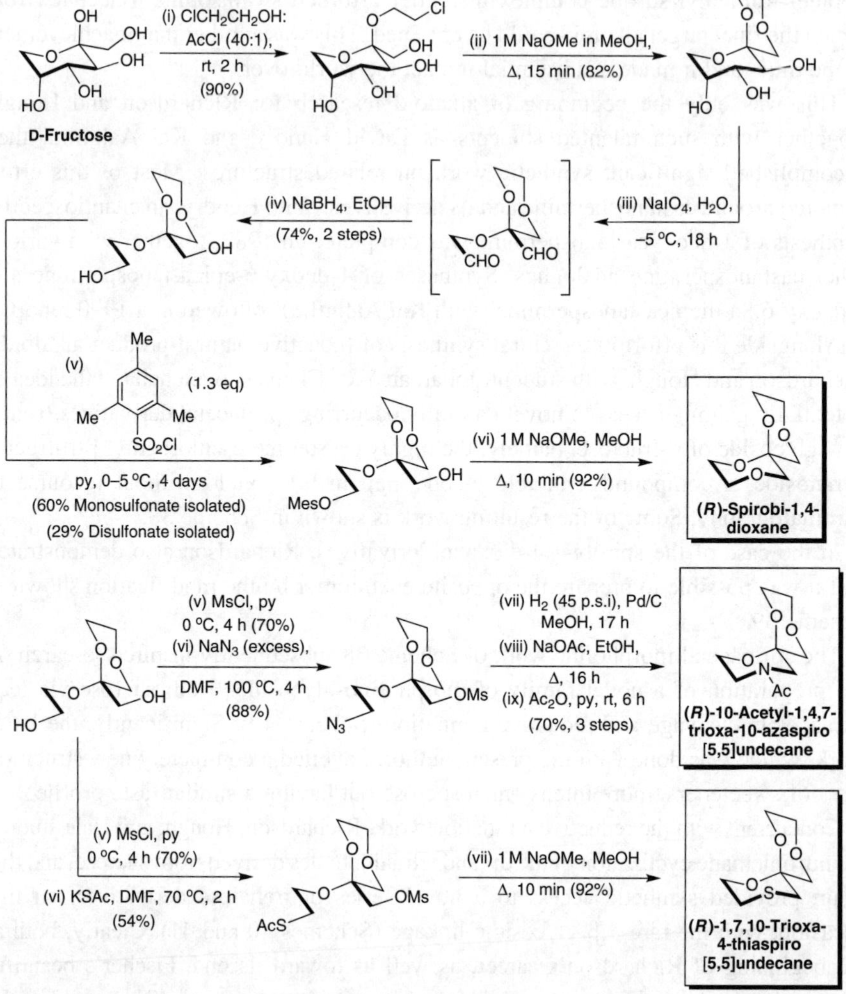

SCHEME 38. The Chan, Hough, and Richardson's enantioselective synthesis of spiroacetal derivatives from the chloroethyl pyranoside derived from D-fructose (1984).

tetraamino-tetradeoxy-monosaccharides with Yousif Ali. These rules were based upon unfavorable steric effects and dipolar interactions operating in the S_N2 transition states of the reaction with certain substrates that possessed neighboring electronegative substituents and permanent dipoles of fixed orientation. According to

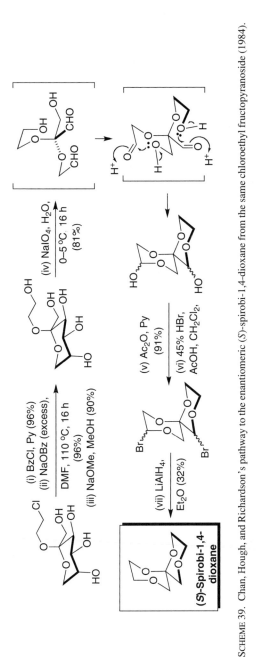

SCHEME 39. Chan, Hough, and Richardson's pathway to the enantiomeric (S)-spirobi-1,4-dioxane from the same chloroethyl fructopyranoside (1984).

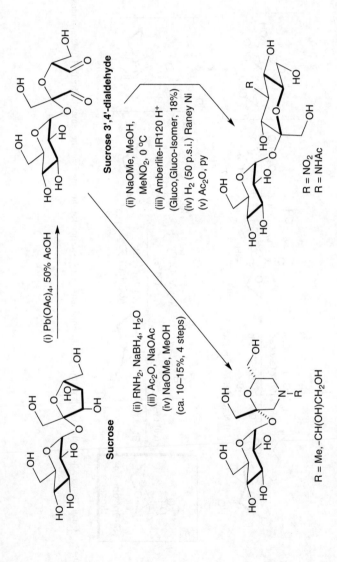

SCHEME 40. Hale, Hough, and Richardson's studies on the reductive amination (1988) and Fischer cyclization (1987) of a 3',4'-dialdehyde derived from lead tetraacetate oxidation of sucrose. The glycosyl morpholines were very sweet!

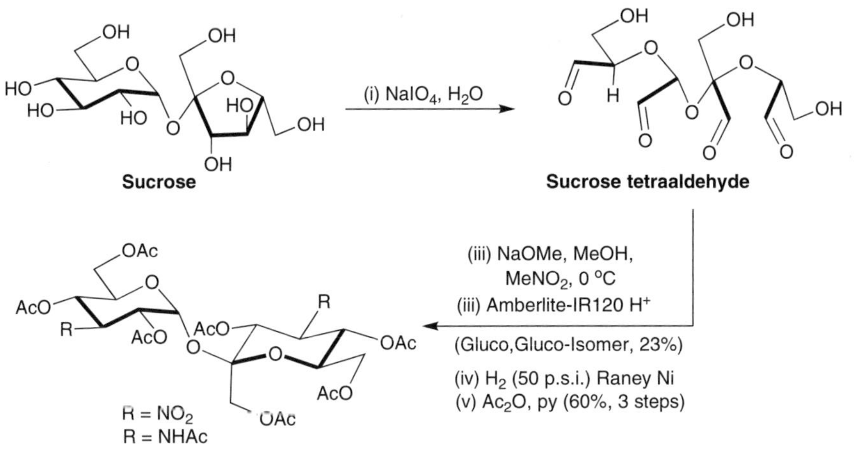

SCHEME 41. Hale, Hough, and Richardson's studies on the Fischer cyclization (1987) of a tetraaldehyde derived from periodate oxidation of sucrose.

Richardson, the extent of dipolar repulsion (and ease of displacement) was dependent upon "(a) the proximity, (b) the angular orientation, and (c) the numbers of permanent dipoles," with the most significant dipolar repulsions occurring when vicinal groups and the developing transitory dipoles in the S_N2 transition state are parallel and aligned (Scheme 42).

Richardson's rules, as enunciated by Richardson himself in updated form (in the MTP International Review of Science article that he wrote in 1973), are as follows:

(i) The presence of three or more vicinal polar groups slows displacement down to the extent that reaction does not proceed at a sufficiently convenient rate. For example, it is difficult to displace a sulfonyloxy group at a position adjacent to an anomeric group.

(ii) Displacements are difficult if the leaving group is adjacent to an axial electronegative group.

(iii) Displacement of a leaving group situated *trans*-axial to a substituent other than hydrogen is severely hindered sterically. This effcct may be overcome by a change in conformation in certain favorable cases (as with 4-sulfonates of altropyranosides).

(iv) Displacements at primary positions are normally facile unless the terminal group (CH_2OR) is attached to either (a) C-5 of a pyranoid ring containing an axial substituent at C-4 (e.g. galactosides, gulosides etc.) or (b) to the anomeric carbon (as in 1-sulfonates or halides of ketoses). The former effect could probably be relieved by a conformational change in certain circumstances.

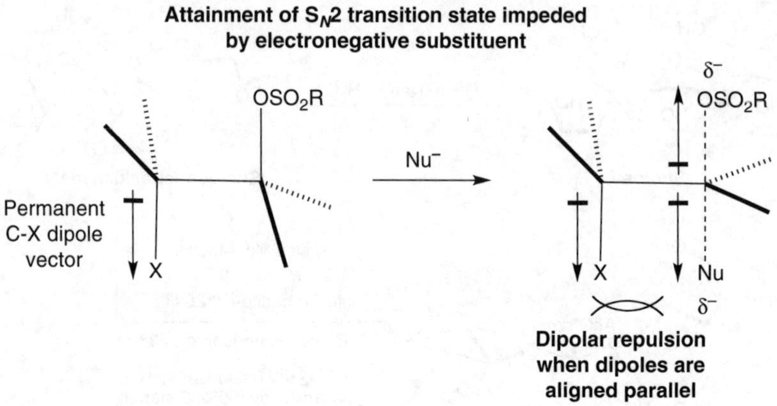

Attainment of S_N2 transition state impeded by electronegative substituent

SCHEME 42. The physical basis of Richardson's rules for nucleophilic displacement of carbohydrate sulfonic esters (1969).

In addition to the foregoing, the rules should be extended in light of more recent experience by the addition of the following:

(v) Attempted displacement of an axial sulfonyloxy or halo group, vicinal to an axial hydrogen, will lead to competing elimination, probably by an E2C mechanism, to give an alkene. The extent of elimination is a function of the basicity of the nucleophilic anion and solvent. The use of acetonitrile is considered to cause less elimination.

(vi) In fused bicyclic systems, only endo groups may be displaced readily by charged nucleophiles; exo-sulfonates are displaced only with difficulty.

(vii) Displacement reactions carried out adjacent to an "activated" [such as –CH(R)-CO_2R'] hydrogen atom always lead to elimination, regardless of the configurations of the leaving group and the activated hydrogen.

(viii) In cases where a strongly nucleophilic group (such as NHCOR, SR, and the like) can participate from the rear, solvolysis products are usually obtained.

The use of neutral nucleophiles, such as ammonia and hydrazine, causes a reversal of polarity of one of the transitory dipoles, and these displacement reactions are therefore more favorable than with azide. On the other hand, the greater basicity of these reagents are more likely to cause elimination.

A pictorial summary of Richardson's main rules is presented in Scheme 43.

Standing back now from the detail, and placing Richardson's entire scientific work in perspective, he made enormous academic contributions as well as doing work having important societal impact. From the standpoint of organic chemistry, his rules for nucleophilic displacement linked together an extensive collection of his own

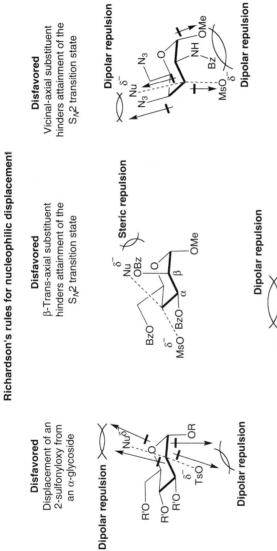

SCHEME 43. A pictorial summary of Richardson's rules.

personal observations with those of many others in the field, and they allowed unbridled empiricism to be replaced by rationality of thought and sound mechanistic understanding. His use of dipolar-repulsion arguments completely changed the way organic chemists thought about organic reactions, especially with regard to the transition states of reactions. Organic chemists could no longer simply ignore the effects of permanent dipoles in substrates undergoing organic reaction. They had to take these physical phenomena into account, and as a consequence, the subject and theory of organic chemistry took an important step forward.

While Richardson was undoubtedly led into this way of thinking by such giants of the carbohydrate field as Ray Lemieux, with his explanations of physical phenomena such as the anomeric effect, Dick took this revolution of thought to a still higher plane. Thus we nowadays see dipolar-repulsion arguments routinely being marshalled to explain the outcome of many organic reactions, notable contemporary examples being the Evans asymmetric aldol reaction and the alkoxy-directed Wittig reaction of α-alkoxymethyl ketones. However, back in 1969, when Richardson first put forward these rules for nucleophilic displacement, no one else was explaining such complex reaction outcomes through the combined operation of dipolar and steric effects. This change in mechanistic thought was rapidly accepted by the general organic chemistry community, in contrast to Hughes and Ingold, whose concepts at an earlier time on the electronic theory of organic chemistry had to break through a long and sustained campaign of opposition before being ultimately accepted. For Dick, having his contributions in organic chemistry so readily adopted by his peers during his lifetime gave him a deep sense of satisfaction.

The fact that his work in carbohydrate chemistry had led to many useful and important commercial products, such as TGS and his various diagnostic kits, also gave him great scientific fulfillment. The same can be said of the valuable synthetic protocols and intermediates that he published throughout his days in chemical research.

Looking now, over a career that spanned some 40 years, one is left with the clear conclusion that Richardson had extraordinary intellect, imagination, and ingenuity as a chemist, and that he was blessed with insights that those who knew him were fortunate to have shared.

In closing, anyone who met A.C. Richardson cannot have failed to have been endeared to him. He was a generous, humorous person with a warmth of spirit and a bonhomie that blessed all who knew him. As an eminent carbohydrate chemist, he made a long-lasting mark on the field. He will be sorely missed by all of his family and friends, and by the broader international scientific community.

Karl J. Hale

Acknowledgments

In the preparation of this biographical memoir, the author thanks Professor Leslie Hough and Caroline Richardson for the detailed information that they provided on A.C. Richardson's early life and subsequent career. Without their kind help, this obituary and tribute could not have been prepared. I also thank Professor Derek Horton for his masterly editing and further improvement of the memoir.

Appendix

Publications of Anthony C. Richardson, in Chronological Order

L. Hough and A. C. Richardson, Oxonium cation intermediates in the nucleophilic degradation of diethylsulphonylglycopyranosylmethane derivatives, *Proc. Chem. Soc.* (1959) 193–194.

A. C. Richardson and H. O. L. Fischer, Cyclizations of dialdehydes with nitromethane, Part V. Preparation of some 3-amino-1,6-anhydro-3-deoxy-β-D-hexoses and the elucidation of their structures, *Proc. Chem. Soc.* (1960) 341–342.

A. C. Richardson and H. O. L. Fischer, Cyclization of dialdehydes with nitromethane, Part VI. Preparation of 3- amino-1,6-anhydro-3-deoxy-β-D-gulose, -β-D-altrose and -β-D-idose derivatives and their characterization by means of inversion of mesyloxy groups, *J. Am. Chem. Soc.*, 83 (1961) 1132–1139.

A. C. Richardson, Derivatives of 3-amino-3,6-dideoxy-L-glucose and -L-talose, *Proc. Chem. Soc.* (1961) 255.

A. C. Richardson, The synthesis and stereochemistry of mycaminose, *Proc. Chem. Soc.* (1961) 430.

L. Hough and A. C. Richardson, Diethylsulphonyl(2-O-methyl-α-D-arabopyranosyl)methane and its behaviour with base, *J. Chem. Soc.* (1961) 5561–5563.

A. C. Richardson, An improved preparation of methyl 3-amino-3-deoxy-α-D-mannopyranoside hydrochloride, *J. Chem. Soc.* (1962) 373–374.

L. Hough and A. C. Richardson, The nucleophilic degradation of diethylsulphonylglycopyranosylmethane derivatives, *J. Chem. Soc.* (1962) 1019–1023.

L. Hough and A. C. Richardson, Conformational and electronic factors affecting the nucleophilic degradation of diethylsulphonylglycopyranosylmethane derivatives, *J. Chem. Soc.* (1962) 1024–1032.

A. C. Richardson and K. A. McLauchlan, 3-Amino-3-deoxy-derivatives of L-glucose, L-galactose and L-talose, *J. Chem. Soc.* (1962) 2499–2506.

A. C. Richardson, The synthesis of D- and L-mycaminose hydrochlorides, *J. Chem. Soc.* (1962) 2758–2760.

R. D. Guthrie, D. Murphy, D. H. Buss, L. Hough, and A. C. Richardson, Aziridino derivatives of carbohydrates, *Proc. Chem. Soc.* (1963) 84.

A. C. Richardson, A stereospecific synthesis of desosamine hydrochloride, *Proc. Chem. Soc.* (1963) 131.

J. Hill, L. Hough, and A. C. Richardson, Replacement of methanesulphonyloxy groups: Conversion of the D-gluco into the D-galacto configuration, *Proc. Chem. Soc.* (1963) 346–347.

D. H. Buss, L. Hough, and A. C. Richardson, Preparation of 2,3-epimino derivatives of pyranosides, *J. Chem. Soc.* (1963) 5295–5301.

A. C. Richardson, The synthesis of desosamine hydrochloride, *J. Chem. Soc.* (1964) 5364–5370.

A. C. Richardson and J. M. Williams, Selective O-acylation of pyranosides, *J. Chem. Soc. Chem. Commun.* (1965) 104–105.

D. H. Buss, L. Hough, and A. C. Richardson, Some ring-opening reactions of 2,3-epimino derivatives, *J. Chem. Soc.* (1965) 2736–2743.

A. C. Richardson, The synthesis of D-daunosamine N-benzoate, *J. Chem. Soc. Chem. Commun.* (1965) 627–628.

C. F. Gibbs, L. Hough, and A. C. Richardson, A new synthesis of methyl 2,3-epimino-α-D-allopyranoside, *Carbohydr. Res.*, 1 (1965) 290–296.

A. C. Richardson, Heterocyclic compounds. Part II. Monosaccharides, *Ann. Rep. Prog. Chem.*, 62 (1965) 368.

J. M. Williams and A. C. Richardson, Selective acylation of pyranosides I. Benzoylation of methyl α-D-glycopyranosides of mannose, glucose and galactose, *Tetrahedron*, 23 (1967) 1369–1378.

A. C. Richardson and J. M. Williams, Selective acylation of pyranosides II. Benzoylation of methyl 6-deoxy-α-D-galactopyranoside and methyl α-D-mannopyranoside, *Tetrahedron*, 23 (1967) 1641–1646.

Y. Ali and A. C. Richardson, The synthesis of derivatives of 2,3,4,6-tetra-amino-2,3,4,6-tetradeoxy-D-galactose and -D-idose, *J. Chem. Soc. Chem. Commun.* (1967) 554–556.

A. D. Barford and A. C. Richardson, A 3,4-epimino-pyranoside, *Carbohydr. Res.*, 4 (1967) 408–414.

A. C. Richardson, Selective acylation of methyl 3-acetamido-3-deoxy-α-L-glucopyranoside: Its conversion into the L-galacto- and L-manno-isomers, *Carbohydr. Res.*, 4 (1967) 415–421.

A. C. Richardson, The synthesis of *N*-benzoyl-D-daunosamine, *Carbohydr. Res.*, 4 (1967) 422–428.

A. C. Richardson, Heterocyclic compounds, Part (ii). Monosaccharides, *Ann. Rep. Prog. Chem.*, 63 (1966) 489.

Y. Ali and A. C. Richardson, The reduction of azides with sodium borohydride. A convenient synthesis of methyl 2-acetamido-2-deoxy-4,6-*O*-benzylidene-α-D-allopyranoside, *Carbohydr. Res.*, 5 (1967) 441–448.

L. Hough and A. C. Richardson, Penta-, hexa- and higher polyhydric alcohols, in D. J. Coffey, (Ed.), *Rodd's Chemistry of Carbon Compounds*, Vol. IF, Elsevier, Amsterdam, 1967, pp. 1–66. Chapter 22.

L. Hough and A. C. Richardson, The monosaccharides: Pentoses, hexoses and higher sugars, in D. J. Coffey, (Ed.), *Rodd's Chemistry of Carbon Compounds*, Vol. IF, Elsevier, Amsterdam, 1967, pp. 67–595. Chapter 23.

Y. Ali, A. C. Richardson, C. F. Gibbs, and L. Hough, Some further ring-opening reactions of methyl 4,6-*O*-benzylidene-2,3-dideoxy-2,3-epimino-α-D-allopyranoside and its derivatives, *Carbohydr. Res.*, 7 (1968) 255–271.

J. Hill, L. Hough, and A. C. Richardson, Nucleophilic replacement reactions of sulphonates. I. The preparation of derivatives of 4,6-diamino-4,6-dideoxy-D-glucose and -D-galactose, *Carbohydr. Res.*, 8 (1968) 7–18.

J. Hill, L. Hough, and A. C. Richardson, Nucleophilic replacement reactions of sulphonates II. The synthesis of derivatives of 4,6-dithio-D-galactose and -D-glucose and their conversion into 4,6-dideoxy-D-*xylo*-hexose, *Carbohydr. Res.*, 8 (1968) 19–28.

Y. Ali and A. C. Richardson, Nucleophilic replacement reactions of sulphonates. III. The synthesis of derivatives of 2,3,4,6-tetra-acetamido-2,3,4,6-tetradeoxy-D-galactose and -D-idose, *J. Chem. Soc., Sect. C* (1968) 1764–1769.

C. F. Gibbs, L. Hough, A. C. Richardson, and J. Tjebbes, Methyl 2,3,6-trideoxy-2,3-epimino-α-D-allopyranoside, *Carbohydr. Res.*, 8 (1968) 405–410.

G. Birch and A. C. Richardson, Chemical modification of trehalose. Part I. Selective sulphonylation of trehalose and the determination of the conformation of hexa-*O*-acetyl-6,6′dideoxy trehalose, *Carbohydr. Res.*, 8 (1968) 411–415.

Y. Ali and A. C. Richardson, Nucleophilic replacement reactions of sulphonates V. The synthesis of derivatives of 2,3,4,6-tetra-amino-2,3,4,6-tetradeoxy-D-glucose, *J. Chem. Soc., Sect. C* (1969) 320–329.

A. C. Richardson, Nucleophilic replacement reactions of sulphonates, Part VI. A summary of steric and polar factors, *Carbohydr. Res.*, 10 (1969) 395–402.

L. Hough and A. C. Richardson, The synthesis of monosaccharides, in W. Pigman and D. Horton, (Eds.) *The Carbohydrates*, Vol. 1A, Academic Press, New York, 1972, pp. 113–164. Chapter 5.

A. C. Richardson, Amino sugars via reduction of azides, *Methods Carbohydr. Chem.*, 6 (1972) 218.

G. Birch and A. C. Richardson, Chemical modification of trehalose. Part II. The synthesis of the galacto analogue of trehalose, *J. Chem. Soc., Sect. C* (1970) 749–752.

M. W. Horner, L. Hough, and A. C. Richardson, Selective benzoylation of methyl 2-benzamido-2-deoxy-α-D-glucopyranoside; A convenient preparation of derivatives of 2-amino-2-deoxy-D-galactose, *J. Chem. Soc., Sect. C* (1970) 1336–1340.

Y. Ali, L. Hough, and A. C. Richardson, Chemical modification of trehalose. Part III. The introduction of azido and amino substituents at the 4,4'- and 6,6'-positions, *Carbohydr. Res.*, 14 (1970) 181–187.

A. D. Barford and A. C. Richardson, Derivatives of 5,6-dideoxy-5,6-epimino-L-iditol, *Carbohydr. Res.*, 14 (1970) 217–230.

A. D. Barford and A. C. Richardson, Derivatives of 3,4-dideoxy-3,4-epimino-L-iditol, *Carbohydr. Res.*, 14 (1970) 231–236.

W. Horner, L. Hough, and A. C. Richardson, Derivatives of 2-amino-2-deoxy-D-gulose: Synthetic and conformational studies, *J. Chem. Soc., Sect. C* (1971) 99–102.

G. Birch, C. K. Lee, and A. C. Richardson, Chemical modification of trehalose. Part IV. The synthesis and conformation of 3,6;3',6'-dianhydro-α, α-trehalose, *Carbohydr. Res.*, 16 (1970) 235–238.

M. W. Horner, L. Hough, and A. C. Richardson, Syntheses of methyl 2,4-dibenzamido-2,4-dideoxy-α-D-glucopyranoside and -galactopyranoside, *Carbohydr. Res.*, 17 (1971) 209–212.

P. L. Gill, M. W. Horner, L. Hough, and A. C. Richardson, Selective benzoylation of methyl (methyl α-D-galactopyranosid)uronate, *Carbohydr. Res.*, 17 (1971) 213–215.

L. Hough, P. A. Munroe, and A. C. Richardson, Chemical modification of trehalose. Part V. The synthesis of mono and diepoxides, *J. Chem. Soc., Sect. C* (1971) 1090–1094.

L. Hough, E. Tarelli, and A. C. Richardson, Chemical modification of trehalose. Part VI. The synthesis of analogues containing 2,2'-dideoxy functions, *J. Chem. Soc., Sect. C* (1971) 1732–1738.

G. Birch, C. K. Lee, and A. C. Richardson, Chemical modification of trehalose. Part VII. Rearrangement of 3,6;3',6'dianhydro-α, α-trehalose into the difuranoid form, *Carbohydr. Res.*, 19 (1971) 119–122.

L. Hough, E. Tarelli, and A. C. Richardson, Chemical modification of trehalose. Part VIII. The synthesis 3,3'-dideoxy analogues, *J. Chem. Soc., Sect. C* (1971) 2122–2127.

C. Bullock, L. Hough, and A. C. Richardson, Rearrangement of α-D-glucopyranose-4-sulphonates to 1,4-anhydro-αβ-D-galactopyranose, *J. Chem. Soc. Chem. Commun.* (1971) 1276–1277.

A. C. Richardson and E. Tarelli, Chemical modification of trehalose. Part IX. The monobenzylidene acetal, *J. Chem. Soc., Sect. C* (1971) 3733–3735.

A. C. Richardson and E. Tarelli, Chemical modification of trehalose. Part X. Some further 3,3'-dideoxy analogues, *J. Chem. Soc., Perkin Trans. 1* (1972) 949–952.

J. M. Ballard, L. Hough, and A. C. Richardson, Sucrochemistry, Part V. A direct preparation of sucrose 2,3,4,6,1',3',4'-hepta-acetate, *Carbohydr. Res.*, 24 (1972) 152–153.

L. Hough, A. K. Palmer, and A. C. Richardson, Chemical modification of trehalose. Part XI. 6,6'-Dideoxy-6,6'-difluoro-α,α-trehalose and its galacto-analogue, *J. Chem. Soc., Perkin Trans. 1* (1972) 2513–2517.

J. M. Ballard, L. Hough, and A. C. Richardson, Reaction of sucrose with sulphuryl chloride, *J. Chem. Soc. Chem. Commun.* (1972) 1097–1098.

L. Hough, P. A. Munroe, A. C. Richardson, Y. Ali, and S. T. K. Bukhari, Chemical modification of trehalose. Part XII. The synthesis of amino- and azido-trehaloses via epoxide derivatives, *J. Chem. Soc., Perkin Trans. 1* (1973) 287–290.

L. Hough, A. K. Palmer, and A. C. Richardson, Chemical modification of trehalose. Part XIII. The synthesis of some 4,4'-difluoro- and 4,4',6,6'-tetrafluoro-analogues, *J. Chem. Soc., Perkin Trans.1* (1973) 784–788.

A. J. Freestone, L. Hough, and A. C. Richardson, A convenient procedure for the synthesis of theophylline nucleosides, *Carbohydr. Res.*, 28 (1973) 378–386.

A. C. Richardson and E. Tarelli, Chemical modification of trehalose. Part XIV. Some tetra- and hexa-deoxy derivatives and their amino derivatives, *J. Chem. Soc., Perkin Trans. 1* (1973) 1520–1523.

J. M. Ballard, L. Hough, A. C. Richardson, and P. H. Fairclough, Sucrochemistry, Part XII. The reaction of sucrose with sulphuryl chloride, *J. Chem. Soc., Perkin Trans. 1* (1973) 1524–1528.

P. L. Durette, L. Hough, and A. C. Richardson, The chemistry of maltose. Part 1. The reaction of methyl α-maltoside with sulphuryl chloride, *Carbohydr. Res.*, 31 (1973) 114–119.

R. Toubiana, M. J. Toubiana, B. C. Das, and A. C. Richardson, Synthese d'analogues du cord factor, Partie 1. Etude de l'esterifaction selective du trehalose par le chlorure de para-toluene sulphonyle; mis en evidence d'un ester dissymetrique Ie 2,6-ditosylate de trehalose, *Biochemie*, 55 (1973) 569–573.

R. G. Edwards, L. Hough, A. C. Richardson, and E. Tarelli, A reappraisal of the selectivity of the DMF-mesyl chloride reagent. Chlorination at secondary positions, *Tetrahedron Lett.* (1973) 2369–2370.

A. C. Richardson, Amino and nitro sugars, *MTP Int. Rev. Sci. (Org. Chem.)*, 7 (1973) 105. Chapter 4.

P. L. Durette, L. Hough, and A. C. Richardson, The chemistry of maltose. Part II. Chemical modification at the reducing unit, *J. Chem. Soc., Perkin Trans. 1* (1974) 88–96.

P. L. Durette, L. Hough, and A. C. Richardson, The chemistry of maltose. Part III. The introduction of azido- and amino-substituents at specific positions of methyl β-maltoside, *J. Chem. Soc. Perkin Trans. 1* (1974) 97–101.

R. S. Bhatt, L. Hough, and A. C. Richardson, Selective benzoylation of methyl β-lactoside, *Carbohydr. Res.*, 32 (1974) C4–C6.

J. M. Ballard, L. Hough, and A. C. Richardson, Sucrochemistry, Part XIV. Further studies on the partial de-O-acetylation of sucrose octa-acetate, *Carbohydr. Res.*, 34 (1974) 184–188.

R. G. Edwards, L. Hough, A. C. Richardson, and E. Tarelli, The stereoselective displacement of hydroxyl by chloride using the mesyl chloride-DMF reagent, *Carbohydr. Res.*, 35 (1974) 111–129.

G. G. Birch, C. K. Lee, and A. C. Richardson, The chemical modification of trehalose. Part XV. The synthesis of 4,4'-dideoxy and 4,4',6,6'-tetradeoxy analogues, *Carbohydr. Res.*, 36 (1974) 97–109.

P. H. Fairclough, L. Hough, and A. C. Richardson, Derivatives of β-D-fructofuranosyl α-D-galactopyranoside, *Carbohydr. Res.*, 40 (1975) 285–298.

R. S. Bhatt, L. Hough, and A. C. Richardson, The synthesis of 4',6'-diacetamido-4',6'-dideoxy-cellobiose hexa-acetate from lactose, *Carbohydr. Res*, 43 (1975) 57–67.

J. Freestone, L. Hough, and A. C. Richardson, Some chemical transformations of 7-β-D-glucopyranosyl-theophylline, *Carbohydr. Res.*, 43 (1975) 239–246.

A. J. Freestone, L. Hough, and A. C. Richardson, Some benzylidene acetal derivatives of theophylline nucleosides, *Carbohydr. Res.*, 45 (1975) 3–9.

A. C. Richardson, The nitro sugars, *MTP Int. Rev. Sci. (Org. Chem.), Ser. 2* (1975) 31. Chapter 4.

R. S. Bhatt, L. Hough, and A. C. Richardson, Selective chlorination of methyl β-lactoside with mesyl chloride in DMF, *Carbohydr. Res.*, 49 (1976) 103–118.

G. G. Birch, C. K. Lee, A. C. Richardson, and Y. Ali, The synthesis of the *allo*-analogue of trehalose, *Carbohydr. Res.*, 49 (1976) 153–161.

R. S. Bhatt, L. Hough, and A. C. Richardson, Transformation of lactose into its 3-epimer, 4-*O*-β-D-galactopyranosyl-D-allopyranose, *Carbohydr. Res.*, 51 (1976) 272–275.

R. G. Edwards, L. Hough, and A. C. Richardson, Transformations of cellobiose derivatives into analogues of lactose, *Carbohydr. Res.*, 55 (1977) 129–148.

A. C. Richardson and L. Hough, Recent aspects of the chemistry of disaccharides, *Pure Appl. Chem.*, 49 (1977) 1069–1084.

R. S. Bhatt, L. Hough, and A. C. Richardson, The chemistry of cellobiose and lactose. Part 7. Selective benzoylation of methyl β-lactoside, *J. Chem. Soc. Perkin Trans. 1*, 1 (1977) 2001–2005.

A. F. Hadfield, L. Hough, and A. C. Richardson, The synthesis of 6,6'-dideoxy-6,6'-difluoro-α,α-trehalose and related analogues, *Carbohydr. Res.*, 63 (1978) 51–60.

C. W. Bird, R. Khan, and A. C. Richardson, Structure of endrin aldehyde, *Chem. Ind.* (1978) 231–232.

L. Hough, M. Chowdhary, and A. C. Richardson, Selective esterification of sucrose using pivaloyl chloride, *J. Chem. Soc. Chem. Commun.* (1978) 664–665.

J. P. H. Verheyden, A. C. Richardson, R. S. Bhatt, B. D. Grant, W. L. Fitch, and J. G. Moffatt, Chiral synthesis of the antibiotics anisomycin and pentenomycin from carbohydrates, *Pure Appl. Chem.*, 54 (1978) 1363–1383.

L. Hough and A. C. Richardson, Monosaccharide chemistry, Comprehensive Organic Chemistry (1979) Pergamon Press, Oxford, Part 26.1, pp. 687–748.

L. Hough and A. C. Richardson, Oligosaccharide chemistry, Comprehensive Organic Chemistry (1979) Pergamon Press, Oxford, Part 26.2, pp. 749–753.

L. Hough, A. C. Richardson, and M. A. Salam, Reaction of raffinose with sulphuryl chloride, *Carbohydr. Res.*, 71 (1979) 85–93.

A. F. Hadfield, L. Hough, and A. C. Richardson, The synthesis of 4,6-dideoxy-4,6-difluoro- and 4-deoxy-4-fluoro-α,α-trehalose, *Carbohydr. Res.*, 71 (1979) 95–102.

L. Hough, A. C. Richardson, and L. A. W. Thelwall, Reaction of lactose with 2,2-dimethoxypropane. A tetra-acetal of novel structure, *Carbohydr. Res.*, 75 (1979) C11–C12.

M. K. Gurjar, L. Hough, and A. C. Richardson, The formation of a 2,1'-anhydro linkage in sucrose, *Carbohydr. Res.*, 78 (1980) C21–C23.

A. F. Hadfield, L. Hough, and A. C. Richardson, The synthesis of further analogues of trehalose containing fluoro and amino substituents, *Carbohydr. Res.*, 80 (1980) 123–130.

L. Hough, A. C. Richardson, and M. A. Salam, 6'-Substituted and 6',6"-disubstituted derivatives of raffinose, *Carbohydr. Res.*, 80 (1980) 117.

J. M. Ballard, L. Hough, and A. C. Richardson, Selective tetratosylation of sucrose. The isolation of the 2,6,1',6'-tetrasulphonate, *Carbohydr. Res.*, 83 (1980) 138–141.

L. Hough, A. A. E. Penglis, and A. C. Richardson, The synthesis of derivatives of 3-amino-2,3-dideoxy-2-fluoro-D-altrose, *Carbohydr. Res.*, 83 (1980) 142–145.

L. Hough, A. A. E. Penglis, and A. C. Richardson, The synthesis of fluoro derivatives of 2-amino-2-deoxy-D-galactose and -D-glucose, *Can. J. Chem.*, 59 (1981) 396–405.

C. T. Yuen, R. G. Price, A. C. Richardson, and P. F. G. Praill, The assay of arylesterase in serum using two new colorimetric substrates. ω-nitrostyrylacetate and propionate, *Clin. Chem. Acta* (1981) 99–105.

C. T. Yuen, R. G. Price, P. F. G. Praill, and A. C. Richardson, A novel substrate for the assay of the lysosomal hydrolase N-acetyl-β-D-glucosaminidase (NAG) in human urine, *Biochem. Soc. Trans.*, 9 (1981) 176P, Tue-S10-27.

A. B. Foster, M. Jarman, R. W. Kinas, J. M. S. van Maanen, G. N. Taylor, J. L. Gaston, A. Parkin, and A. C. Richardson, 5-Fluoro- and 5-chloro-cyclophosphamide: Synthesis, metabolism and antitumour activity of the cis and trans isomers, *J. Med. Chem.*, 24 (1981) 1399–1403.

A. K. B. Chiu, L. Hough, A. C. Richardson, and L. V. Sincharoenkul, Two unequivocal syntheses of 1',2:3,6:3',6'-trianhydrosucrose, *Tetrahedron Lett.*, 22 (1981) 4345–4346.

R. Cortes-Garcia, L. Hough, and A. C. Richardson, Acetalation of sucrose with concomitant fission of the glycosidic bond. Some new acetals of D-glucose and methyl α-D-fructofuranoside, *J. Chem. Soc., Perkin Trans. 1* (1981) 3176–3181.

C. T. Yuen, R. G. Price, P. F. G. Praill, and A. C. Richardson, A new and simple colorimetric assay procedure for N-acetyl-β-D-glucosaminidase in urine, *J. Clin. Chem. Biochem.*, 19 (1981) 884. (Abstr. XI Int. Congr. Clin. Chem.).

R. G. Price, C. R. R. Corbett, C. T. Yuen, P. F. G. Praill, A. C. Richardson, A. E. Thompson, and P. R. N. Kind, The assay of N-acetyl-β-D-glucosaminidase (NAG) in urine of transplant recipients using a new colorimetric substrate, *Ber. Oster. Gesellschaft Klin. Chem.*, 4 (1981) 149. (Abstr. III Int. Congr. Clin. Enzymol.).

A. K. B. Chiu, M. K. Gurjar, L. Hough, L. V. Sincharoenkul, and A. C. Richardson, The synthesis of 2,1'-anhydro-2,1':3,6-dianhydro- and 2,1':3,6:3',6'-trianhydro-sucrose, *Carbohydr. Res.*, 100 (1982) 247–261.

C. T. Yuen, R. G. Price, L. Chattagoon, A. C. Richardson, and P. F. G. Praill, Colorimetric assays for N-acetyl-β-D-glucosaminidase and β-D-galactosidase in human urine using newly developed ω-nitrostyryl substrates, *Clin. Chem. Acta*, 124 (1981) 195–204.

J. Y. C. Chan, L. Hough, and A. C. Richardson, Chiral synthesis of (R)-spirobi-1,4-dioxan and related compounds from D-fructose, *J. Chem. Soc. Chem. Commun.* (1982) 1151–1152.

B. E. Davison, R. J. Ferrier, A. C. Richardson, and N. R. Williams, Mono-, di- and tri-saccharides and their derivatives, *Specialist Periodical Reports: Carbohydrate Chemistry*, Vol. 13, (1982) Royal Society of Chemistry, London, Part 1.

A. C. Richardson, R. G. Price, and P. F. G. Praill, Substrates for the assay of enzymes, (1982) UK Patent, 2,008,103.

P. F. G. Praill, R. G. Price, and A. C. Richardson, Enzyme assay methods and apparatus, (1982) UK Patent, 2,079,450.

B. E. Davison, R. J. Ferrier, A. C. Richardson, and N. R. Williams, Mono-, di- and tri-saccharides and their derivatives, *Specialist Periodical Reports: Carbohydrate Chemistry*, Vol. 14, (1983) Royal Society of Chemistry, London, Part 1.

L. Hough, A. K. M. S. Kabir, and A. C. Richardson, The synthesis of some fluoro derivatives of sucrose, *Carbohydr. Res.*, 125 (1984) 247–252.

M. S. Chowdhary, L. Hough, and A. C. Richardson, Sucrochemistry, Part 33. The selective pivaloylation of sucrose, *J. Chem. Soc., Perkin Trans. 1* (1984) 419–427.

M. H. Ali, L. Hough, and A. C. Richardson, A chiral synthesis of swainsonine from D-glucose, *J. Chem. Soc., Chem. Commun.* (1984) 447–448.

C. T. Yuen, P. R. N. Kind, R. G. Price, P. F. G. Praill, and A. C. Richardson, Colorimetric assay for N-acetyl-β-D-glucosaminidase (NAG) in pathological urine using the oo-nitrostyryl substrate: the development of a kit and the comparison of the manual procedure with the automated fluorimetric method, *Ann. Clin. Biochem.*, 21 (1984) 295–300.

L. Hough, A. K. M. S. Kabir, and A. C. Richardson, The synthesis of some 4-deoxy-4-fluoro and 4,6-dideoxy-4,6-difluoro derivatives of sucrose, *Carbohydr. Res.*, 131 (1984) 335–340.

M. H. Ali, L. Hough, and A. C. Richardson, Synthesis of the indolizidine alkaloid swainsonine from D-glucose, *Carbohydr. Res.*, 136 (1985) 225–240.

J. Y. C. Chan, P. P. L. Cheong, L. Hough, and A. C. Richardson, The preparation and reactions of a new glycoside: 2′-chloroethyl β-D-fructopyranoside, *J. Chem. Soc., Perkin Trans. 1* (1985) 1447–1455.

J. Y. C. Chan, L. Hough, and A. C. Richardson, The synthesis of (R)- and (S)-spirobi-1,4-dioxan and other related spirobicycles from D-fructose, *J. Chem. Soc., Perkin Trans. 1* (1985) 1457–1462.

A. Patel and A. C. Richardson, 3-Methoxy-4-(2′-nitrovinyl)-phenyl glycosides as potential chromogenic substrates for the assay of glycosidases, *Carbohydr. Res.*, 146 (1986) 241–249.

M. S. Chowdhary, L. Hough, and A. C. Richardson, The use of pivaloyl esters of sucrose for the synthesis of chloro-, azido-, and anhydro derivatives, *Carbohydr. Res.*, 147 (1986) 49–58.

C. Bullock, L. Hough, and A. C. Richardson, The synthesis of some substituted D-glucose 1-phosphate derivatives and their reactivity towards nucleophilic substitution, *Carbohydr. Res.*, 147 (1986) 330–336.

M. K. Gurjar, L. Hough, A. C. Richardson, and L. V. Sincharoenkul, Preparation and ring-opening reactions of a 2,3-anhydride derived from sucrose, *Carbohydr. Res.*, 150 (1986) 53–61.

K. J. Hale, L. Hough, and A. C. Richardson, The cyclisation of di- and tetra-aldehydes derived from sucrose with nitroalkanes, *Tetrahedron Lett.*, 28 (1987) 891–894.

K. H. Aamlid, L. Hough, A. C. Richardson, and (in part) D. Hendry, An enantiospecific synthesis of (R)-1,4,7-trioxaspiro[5.5]undecane from D-fructose, *Carbohydr. Res.*, 164 (1987) 373–390.

A. K. B. Chiu, L. Hough, A. C. Richardson, I. A. Toufeili, and S. Z. Dziedzic, Synthesis of 4-C-methyl- and 4-C-allyl- derivatives of sucrose, *Carbohydr. Res.*, 162 (1987) 316–322.

D. Hendry, L. Hough, and A. C. Richardson, Enantiospecific synthesis of I-deoxy-castanospermine, (6S,7R,8R,8aR)-trihydroxyindolizidine, from D-glucose, *Tetrahedron Lett.*, 28 (1987) 4597–4600.

D. Hendry, L. Hough, and A. C. Richardson, Enantiospecific synthesis of, (6S,7R,8R,8aR)-6,7-dihydroxy-indolizidine and (6R,7R,8S,8aR)-6,7,8-trihydroxyindolizidine, from D-glucose, *Tetrahedron Lett.*, 28 (1987) 4601–4604.

L. Hough, L. V. Sincharoenkul, A. C. Richardson, F. Akhtar, and M. J. Drew, Bridged derivatives of sucrose. The synthesis of 6,6'-dithiosucrose, 6,6'-epidithiosucrose and 6,6'-epithiosucrose, *Carbohydr. Res.*, 174 (1988) 145–160.

K. J. Hale, L. Hough, and A. C. Richardson, The synthesis of some novel Group IVa organometallic derivatives of carbohydrates, *Carbohydr. Res.*, 177 (1988) 259–264.

K. J. Hale, L. Hough, and A. C. Richardson, Morpholino-glucosides: New potential sweeteners derived from sucrose, *Chem. Ind. (Lond.)* (1988) 268–269.

D. Hendry, L. Hough, and A. C. Richardson, Enantiospecific synthesis of polyhydroxylated indolizidines related to castanospermine: 1-Deoxy-castanospermine, *Tetrahedron*, 44 (1988) 6143–6152.

D. Hendry, L. Hough, and A. C. Richardson, Enantiospecific synthesis of polyhydroxylated indolizidines related to castanospermine: (6R,7S,8aR)-6,7-dihydroxy-indolizidines and (6R, 7R,8S,8aR)-6,7,8-trihy-droxy-indolizidine, *Tetrahedron*, 44 (1988) 6153–6168.

K. H. Aamlid, G. Lee, R. G. Price, A. C. Richardson, B. V. Smith, and S. A. Taylor, Development of improved chromogenic substrates for the detection and assay of hydrolytic enzymes, *Chem. Ind. (Lond.)* (1989) 106–108.

C. Bullock, L. Hough, and A. C. Richardson, A novel route to 1,4-anhydro derivatives of β-D-galactopyr-anoside, *Carbohydr. Res.*, 197 (1990) 131–138.

L. Hough and A. C. Richardson, *Carbohydr. Res.*, 200 (1990) ix–x.

R. C. Garcia, L. Hough, and A. C. Richardson, The synthesis of hepta-, hexa- and penta-pivalates of trehalose by selective pivaloylation, *Carbohydr. Res.*, 200 (1990) 307–317.

K. H. Aamlid, L. Hough, and A. C. Richardson, Synthesis of 1-deoxy-6-epi-castanospermine and 1-deoxy-6,8a-di-epicastanospermine, *Carbohydr. Res.*, 202 (1990) 117–129.

S. A. Taylor, A. C. Richardson, P. F. G. Praill, and R. G. Price, A new colorimetric kit for the assay of β-D-glucosaminidase in urine, *Clin. Chem. News*, 16(7), (1990) 24.

R. G. Price, I. Pocsi, A. C. Richardson, and P. H. Whiting, Rate assays of N-acetyl-β-D-glucosaminidase; some problems and fundamental principles, *Clin. Chem.*, 36 (1990) 1259–1260.

B. G. Winchester, I. Cenci de Bello, A. C. Richardson, R. J. Nash, L. A. Fellows, N. G. Ramsden, and G. Fleet, The structural basis of the inhibition of human glycosidases by castanospermine analogues, *Biochem. J.*, 269 (1990) 227–231.

A. C. Richardson, Dedication: Professor Leslie Hough, *Carbohydr. Res.*, 202 (1990) xi–xiv.

K. H. Aamlid, G. Lee, B. V. Smith, A. C. Richardson, and R. G. Price, New colorimetric substrates for the assay of glycosidases, *Carbohydr. Res.*, 205 (1990) C5–C9.

I. Pocsi, S. A. Taylor, A. C. Richardson, K. H. Aamlid, B. V. Smith, and R. G. Price, VRA-GlcNAc: Novel substrate for N-acetyl-β-D-glucosaminidase applied to assay of this enzyme in human urine, *Clin. Chem.*, 36 (1990) 1884–1888.

B. W. Bainbridge, N. Mathias, R. G. Price, A. C. Richardson, J. Sandhu, and B. V. Smith, Improved methods for the detection of β-galactosidase activity in colonies of *Escherichia coli* using a new chromogenic substrate: VBzTM-gal (2-[2-(4-β-D-galactopyranosyl-3-methoxyphenyl)-vinyl]-3-methyl-benzothiazolium toluene- 4-sulphonate), *FEMS Microbiol. Lett.*, 80 (1991) 319–324.

M. A. Ali, L. Hough, and A. C. Richardson, Thio and epidithio derivatives of methyl β-lactoside, *Carbohydr. Res.*, 216 (1991) 271–287.

L. Hough, K. C. McCarthy, and A. C. Richardson, The preparation and reactions of 2-halogenoethyl β-L-arabinopyranosides, *Recl. Trav. Chim. Pays-Bas*, 110 (1991) 450–458.

R. J. Miles, E. L. T. Sui, C. Carrington, A. C. Richardson, B. V. Smith, and R. G. Price, The detection of lipase activity in bacteria using novel chromogenic substates, *FEMS Microbiol. Lett.*, 90 (1992) 283–288.

L. Hough and A. C. Richardson, From sugars to morpholines, spiro-acetals and alkaloids, in H. Ogura, A. Hasegawa, and T. Suami, (Eds.), *Carbohydrates: Synthetic Methods and Applications in Medicinal Chemistry,* Kodansha, Tokyo, 1992, pp. 108–119.

I. Izquierdo-Cubero, M. T. P. Lopez-Espinosa, M. D. Suarez-Ortega, and A. C. Richardson, Enantiospecific synthesis of 1-deoxythiomannojirimycin from a deriviative of D-glucose, *Carbohydr. Res.*, 242 (1993) 109–118.

I. Izquierdo-Cubero, M. T. P. Lopez-Espinosa, A. C. Richardson, and K. H. Aamlid, Enantiospecific synthesis of (*R*)-1,6-dioxaspiro[4.5]decane from a derivative of D-fructose, *Carbohydr. Res.*, 242 (1993) 281–286.

I. Izquierdo Cubero, M. T. P. Lopez Espinosa, and A. C. Richardson, Enantiospecific synthesis from D-fructose of (2*S*,5*R*)- and (2*R*,5*R*)-2-methyl-1,6-dioxaspiro[4.5]decane [The odour bouquet, minor components, of *Paravespula vulgaris* (L.)], *J. Chem. Ecol.*, 19 (1993) 1265–1283.

I. Pocsi, S. A. Taylor, A. C. Richardson, B. V. Smith, and R. G. Price, Comparison of several new chromogenic galactosides as substrates for various β-D-galactosidases, *Biochim. Biophys. Acta*, 1163 (1993) 54–60.

A. C. Richardson and K. J. Hale, Carbohydrates, in R. H. Thompson, (Ed.), *The Chemistry of Natural Products,* Blackie Academic and Professional, Edinburgh, 1993, pp. 1–59. Chapter 1.

R. G. Price, A. C. Richardson, I. Pocsi, L. Csathy, and V. A. Olah, Reply to letter of K. Jung and F. Preim, What are the criteria to introduce new methods for the determination of urinary *N*-acetyl-β-D-glucosaminidase? *Ann. Clin. Biochem.*, 30 (1993) 503–504. (pp. 501–503).

V. M. Cooke, R. J. Miles, R. G. Price, and A. C. Richardson, A novel chromogenic ester agar medium for detection of *Salmonellae*, *Appl. Environ. Microbiol.*, 65 (1999) 807–812.

K. J. Hale and A. C. Richardson, Chemical synthesis of monosaccharides, in P. Finch, (Ed.), *Carbohydrates: Structures, Syntheses and Dynamics,* Kluwer Academic Publishers, Dordrecht, The Netherlands, 1999, pp. 47–106. Chapter 2.

V. M. Cooke, R. J. Miles, R. G. Price, G. Midgely, W. Khamri, and A. C. Richardson, A novel chromogenic ester agar medium for detection of *Candida* spp., *Appl. Environ. Microbiol.*, 68 (2002) 3622–3627.

DETECTION AND STRUCTURAL CHARACTERIZATION OF OXO-CHROMIUM(V)–SUGAR COMPLEXES BY ELECTRON PARAMAGNETIC RESONANCE

Luis F. Sala[a], Juan C. González[a], Silvia I. García[a],
María I. Frascaroli[a] and Sabine Van Doorslaer[b]

[a]Departamento de Química Física—Área Química General, Facultad de Ciencias Bioquímicas y Farmacéuticas, Universidad Nacional de Rosario, Rosario, Santa Fe, Argentina
[b]Physics Department, University of Antwerp, Antwerp-Wilrijk, Belgium

I. Introduction	70
1. Introduction to Chromium (Bio)chemistry	70
2. State-of-the-Art EPR Techniques	72
II. General Aspects of EPR of Oxo-Cr(V) Complexes	73
III. Systematic Structural Characterization of Oxo-Cr(V)–Sugar Complexes Using EPR	77
1. Complexes with Alditols	77
2. Complexes with Aldohexoses and Aldopentoses	86
3. Complexes with Methyl Glycosides	94
4. Oxo-Cr(V) Complexes with Other Sugars	98
5. Complexes with Hydroxy Acids: Alduronic and Glycaric Acids	100
6. Complexes with Oligosaccharides and Polysaccharides	108
IV. Conclusions	114
Acknowledgments	115
References	115

Abbreviations

CW, continuous wave; Cys, cysteine; DFT, density functional theory; ENDOR, electron nuclear double resonance; ehba, 2-ethyl-2-hydroxybutanoate^{2-}; EPR, electron paramagnetic resonance; Glc6P, D-glucose 6-phosphate; GSH, reduced glutathione; HEPES, 4-(2-hydroxyethyl)-1-piperazineethanesulfonic acid; Hex, aldohexose;

HexOMe, methyl glycoside; Hex-onic, aldohexonic acid; hmba, 2-hydroxy-2-methylbutanoate^{2-}; HYSCORE, hyperfine sublevel correlation spectroscopy; mod. ampl., modulation amplitude; polyGalA, galacturonan; Qa, quinic acid; XAFS, X-ray absorption fine structure spectroscopy

I. INTRODUCTION

1. Introduction to Chromium (Bio)chemistry

Chromium, widely used in various forms in industry, has been shown to have toxic and carcinogenic effects in humans and animals. Specifically, chromates have been implicated as chemical compounds that produce malignancies in individuals involved in welding and chemical processing of this metal.[1] Electroplaters, for example, who are exposed to mists of chromic acid, suffer from nasal septum lesions.[2] Chromium (VI) causes significant DNA damage in human nasal or gastric mucosal cells, which are rich in glycoproteins.[3] Scanning electron microscopy has revealed that the catfish, *Sacobranchus fossilis*, kept in test water containing 0.1 mM Cr(VI) for 7 days, undergo severe changes of the mucous goblet cells of their skin, changing from normal flat, hexagonal, or polygonal disks with microvilli-like structures to cylindrical shapes with needle-shaped forms at their tips.[4] *In vivo* administration of sodium dichromate onto the inner-shell membrane of 14-day-old chick embryos resulted in the formation of a persistent chromium(V) species in the liver cell.[5] From these few examples, it is already clear that there is a strong need to understand chromium biochemistry and to remediate chromium pollution.

A key question in chromium chemistry involves understanding the conversion pathways from one oxidation state of chromium to another. The anionic hexavalent form of chromium can readily cross the cell membranes, and being a strong oxidizing agent, it is reduced inside the cell to its trivalent form and thus may become clastogenic.[6] This expectation is supported by findings that trivalent chromium reacts avidly with DNA[7] and, when incubated with isolated nuclei, causes cross-linking of nuclear proteins to DNA.[8]

Already in 1982, it was suggested that the intermediate chromium(V) state is involved in the carcinogenic process.[9] Reactive Cr(V) and Cr(IV) intermediates may be harmful in many ways: acting as tyrosine phosphatase inhibitors, or by forming organic radicals upon reaction with cellular reductants, which in turn can react with O_2 and lead to reactive oxygen species.[10] Reaction of chromium(VI) with

NADH or NAD(P)H in the presence of either rat-liver cytosolic or microsomal fractions led to the formation of stable chromium(V)–NAD(P)H complexes. When D-glucose 6-phosphate (Glc6P) was present in the reaction as part of an NADPH-generating system, stable chromium(V)–Glc6P complexes were formed in addition to the chromium(V)–NAD(P)H complexes.[11] Cr(V) was also found to be a stable intermediate in the interaction of Cr(VI) with various biological substances, such as soil fulvic acid[12] and milk.[13] Relatively long-lived Cr(V) species are formed following Cr(VI) uptake by such plants as garlic roots[14] and algae.[15] Plants may thus play a role in the environmental genotoxicity of chromium with respect to animals that graze on chromium-contaminated sites.[13]

Bacteria indigenous to Cr(VI)-polluted areas are Cr(VI) tolerant and/or resistant and have been considered as potential candidates for bioremediation of Cr(VI)-contaminated sites.[16] However, the ability of bacteria to reduce Cr(VI) to the less-toxic Cr(III) compounds may produce reactive intermediates (such as Cr(V), Cr(IV), radicals), which are known to be active genotoxins and are likely to be carcinogenic.[17] Therefore, the formation and lifetimes of Cr(V) intermediates, produced via bacterial reduction of Cr(VI), need to be evaluated carefully if microorganisms are to be employed as a means for remediation of chromium-polluted subsurface environments. Similarly, Cr(V) accumulation should first be monitored when considering plants and algae as biosorption materials for the bioremediation in the event of chromium pollution.[18]

More details on the biochemistry of chromium and its carcinogenicity can be found in excellent reviews by the group of Lay.[10,19,20]

Gaining an insight into Cr(V) chemistry is, however, important not just for understanding the mechanism of biochemical Cr(VI) oxidation; Cr(V) complexes play an important role in many other applications, such as target complexes for polarized deuterons and protons in high-energy physics applications[21] and as key molecules in a variety of synthetic catalytic processes.[22–24]

In this article, we focus on the interaction of Cr(V) with carbohydrates, producing oxo-chromium(V) complexes of the form [Cr(V)O(Sug$_n$X$_m$)] with Sug being the sugar, X a solvent molecule and n,m the number of ligands. Although several Cr(V) species with nitrogen- and/or sulfur-containing ligands (such as amino acids, peptides, imines, and thiols) have been characterized,[25–30] oxygen-containing ligands dominate in Cr(V) coordination chemistry. This can be rationalized in terms of hard–soft acid–base theory, which predicts that hard metal ions (such as Cr(V)) are preferentially stabilized by hard donor atoms (such as oxygen). The preference of Cr(V) for oxygen-containing ligands is reflected by the significant number of oxo-Cr(V)–carbohydrate complexes that have been characterized over recent years (see Section III).

Characteristic to the whole field of transition-metal chemistry, a multitude of spectroscopic techniques have been employed to study Cr(V) complexes, including optical absorption spectroscopy,[31] X-ray absorption spectroscopy,[29,32,33] X-ray crystallography,[29,34] and electron paramagnetic resonance (EPR).[10,29,35] Although many Cr(V) complexes with biologically relevant ligands have been identified in solution, a Cr(V)–glutathione complex is the only one that has been isolated in the solid state, and even for this complex the structure is uncertain.[36] Therefore, EPR spectroscopy has been one of the key spectroscopic techniques for gaining insight into the structure of Cr(V) complexes of biologically relevant molecules, particularly because of its versatility in detecting Cr(V) in liquid[20,37] and solid[18] samples, and even in cells.[5] For this reason, we focus here mainly on the EPR results reported for oxo-Cr(V)–sugar complexes, while comparing some of the results to findings obtained by other analytical methods when appropriate. Before focusing on the different aspects of EPR of oxo-chromium(V) complexes, and in particular oxo-Cr(V)–sugar complexes, we first give a brief overview of the status of modern-day EPR.

2. State-of-the-Art EPR Techniques

Since its first introduction, EPR spectroscopy, and its sister technique ENDOR (electron nuclear double resonance), has been used extensively for tackling problems in biology, medicine, physics, chemistry, and material sciences.[38–42] The acronyms EPR and ENDOR denote a group of versatile magnetic-resonance experiments that enable analysis of the structure and dynamics of paramagnetic systems in both the liquid and the solid state. By using spin probes[43] or spin labels,[44] these techniques can be extended even to diamagnetic materials.

Many parameters can be determined from a set of EPR and ENDOR experiments,[45,46] among which are the so-called g tensor that is a measure for the electronic structure and local symmetry experienced by the electron spin, the hyperfine tensor that reflects the interaction between the unpaired electron and the surrounding magnetic nuclei, and the nuclear quadrupole tensor that quantifies the interaction between a high nuclear spin ($I > 1/2$) with the electric field gradient of the molecule. From the EPR parameters, valuable information can be obtained concerning the geometric and electronic structure of a paramagnetic compound to a point where they can be used as fingerprints of specific paramagnetic molecules. In the past few years, the interpretation of EPR data has become more and more coupled to quantum-chemical computations, such as DFT (density functional theory).[47] This trend parallels the instrumental and methodological developments that have revolutionized the field of EPR over the past decades. For a long time, most EPR applications involved continuous-wave (CW) experiments at the standard X-band (~9.5 GHz) microwave frequencies. Since the late 1980s, the introduction of fast electronics and the advent of (commercial) pulsed and/or high-field EPR/ENDOR spectrometers have increased enormously the set of parameters that can be obtained.[41,48,49] In this way, the true complexity of the systems under study is often revealed, a complexity that may not be apparent when only a limited set of parameters is known. This development explains the need for combining experimental results with quantum-chemical computations.

II. General Aspects of EPR of Oxo-Cr(V) Complexes

Analogous to the V(IV) and oxo-vanadium(IV) complexes, the Cr(V) and oxo-chromium(V) complexes are characterized by a paramagnetic d^1 ground state that can be detected with EPR.[50] Most of the EPR data reported in the literature on oxo-chromium(V) complexes concerns room-temperature X-band CW-EPR experiments of these complexes in solution. Room-temperature CW-EPR has the advantage that it is a fast and easy technique, but such an experiment has the drawback that only a limited amount of information can be retrieved from it. Since the paramagnetic complexes are fast rotating in solution, the tensorial character of the EPR parameters is averaged out and only the "isotropic" g and hyperfine values can be determined. These parameters represent the sum of the principal values of the corresponding tensors. This leads to relatively simple spectra that may be used as molecular fingerprints, but care should be taken not to overinterpret these features, since their information content is relatively limited.

Figure 1 shows a typical room-temperature X-band CW-EPR spectrum of an oxo-Cr(V) complex. Besides the central line due to the ^{52}Cr-containing complexes, four side-bands are visible (Fig. 1A) stemming from the ^{53}Cr-containing complexes (^{53}Cr: $I=3/2$, natural abundance 9.5%). The splitting between these lines allows facile determination of the isotropic ^{53}Cr hyperfine interaction ($^{Cr}a_{iso}$). The magnitude of this value is characteristic of the coordination number of the Cr(V) species (penta- or hexa-coordination).[35] When zooming in on the central line, further substructure can be observed that stems from the isotropic hyperfine interaction of the unpaired electron with the nearby hydrogen atoms of the ligand (in the example presented, the interaction with two protons is observed) (Fig. 1b). This fine structure evidently depends strongly on the type of the ligand, and it can thus be used as fingerprint for a particular complex.[20]

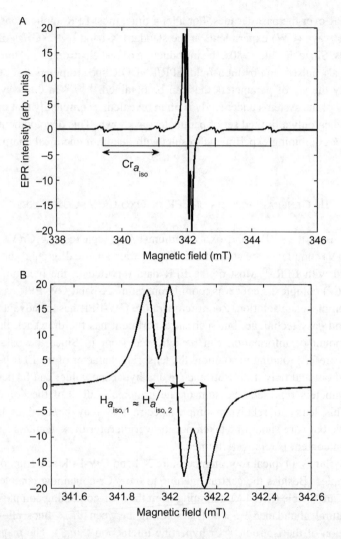

FIG. 1. Simulated room-temperature CW-EPR spectrum for an oxo-chromium complex with $g_{iso} = 1.9803$, $^{Cr}a_{iso} = 15.9 \times 10^{-4}$ cm^{-1} and $^{H}a_{iso,1} = 0.84 \times 10^{-4}$ cm^{-1} and $^{H}a_{iso,2} = 0.87 \times 10^{-4}$ cm^{-1}. (A) Full-field scan and (B) central detail of spectrum. The microwave frequency was set to 9.48 GHz.

Although these room-temperature CW-EPR spectra can be useful for making a rough classification of the types of ligations occurring at the Cr(V) center,[10–20] they do not allow derivation of a full geometric model. As illustrated in Fig. 1, the method picks up only the strongest isotropic hyperfine couplings. The isotropic hyperfine coupling, also termed the Fermi-contact term, reflects the spin density in the s-orbitals but does not give any information about the dipolar interactions.[45,46] While the s-spin density depends on the orbital overlap, the dipolar interactions are determined by the inter-spin distances. A hydrogen atom close to the Cr(V) center may have a quasi zero isotropic hyperfine coupling but will have a very large anisotropic hyperfine coupling. If analysis were based only on the room-temperature CW-EPR spectrum, it would be concluded that there is no proton close to the Cr(V) center, as no resolved isotropic hyperfine splitting is seen. Only when applying ENDOR or pulsed EPR techniques to the solid phase, namely on the frozen solution, is it possible to resolve the dipolar contribution and arrive at the correct conclusion. It should be noted that, in many instances, no hyperfine coupling is resolved in the room-temperature CW-EPR spectra, and this factor emphasizes the need to apply more advanced techniques.

Figure 2A shows a pulsed ENDOR spectrum of an oxo-Cr(V) complex where the unpaired electron is interacting with a ^{1}H nucleus with principal hyperfine values $^Ha_x = {}^Ha_y = -2$ MHz and $^Ha_z = 5$ MHz. In this case, the isotropic hyperfine value is $^Ha_{iso} = ({}^Ha_x + {}^Ha_y + {}^Ha_z)/3 = 0.33$ MHz ($= 0.11 \times 10^{-4}$ cm^{-1}), a value that is not resolved in the CW-EPR spectrum. The hyperfine contribution of this proton is, however, clearly resolved in the ENDOR spectrum. The ENDOR spectrum is centered around the proton Larmor frequency (v_H), identifying the contribution as stemming from an interaction with a ^{1}H nucleus. The principal values can be read directly from the spectrum, as indicated in Fig. 2A.

Figure 2B shows a HYSCORE (hyperfine sublevel correlation[51]) spectrum of an oxo-Cr(V) complex, with the unpaired electron interacting with a ^{1}H nucleus with principal hyperfine values $^Ha_x = {}^Ha_y = -2$ MHz and $^Ha_z = 10$ MHz and $^Ha_{iso} = 2$ MHz ($= 0.67 \times 10^{-4}$ cm^{-1}). The HYSCORE experiment consists of a four-microwave-pulse sequence, $\pi/2 - \tau - \pi/2 - t_1 - \pi - t_2 - \pi/2 - \tau - echo$, in which the spin echo is detected as a function of t_1 and t_2. After Fourier transformation of this two-dimensional time-domain signal, cross-peaks are found linking different nuclear frequencies. As indicated in Fig. 2B, the cross-ridges again permit determination of the nature of the contributing nucleus (here, ridge centered around (v_H, v_H)) and the principal hyperfine values. The observation that the ridge crosses the diagonal is a direct proof of the fact that the hyperfine values ^Ha_z and $^Ha_x, {}^Ha_y$ have opposite signs. The two-dimensionality of the HYSCORE spectrum allows the unraveling of different spectral components that may overlap in a one-dimensional spectrum, such as the ENDOR spectrum shown in Fig. 2A. Several other pulsed EPR and ENDOR techniques are known that can elucidate complex hyperfine spectra. For further details, the specialized literature[49] should be consulted.

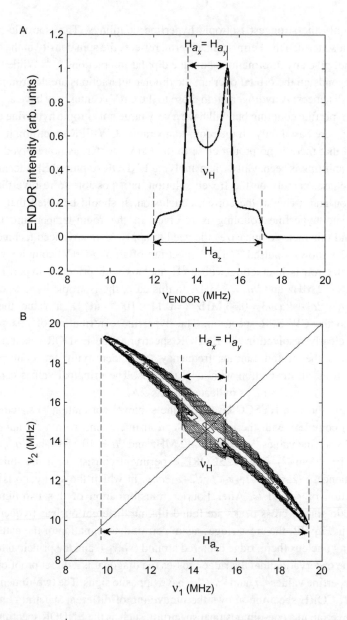

FIG. 2. (A) Simulated ^{1}H ENDOR spectrum for an oxo-chromium(V) complex with $g = 1.9803$, and $^Ha_x = {^Ha_y} = -2$ MHz, $^Ha_z = 5$ MHz. (B) Simulated HYSCORE spectrum for an oxo-chromium complex with $g = 1.9803$, and $^Ha_x = {^Ha_y} = -2$ MHz, $^Ha_z = 10$ MHz. The microwave frequency was set to 9.48 GHz, the observer magnetic field was 342 mT.

Furthermore, room-temperature CW EPR only picks up the isotropic g value that represents the average of the three principal g values. Two complexes with entirely different surroundings, and hence entirely different $\mathbf{g}$ tensors, may accidentally have the same isotropic g value. For instance, the example in Fig. 1 may stem from a complex with $g_x = 1.9825$, $g_y = g_z = 1.9792$ or from a complex with $g_x = 1.9825$, $g_y = 1.9801$, $g_z = 1.9784$. The first concerns a molecule where the unpaired electron senses a local axial symmetry, the latter is due to a Cr(V) site with rhombic surroundings. From the room-temperature CW-EPR spectrum, both cases would be considered identical if no additional proton hyperfine value is resolved. The full $\mathbf{g}$ tensor can only be obtained from low-temperature CW-EPR, preferably at higher microwave frequencies.[42,48]

Amazingly, low-temperature CW EPR has only been sparsely applied for the study of oxo-Cr(V) complexes,[29,52–55] and even fewer studies involve CW ENDOR[55,56] or pulsed EPR/ENDOR experiments,[52] or attempt a combination of EPR and quantum-chemical computations.[29,52,54] A great potential evidently still exists for further research along this line.

One of the additional assets of EPR in the study of chromium (bio)chemistry is the fact that, besides the Cr(V) state already discussed, the biologically relevant Cr(III) state[57] can also be detected readily with this technique. Even EPR spectra of highly unusual Cr(I) complexes[58] have been reported, and high-field EPR allows detection of the Cr(II) state that is "EPR-silent" at conventional microwave frequencies.[59] Furthermore, organic radicals that may be formed during the reduction of Cr(VI) to Cr(III) can be detected with EPR.

III. Systematic Structural Characterization of Oxo-Cr(V)–Sugar Complexes Using EPR

1. Complexes with Alditols

1: R = R' = H
2: R = H; R' = Me
3: R = R' = Me

Branca et al. performed pioneering work in analyzing the room- and low-temperature CW-EPR and CW-ENDOR spectra of oxo-Cr(V) complexes of a number of 1,2-diols (**1–3**) in methanol.[55] They revealed a clear anisotropy of the g and ^{Cr}A tensors (Table I) and determined the full proton hyperfine tensors. They were able to identify the presence of two classes of protons in the chelated ethylene bridge of the complexes. For complex **1**, a fast interconversion between these two classes was found at room temperature.

The same authors also showed that, in systems containing GSH, chromate, and sugars, the Cr(VI) species interacts with the sugar ligand to yield esters that are reduced by GSH. The oxo-Cr(V) ions thus formed are then stabilized by the sugar ligands, where those sugars that have pairs of *cis*-disposed hydroxyl groups are found to be the most effective chelators.[60]

It is known that five-membered Cr(V) chelates are favored over six-membered ones.[19,50,61] For Cr(V)–diolato complexes formed with linear diols, it was observed that all of the protons are equivalent in the isotropic EPR spectra,[62] although the strain of a six-membered ring imparts inequivalence to the magnetic environment of the protons in the second coordination sphere.[63] This observation at room temperature again points to rotational flexibility in the chelate ring (namely, puckering in the δ or λ configuration),[20] similar to that observed for **1**.

The redox and complexation chemistry of alditol/Cr(VI) systems[64] and the *myo*-inositol/Cr(VI) system[65] has been reported. In the first case, when an excess of the alditol over Cr(VI) is used, the secondary OH groups are inert to oxidation, and alditols are selectively oxidized at the primary OH group to yield the aldonic acid as the only oxidation product. The corresponding reaction involves a Cr(VI) → Cr(V) → Cr(III) reduction pathway, and the relative rate of each step depends on [H$^+$]: at

TABLE I
Principal g and ^{53}Cr Hyperfine Values of Complexes 1–3[55]

Complex	g_x	g_y	g_z	g_{iso}	$^{Cr}a_x$	$^{Cr}a_y$	$^{Cr}a_z$	$^{Cr}a_{iso}$
1	1.9840	1.9800	1.9770	1.9803	20.2	105.2	24.2	49.9
					(6.7)	(35.1)	(8.1)	(16.6)
2	1.9836	1.9796	1.9766	1.9799	20.2	105.2	24.2	49.9
					(6.7)	(35.1)	(8.1)	(16.6)
3	1.9836	1.9796	1.9761	1.9798	20.2	105.2	24.2	49.9
					(6.7)	(35.1)	(8.1)	(16.6)

The hyperfine values are given in megahertz (10^{-4} cm^{-1}).

[H$^+$] > 0.1 M, the rate is determined by the first step [Cr(VI) → Cr(V)], whereas at pH > 1, the second step [Cr(V) → Cr(III)] becomes rate determining.[50]

4

5

EPR spectra of mixtures of D-glucitol (**4**) or D-mannitol (**5**) and Cr(VI) in HClO$_4$ (0.1–0.8 M) show the formation of several intermediate oxo-Cr(V) species.[64] At these pH values, the EPR spectra consist of a minority contribution (g_{iso} = 1.9793) and a dominant contribution (g_{iso} = 1.9718).[64] The latter signal can no longer be observed at pH > 1, where the reaction becomes extremely slow, while the former contribution remains observable for a long time in solution. The Cr(V) species is stabilized by an excess of the alditols or their oxidation products, and an estimation of the Cr(V) species formed in the redox process can be evaluated by using room-temperature CW EPR for different pH values, [Cr(VI)], and [substrate]. The experimental g_{iso} and $^{Cr}a_{iso}$ values show that five- and six-coordinated oxo-Cr(V) complexes are formed depending on [H$^+$]. At pH > 2, only five-coordinated Cr(V) species are observed [g_{iso} = 1.9799, $^{Cr}a_{iso}$ = 16.5 × 10^{-4} cm^{-1} (= 49.5 MHz)]. Under these experimental conditions, the intramolecular redox reactions are very slow and the Cr(V) complexes formed remain in solution for several days or weeks (Fig. 3).

The hyperfine substructure (quintet) of the signal at g_{iso} = 1.9799 (Fig. 3A) stems from the interaction with four nearby protons that bear substantial spin density on the ^{1}H nucleus. This suggests that the five-coordinated Cr(V) center ligates to four secondary hydroxyl groups, probably stemming from the *vic*-diols donor sites of two bidentate-ligating alditols, which are schematically represented in Scheme 1 (type A). Actually, those oxo-Cr(V) complexes that are more stable do not involve Cr(V) coordination to the primary hydroxyl group. In addition to this type of complex, a second type of oxo-Cr(V) complex (g_{iso} = 1.9787) appears at pH 1–2 (Fig. 3B), and its EPR parameters suggest that Cr(V) now ligates to the oxidation product with coordination via the 1-carboxylate-2-hydroxy donor group (type B in Scheme 1). Moreover, at [H$^+$] ≥ 0.1 M, a positively charged six-coordinated Cr(V) monochelate is formed (g_{iso} = 1.9719, $^{Cr}a_{iso}$ = 20.0 × 10^{-4} cm^{-1}, type C in Scheme 1), dominating the Cr(V) species in solution (Fig. 3C).[50]

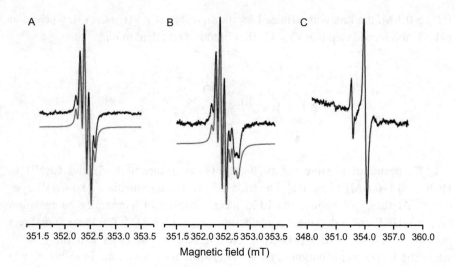

FIG. 3. X-band room-temperature (25 °C) CW-EPR spectra of solutions of **4** and Cr(VI). (A) **4**:Cr(VI) = 10:1, pH 5.0; (B) **4**:Cr(VI) = 20:1, pH 2.0, mod. ampl. = 0.04 mT; (C) **4**:Cr(VI) = 10:1, pH 1.0, mod. ampl. 0.1 mT. Microwave frequency = 9.7815 GHz. [Cr(VI)] = 2.5 mM. (A and B) Experimental (top) and simulated (bottom) spectra. Figure adapted from ref. 64.

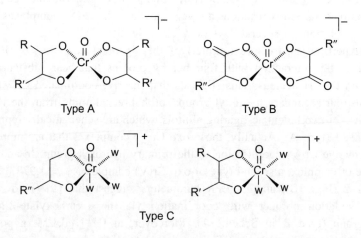

SCHEME 1. Schematic classification of the different oxo-Cr(V) species formed in the reaction between Cr(VI) and alditols (w denotes water).

EPR CHARACTERIZATION OF OXO-CHROMIUM(V)–SUGAR COMPLEXES

6   **7**

For cyclic 1,2-diol–Cr(V) complexes, the multiplicity of the room-temperature CW-EPR signal depends upon the cyclic strain in the ligand.[50] For these systems, the observed EPR hyperfine pattern of the Cr(V)-diolate$_2$ formed by reduction of Cr(VI) with GSH in the presence of cis-1,2-cyclohexanediol (**6**) or trans-1,2-cyclohexanediol (**7**)[7,19,50] shows that the strain of the six-membered ring renders the protons in the second coordination sphere magnetically nonequivalent. Thus, a mixture of **7** with Cr(VI) leads to a single EPR line (due to [CrO(trans-1,2-cyclohexanediolate)$_2$]$^-$, $g_{iso} = 1.9800$), whereas the cis-isomer **6** leads to two triplet EPR contributions agreeing with two geometric isomers of [CrO(cis-1,2-cyclohexanediolate)$_2$]$^-$. For each of these configurations, two equivalent protons were coupled to the electronic spin ($g_{iso} = 1.9799$, $^Ha_{iso} = 0.97 \times 10^{-4}$ cm^{-1} (=2.9 MHz); $g_{iso} = 1.9796$, $^Ha_{iso} = 0.89 \times 10^{-4}$ cm^{-1} (=2.67 MHz)) (Fig. 4).

This difference in the EPR parameters can be explained as follows. When the protons lie in the Cr(V)-ligand plane, there is maximal overlap between their orbitals and the Cr(V) d$_{xy}$ orbital in which the unpaired electron resides.[63] This condition is reached for ligation with the cis-isomer **6** (Fig. 5, left). As the proton moves away from Cr(V)-ligand plane, the magnitude of the coupling diminishes until reaching a minimum, as is the case for the trans-isomer **7** (Fig. 5, right).

Because of the fact that, in [CrO(cis-O^1,O^2-cyclohexanediolate)$_2$]$^-$ (situation as in Fig. 5, left), only one of the two carbinolic protons of each diol is in the Cr(V)-ligand plane, a triplet is observed in the room-temperature CW-EPR spectrum (two equivalent protons, Fig. 4B). On the other hand, the [CrO(trans-O^1,O^2-cyclohexanediolate)$_2$]$^-$ species has both protons pointing above the Cr(V)-ligand plane (situation as in Fig. 5, right), and hence the CW-EPR spectrum shows no resolved hyperfine splitting (Fig. 4A).

While the [CrO(cis-O^1,O^2-cyclohexanediolate)$_2$]$^-$ species can in principle assume three geometrical isomeric forms (Scheme 2A), only two are possible for [CrO(trans-O^1,O^2-cyclohexanediolate)$_2$]$^-$ (Scheme 2B).

8   **9**

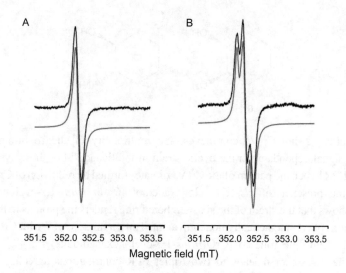

FIG. 4. Experimental (top) and simulated (bottom) X-band room-temperature CW-EPR spectra of mixtures of (A) GSH:Cr(VI):**7** = 1:1:10,000, microwave frequency = 9.7743 GHz, mod. ampl. = 0.04 mT; (B) GSH:Cr(VI):**6** = 1:1:10,000, microwave frequency = 9.7739 GHz, mod. ampl. = 0.02 mT. [Cr(VI)] = 2.5 mM, pH 7.5.

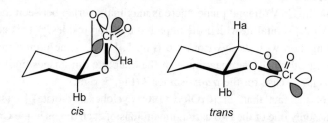

FIG. 5. Schematic representation of the orbital overlap between the chromium d_{xy} orbital and the ligand protons Ha and Hb for oxo-Cr(V) complexes of **6** (left) and **7** (right).

The study of oxo-Cr(V)–diols complexes formed by reduction of Cr(VI) by GSH in the presence of *cis*-cyclopentanediol (**8**) or *trans*-cyclopentanediol (**9**) expanded the EPR analysis of oxo-Cr(V)–diols species to five-membered ring 1,2-diols.[66] Unlike the Cr(V) complexes formed with **6** and **7**, the 1,2-cyclopentanediols **8** and **9** lead to oxo-Cr(V) complexes having very similar EPR spectral patterns (Fig. 6). The spectra are in both instances the sum of two EPR components, both split by the hyperfine interaction with nonequivalent carbinolic protons (Table II).

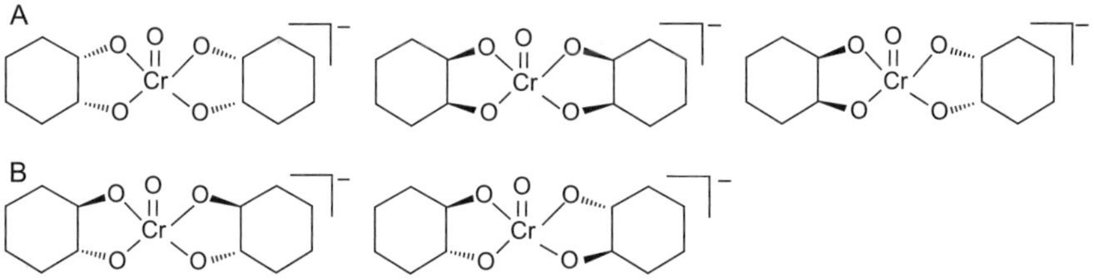

SCHEME 2. Possible isomers formed upon reaction of Cr(VI) with (A) **6** and (B) **7** in the presence of GSH.

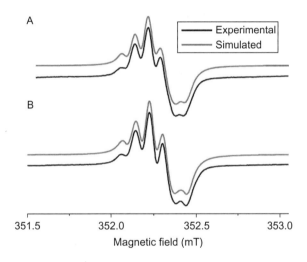

FIG. 6. Experimental and simulated room-temperature X-band EPR spectra of mixtures of (A) GSH:Cr(VI):**9** = 1:1:5000, [Cr(VI)] = 1.0 mM, microwave frequency = 9.7620 GHz, mod. ampl. = 0.04 mT; (B) GSH:Cr(VI):**8** = 1:1:5000, [Cr(VI)] = 0.5 mM, microwave frequency = 9.7634 GHz, mod. ampl. = 0.04 mT. Figure adapted from ref. 66.

As for the six-membered ring case (Scheme 2), three different geometric isomers are possible for [CrO(*cis*-1,2-cyclopentanediolato)$_2$]$^-$, whereas only two can exist for the [CrO(*trans*-1,2-cyclopentanediolato)$_2$]$^-$ (Scheme 3). However, in both situations, only two species are required to simulate the EPR signal (Table II). It seems therefore reasonable to assume that the angles between the carbinolic protons and the Cr(V)-ligand plane in two of the geometric isomers of [CrO(*cis*-1,2-cyclopentanediolato)$_2$]$^-$ are very similar.

TABLE II
Isotropic g, and ^{53}Cr and ^{1}H Hyperfine Values of Oxo-Cr(V) Complexes of 8 and 9

Cr(V)/diol	g_{iso}	$^{Cr}a_{iso}$	$^{H}a_{iso}$
Cr(V)/8	1.9803	47.7	2.52, 2.43, 2.22, 1.92
		(15.9)	(0.84, 0.81, 0.74, 0.64)
	1.9798	47.7	2.28, 1.95, 1.23
		(15.9)	(0.76, 0.65, 0.41)
Cr(V)/9	1.9804	47.7	2.52, 2.28, 2.16, 1.86
		(15.9)	(0.84, 0.76, 0.72, 0.62)
	1.9799	47.7	2.25, 1.92, 1.41
		(15.9)	(0.75, 0.64, 0.47)

The hyperfine values are given in megahertz (10^{-4} cm^{-1}).

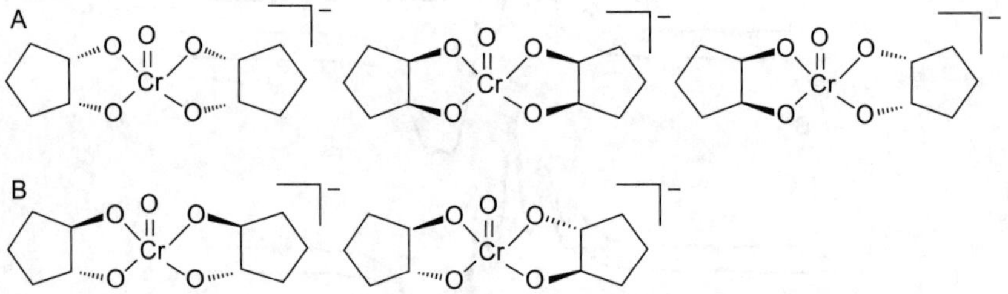

SCHEME 3. Possible isomers formed upon reaction of Cr(VI) with (A) 8 and (B) 9 in the presence of GSH.

While for [CrO(*cis*-O^1,O^2-cyclohexanediolate)$_2$]$^-$ the carbinolic protons are pairwise magnetically equivalent (Fig. 4), the strained bicyclic structure in the [CrO(1,2-cyclopentanediolate)$_2$]$^-$ complex leads to a local inequivalence of all carbinolic protons (Table II).

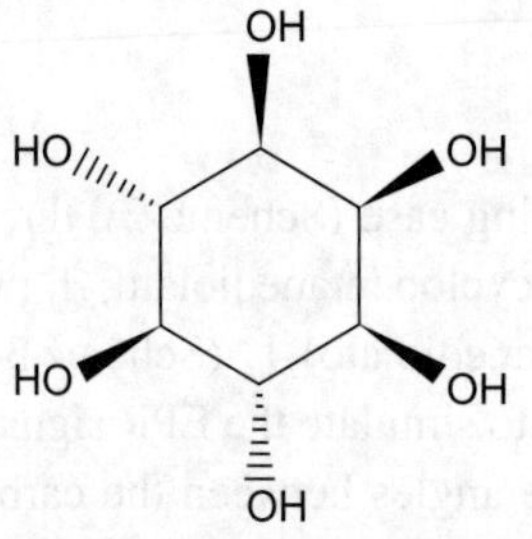

10

Room-temperature EPR spectroscopy was also used to detect intermediate Cr(V) species formed in the reaction of *myo*-inositol (**10**) with Cr(VI). The EPR spectra of mixtures of **10** and Cr(VI) in $HClO_4$ show one dominant signal ($g_{iso} = 1.9800$) and a weak contribution (<10% of total Cr(V), $g_{iso} = 1.972$).[65] At these pH values, a large modulation amplitude of 0.4 mT was required to observe the Cr(V) species. Therefore, the ^{1}H-hyperfine pattern of the EPR signal could not be resolved [^{1}H hyperfine splitting for Cr(V)–alkoxide species is usually <0.1 mT]. Accordingly, the structure of the Cr(V) intermediates was derived from the EPR spectral features obtained after the addition of a 25- to 500-times excess of **10** to a 1:1 mixture of chromate with GSH at pH values of 4.4 and 7.4. Under these conditions, Cr(V) is sufficiently stabilized by **10** to yield oxo-Cr(V)–**10** species that remain in solution over a long period of time. At pH 7.4, the corresponding room-temperature CW-EPR spectra were composed of one triplet [$g_{iso} = 1.9801$, $^Ha_{iso} = 1.03 \times 10^{-4}$ cm^{-1} ($= 3.1$ MHz)] and one doublet [$g_{iso} = 1.9799$, $^Ha_{iso} = 0.95 \times 10^{-4}$ cm^{-1} ($= 2.85$ MHz)]. These two spectra can be correlated with five-coordinated oxochromate(V) complexes with four hydroxyl group donors, similar to the aforementioned examples. The ^{1}H hyperfine splittings found for the two components of the EPR signal are, respectively, in agreement with those expected for oxo-Cr(V)-diolato$_2$ species with two carbinolic protons (one from each chelate ring), and with one carbinolic proton coupled to the unpaired electron.[50] The triplet EPR signal can be attributed to a bis-chelate with oxo-Cr(V) bound to the *cis*-diol moiety of two molecules of **10** (Scheme 4, complex I), while the doublet most

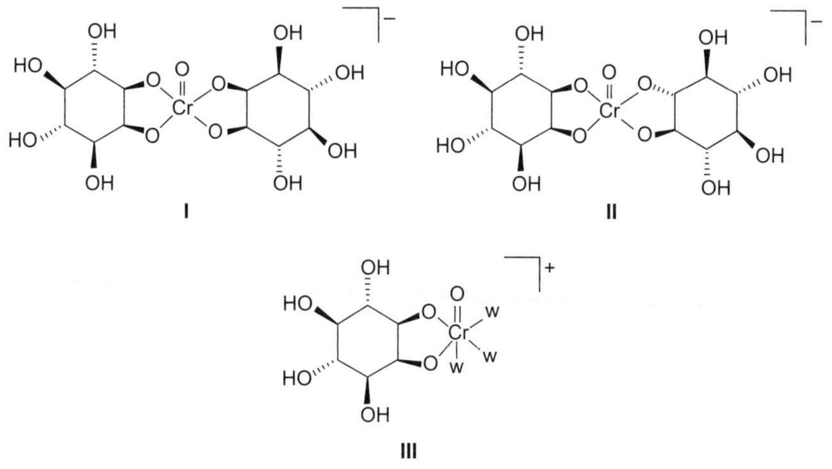

SCHEME 4. Proposed geometric structures of Cr(V)–**10** complexes.

probably corresponds to the bis-chelate **II** (Scheme 4). The ratio of the EPR triplet [Cr(V)-*cis*/*cis*-diol$_2$]$^-$ to the EPR doublet [Cr(V)-*cis*/*trans*-diol$_2$]$^-$ was 30:70, which means that 70% of the ligand is bound to Cr(V) through the *cis*-diol moiety. However, the ratio of *cis*:*trans* diol sites in **10** is 2:4. The marked preference of the Cr(V) ion for binding *cis*- rather than *trans*-diol groups of cyclic diols explains the higher proportion of the Cr(V)-*cis*-diol binding-mode in the mixture.[50]

The same two components (30:70 ratio) were observed at pH 4.4, but a new weak signal was observed having $g_{iso} = 1.979$. Despite the lack of resolution of this EPR signal, the decrease of its proportion with increasing concentration of **10** suggests that this species corresponds to a Cr(V) monochelate (Scheme 4, complex **III**) that transforms into the bis-chelate as the concentration of **10** increases.

2. Complexes with Aldohexoses and Aldopentoses

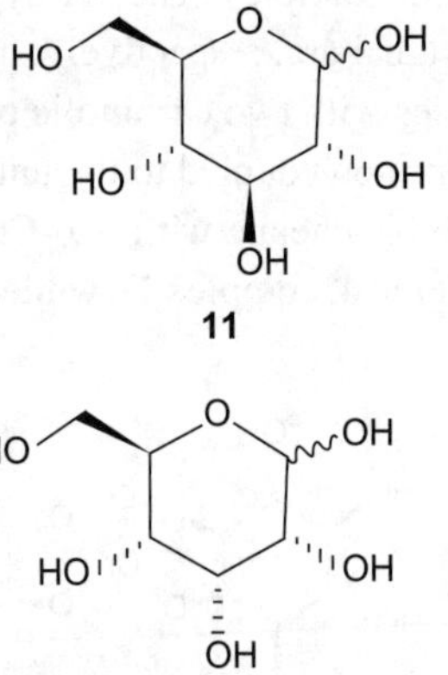

11 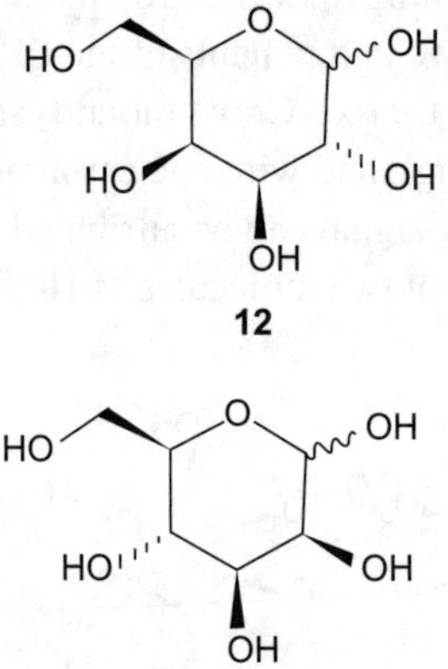   12

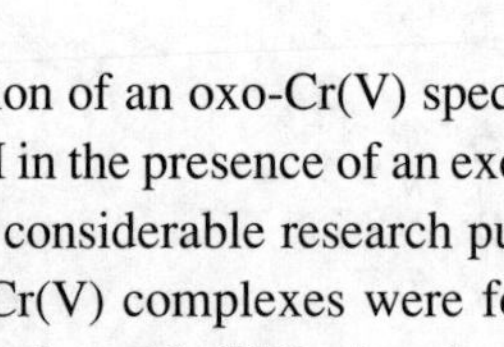

13 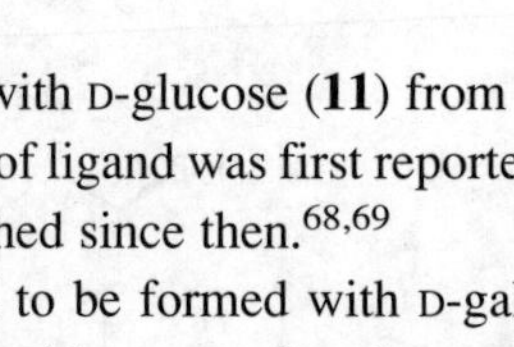   14

The generation of an oxo-Cr(V) species with D-glucose (**11**) from the reduction of Cr(VI) by GSH in the presence of an excess of ligand was first reported[67] in 1992, and there has been considerable research published since then.[68,69]

Stable oxo-Cr(V) complexes were found to be formed with D-galactose (**12**) and D-fructose ligands, with EPR signals detectable even after 1 month.[31,50,70] It was therefore concluded that periods of weeks are necessary to convert the Cr(VI) to Cr(III) complexes having saccharides as ligands, and this poses many environmental implications. In the studies of Cr(VI) oxidation of sugars conducted by Sala *et al.*, the existence of intermediate Cr(V) complexes has indeed been firmly established.[71]

The ternary systems, Cr(VI)/GSH/sugars, have been studied in detail by Kozlowski and coworkers.[60] EPR signals were observed for the reactions of Cr(V) with biological reductants (such as GSH) in the presence of various sugars that possess 1,2-diol moieties [such as D-glucose (**11**) or D-ribose].[72] Compound **11** is a probable candidate ligand responsible for Cr(V) coordination in the $g_{iso} = 1.979$ species in Cr (VI)-exposed animal tissues.[19]

As mentioned before, chromium(VI) reacts with NADH or NAD(P)H in the presence of rat-liver cytosolic or microsomal fractions, forming stable Cr(V)–NAD (P)H complexes, while addition of Glc6P led to stable Cr(V)–Glc6P complexes together with the Cr(V)–NAD(P)H complexes.[11] The chromium(V) complexes have isotropic g values of 1.980 or 1.982, and ^{1}H hyperfine splitting constants of $0.8–0.9 \times 10^{-4}$ cm^{-1} (2.4–2.7 MHz), characteristic of bis(diol)oxo-chromium(V) complexes.

In acidic media, all of the aldohexoses studied are selectively oxidized by Cr(VI) at the anomeric hemiacetal function, yielding the corresponding aldonic acid and Cr(III).[68,73–76] As with other substrates, the redox reaction was found to occur via two pathways: Cr(VI) → Cr(IV) → Cr(III) and Cr(VI) → Cr(V) → Cr(III).[68,74,75] When a H$^+$ concentration of 0.75 M is used, the rates of oxidation by Cr(VI) and Cr(V) of D-glucose (**11**), D-allose (**13**), D-mannose (**14**), and D-galactose (**12**) are comparable.[68] It has been noted that the observed rate constants, k_5 and k_6, for the Cr(VI) and Cr(V) reduction by aldohexoses depend on the [H$^+$]. Moreover, if a large excess of aldohexoses over Cr(VI) is used, then k_6 is lower than k_5 at low pH, but at high pH this trend reverses. The rate of reduction of Cr(V) at pH 2 becomes extremely low and the Cr(V) species generated in the reaction remains in solution for quite a long time.[35,68,76] In these cases, the EPR spectra of the Cr(VI)/aldohexose mixtures show different contributions of five-coordinated oxo-Cr(V)-bischelates ($g_{iso} = 1.979$, $^{Cr}a_{iso} = 16.4(3) \times 10^{-4}$ cm^{-1} ($\equiv 49.2$ MHz)), with the relative proportion of each species depending on the reaction time and on the type of sugar,[68] as illustrated in the following examples. The oxo-Cr(V) complexes of D-glucose (**11**) give rise to an EPR-triplet ($g_{iso} = 1.9789$) and a quartet ($g_{iso} = 1.9793$), consistent with the spectral parameters of two isomeric complexes [CrO(O^1,O^2-Glc-onic)(*cis*-O^1,O^2-Glc)]$^-$ and [CrO(*cis*-O^1,O^2-Glc)(O$^{2(3)}$,O$^{3(4)}$-Glc-onic)]$^-$, where Glc = **11** and Glc-onic = D-gluconic acid. The room-temperature CW-EPR spectrum of the oxo-Cr(V) complexes of D-galactose (**12**) consists of a quartet ($g_{iso} = 1.9789$) and a quintet ($g_{iso} = 1.9796$), most probably corresponding to [CrO(*cis*-O^3,O^4-Gal)(O$^{2(4)}$,O$^{3(5)}$-GalA)]$^-$ and [CrO(O$^{2(4)}$,O$^{3(5)}$-GalA)$_2$]$^-$ (Gal = **12**, GalA = D-galacturonic acid). The spectrum of the oxo-Cr(V) complexes of D-allose (**13**) shows two triplets ($g_{iso} = 1.9794$ and 1.9785), which may be assigned, respectively, to [CrO(*cis*-O$^{1(2,3)}$,

$O^{2(3,4)}$-All)$_2$]$^-$ and [CrO(O^1,O^2-All-onic)$_2$]$^-$ (All = **13**, All-onic = D-allonic acid). Addition of D-mannose (**14**) gives rise to a triplet (g_{iso} = 1.9791) and a quartet (g_{iso} = 1.9795), which can be, respectively, assigned to two isomeric complexes: [CrO(*cis*-O$^{1(2)}$,O$^{2(3)}$-Man)(O^1,O^2-Man-onic)]$^-$ and [CrO(*cis*-O$^{1(2)}$,O$^{2(3)}$-Man)(O^3, O^4-Man-onic)]$^-$ (Man = **14**, Man-onic = D-mannonic acid).

Interestingly, for all of the substrates, except **13**, there was observed a mixture of Cr(V)–aldohexose/aldohexonic acid (Hex-onic) and/or Cr(V)-Hex-onic$_2$ species with an open-chain Hex-onic bound to Cr(V) via the *vic*-diol or 2-hydroxy-1-carboxylate group.

In contrast, at [H$^+$] ≥ 0.1 M, the same reaction results in two or three new EPR signals (g_{iso} = 1.974, 1.971, and 1.966) in addition to the one already mentioned (g_{iso} = 1.979).[68,75] These EPR signals turn out to be consistent with six-coordinated oxo-Cr(V) species. In this situation, the relative intensity of the EPR signal is pH dependent but is independent of the aldohexose/Cr(VI) ratio. In fact, six-coordinated species are dominant at [H$^+$] > 0.75 M. In addition, both species [six- and five-coordinated oxo-Cr(V) complexes] decay at the same rate, meaning that they are in a rapid equilibrium. Scheme 5 shows the complexation chemistry and the observed Cr(V)–sugar redox processes.

In Scheme 5, the first step is formation of the six-coordinated oxo-Cr(V) complex, which can then yield the redox product or the Cr(V)-bischelate, depending on the [H$^+$]. At high [H$^+$], the acid-catalyzed redox reaction turns out to be faster than the complexation reaction by a second molecule of substrate. However, when [H$^+$] decreases, both reactions start to compete and finally at low [H$^+$], the complexation reaction becomes more favored, and five-coordinated Cr(V)-bischelates are formed and become stabilized.

Five and six-coordinated oxo-Cr(V) species were also formed with L-rhamnose and Cr(VI) (ratio 10:1) in 50–100% acidic solution media, presenting EPR signals with g_{iso} 1.978 and 1.973, respectively.[77] In aqueous solution, the corresponding EPR signals appear at g_{iso} = 1.977 [$^{Cr}a_{iso}$ = 16.4 × 10^{-4} cm^{-1} (=49.2 MHz)] for the

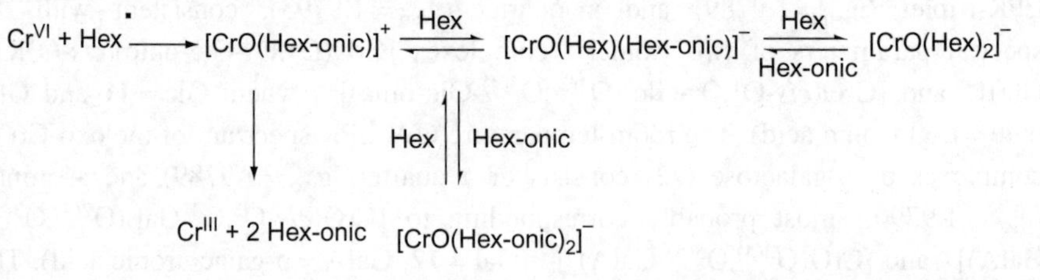

$$Cr^{VI} + Hex \longrightarrow [CrO(Hex\text{-}onic)]^+ \xrightleftharpoons{Hex} [CrO(Hex)(Hex\text{-}onic)]^- \xrightleftharpoons[Hex\text{-}onic]{Hex} [CrO(Hex)_2]^-$$

$$\downarrow \qquad\qquad Hex \updownarrow Hex\text{-}onic$$

$$Cr^{III} + 2\ Hex\text{-}onic \qquad [CrO(Hex\text{-}onic)_2]^-$$

SCHEME 5. Redox and complexation chemistry in a Cr(VI)/aldohexose system. Hex, aldohexose; Hex-onic, aldohexonic acid.

five-coordinated species prevailing at pH 3–7, and at $g_{iso} = 1.970$ for the six-coordinated species, which dominates the spectra at pH < 1.

In addition, the reduction of α-**11** and β-**11** was studied in dimethyl sulfoxide as solvent, in the presence of pyridinium *p*-toluensulfonate. This medium makes mutarotation slower than the redox process.[73] The two anomeric forms could reduce Cr(VI) and Cr(V) by formation of a Cr(VI) and Cr(V) ester intermediate. The equilibrium constant for this step and the rate of the redox step were different for each anomer: for α-**11**, the equilibrium constant for ester formation is higher than for β-**11**, but the redox process within this complex is faster for the latter anomer. These differences can be explained by the better chelating capacity of the 1,2-*cis*-diol moiety of α-**11**. Room-temperature CW-EPR spectra of these mixtures revealed for the α anomer several five-coordinated Cr(V)-bischelates ($g_{iso} = 1.9820$ [$^{Cr}a_{iso} = 15.9 \times 10^{-4}$ cm^{-1} (=47.7 MHz)], $g_{iso} = 1.9785$ and 1.9756) and, for the β anomer, a mixture of six-coordinated oxo-Cr(V) monochelates ($g_{iso} = 1.9700$) and five-coordinated oxo-Cr(V)-bischelates ($g_{iso} = 1.9820$ [$^{Cr}a_{iso} = 15.8 \times 10^{-4}$ cm^{-1} (=47.4 MHz)], $g_{iso} = 1.9789$ and $g_{iso} = 1.9767$). The fact that six-coordinated oxo-Cr(V) monochelate species are observed only with the β anomer indicates the binding of the *trans*-1,2-diol moiety rather than the *cis*-1,2-diol of the α anomer. The latter forms a bis-chelate more easily than the former, because of the enhanced ability of the *cis*-1,2-diol moiety to bind Cr(V).

Complex EPR signals have been obtained during ligand-exchange reactions between D-glucose (**11**) and [CrO(ehba)$_2$]$^-$, which were analyzed with Q- (35 GHz), X-, and S-band (2–4 GHz) CW-EPR spectroscopy, using partially or fully deuterated **11**. The Q-band EPR spectra of Cr(V)/^{2}H$_7$-**11** solutions were shown to contain at least six Cr(V) species.[35]

The observation of long-lived oxo-Cr(V)/lactose species generated by direct reaction of Cr(VI) with lactose (**15**) in milk that is contaminated with chromates points out the capability of disaccharides to stabilize Cr(V).[13] Further work revealed that disaccharides form stable hypervalent chromium chelates at pH > 2, while at higher [H$^+$], their oxidation by Cr(VI)/Cr(V) is favored.[70] When an excess of disaccharide over Cr is used, the oxidation occurs selectively at the hemiacetal group, with first-order kinetics both in [Cr(VI)] and in [disaccharide].[70] The redox and complexation chemistry observed for the disaccharide–Cr(VI) systems parallels that observed with aldoses. However, the (nonreducing) glycoside moiety can delay the redox rate of the aglycone (hemiacetal component) when additional binding to Cr(VI) by the glycoside moiety is possible. In the case of maltose (**16**), this effect decreases the observed rate constant to one half of the value obtained for D-glucose (**11**).[70] The relative ability of the four disaccharides **15–18** to reduce Cr(VI) is found to be melibiose (**17**) > **15** > cellobiose (**18**) > **16**. This is attributed to the increasing stability of the intermediate disaccharide–Cr(VI) chelates that are formed. Room-temperature CW-EPR spectroscopy was used to determine the Cr(V) species formed in the reaction of Cr(VI) with **15–18** or in the reaction of Cr(VI) with GSH in the presence of excess disaccharide over the 5–7 pH range.[70] The EPR spectral features of the disaccharide/Cr(V) mixtures were interpreted in terms of Cr(V)–diolato complex species involving mainly Cr(V) binding through the cis-O^1,O^2-diol groups of **16** and **18** and the cis-O^1, O^2-diol and cis-O$^{3'}$,O$^{4'}$-diol groups of **15** and **17**.

In 0.2 M HClO$_4$, the room-temperature CW-EPR spectra of the reaction between lactose (**15**) and Cr(VI) show three signals at g_{iso} = 1.9797, 1.9743, and 1.9715. The signal at g_{iso} = 1.9715 is attributed to the positively charged, six-coordinated oxo-Cr(V) monochelate [CrO(cis-O,O-lactose)(H$_2$O)$_3$]$^+$ (Scheme 6, complex **I**) and the one at g_{iso} = 1.9743 to the [CrO(O^1,O^2-lactobionato)(H$_2$O)$_3$]$^+$ complex (Scheme 6, complex **II**). The positive charge of these species is consistent with their appearance at high [H$^+$], and coordination to the oxidized ligand agrees with the fact that Cr(VI) is a stronger oxidant in more acidic medium. Only the signal at g_{iso} = 1.9797 persists at higher pH, and it could be deconvoluted into two triplets at $g_{iso,1}$ = 1.9800 and $g_{iso,2}$ = 1.9794 ($^{Cr}a_{iso}$ = 16.5(3) × 10^{-4} cm^{-1} (= 49.5 MHz), ~50:50 $g_{iso,1}/g_{iso,2}$ ratio) (Fig. 7).[70] These EPR spectral parameters are consistent with those expected for any of the three linkage isomers of the [CrO(cis-O,O-lactose)$_2$]$^-$ chelate (complexes **III–V**, Scheme 6). In the reaction of Cr(VI) with maltose (**16**), the EPR spectra consist of two triplets at g_{iso} = 1.9795 and 1.9794, which are attributed to the two geometric isomers of the [CrO(O^1,O^2-maltose)$_2$]$^-$ chelate (complexes **IV**, Scheme 6), with **16** bound to Cr(V) via the 1,2-cis-diol moiety. The room-temperature CW-EPR spectra of **18**/Cr(VI) mixtures consist of one triplet at g_{iso} = 1.9793, which

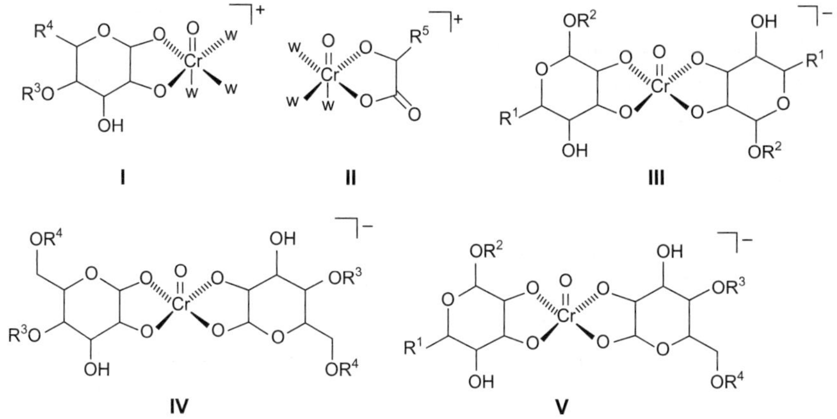

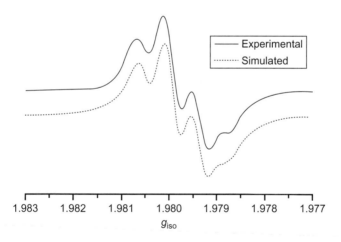

SCHEME 6. Oxo-Cr(V)-mono- and bis-chelates formed in the reaction between Cr(VI) and disaccharides. $R^1 = CH_2OH$. For oxo-Cr(V)-15_2: $R^2 = O^4$-**11**; for oxo-Cr(V)-17_2: $R^2 = O^6$-**11**; for oxo-Cr(V)-15_2 (oxo-Cr(V)-16_2, oxo-Cr(V)-18_2): R^3 = glycoside, R^4 = H; for oxo-Cr(V)-17_2: R^3 = H, R^4 = glycoside. w = water. $R^5 = C_{10}O_9H_{19}$.

FIG. 7. Experimental and simulated room-temperature (20 °C) X-band CW-EPR spectra of a mixtures of Cr(VI) and lactose (**15**), pH 5.0. Spectrum recorded 24 h after mixing, mod. ampl. = 0.04 mT, microwave frequency ≈ 9.7 GHz. Figure is adapted from ref. 70.

corresponds to the [CrO(O^1,O^2-cellobiose)$_2$]$^-$ chelate (Scheme 6, complex **IV**). In the reaction of Cr(VI) with melibiose (**17**), the EPR spectra show two triplets at $g_{iso,1} = 1.9802$ and $g_{iso,2} = 1.9798$ for ∼50:50 $g_{iso,1}/g_{iso,2}$ ratio, corresponding to the different linkage isomers of the five-coordinated oxo-Cr(V) complex [CrO(O,O-melibiose)$_2$]$^-$ chelate (complexes **III–V**, Scheme 6).

Species **III–V** in Scheme 6 are also formed when Cr(V) is generated by reaction of Cr(VI) with GSH (1:1 ratio) at pH 5.0 and stabilized by the presence of a 10-fold excess of the disaccharide over Cr(VI).[70] This result provides evidence that the Cr(V) species observed at this pH correspond effectively to Cr(V) chelates formed with the disaccharide and not with the oxidized product (the aldobionic acid).

The fact that only Cr(V)–disaccharide species are formed in the Cr(VI)/disaccharide reaction at pH 3–7 indicates that the disaccharides are better chelating agents for Cr(V) than are aldoses. The latter are poorer complexation agents than the corresponding aldonic acids so that oxo-Cr(V)–aldose–aldonic acid and oxo-Cr(V)-(aldonic acid)$_2$ bis-chelates are the major species observed in the EPR spectra of the Cr(VI)/aldose reaction mixtures.

19   **20**

The biological relevance of the interaction of Cr(V) with diol groups of furanose sugars was first pointed out in a paper where it was shown that Cr(V) complexation occurred with single ribonucleotide units but not with the analogous deoxyribonucleotides.[25] This suggested that the *cis*-diol group of the ribose component was involved in coordination with Cr(V). Complexes of Cr(V) formed with ribonucleotides and with D-ribose 5′-monophosphate (**20**) have been characterized, and yield EPR spectroscopic parameters ($g_{iso} \approx 1.979$ and $^{Cr}a_{iso} \approx 16.3 \times 10^{-4}$ cm^{-1} (=48.9 MHz)) indicative of Cr(V)–diolato binding.[78] The redox chemistry of the D-ribose(**19**)/Cr(VI) system has also been studied.[79] It was found that the reduction of Cr(VI) by **19** is six times faster than that of aldohexoses having the same configuration at C-2 and C-3. The fact that the furanose ring favors oxidation processes versus the complexation mode was interpreted as resulting from the strain-induced instability of the chromate ester on the redox rate. This effect is enhanced in the reaction between Cr(V) and the furanose derivative and can be evidenced when the observed rate constants for the reduction of Cr(V) (k_5) and Cr(VI) (k_6) by the aldose are compared with those for the pyranose and furanose forms. In the chromic oxidation of aldohexoses (pyranose form) in 0.75 M $HClO_4$, the k_5 values are no more than 10 times higher[68] than k_6, whereas[79] for **19**, $k_5 > 100\ k_6$. The higher $k_{oxidation}/k_{complexation}$ ratio for the **19**/Cr(VI) system means that intramolecular electron-transfer paths proceed faster within the less stable, intermediate hypervalent chromium complexes.

The reaction of Cr(VI) with an excess of D-ribose (**19**) in the pH range 4–6 yields a room-temperature CW-EPR spectrum dominated by a single detectable signal [$g_{iso} = 1.9787$, $^{Cr}a_{iso} = 15.7 \times 10^{-4}\ cm^{-1}$ ($=47.1$ MHz)], corresponding to five-coordinated oxo-Cr(V) complexes. In this pH range, Cr(V) remains in solution for several days. At low modulation amplitude, the hyperfine pattern can be partially resolved in the EPR spectrum, which is composed of contributions from several Cr(V) species. Based on the EPR hyperfine spectral pattern found for the cyclopentanediol/Cr(V) system (Fig. 6), the present spectra could be deconvoluted into three signals having EPR parameters consistent with the presence of [CrO($O^{1(2)},O^{2(3)}$-**19**)(O^1,O^2-ribonic acid)]$^-$ ($g_{iso} = 1.9791$ and 1.9792) and [CrO(O^1,O^2-ribonic acid)$_2$]$^-$ ($g_{iso} = 1.9785$).[79]

In [$HClO_4$] ≥ 0.1 M, the EPR spectra of the reaction of Cr(VI) with an excess of **19** show that six-coordinated oxo-Cr(V) monochelates ($g_{iso} = 1.9711$) are formed as well as the five-coordinated oxo-Cr(V) species at $g_{iso} = 1.9787$, and they are the major Cr(V) species in 0.15–1.0 M $HClO_4$.[79] These two Cr(V) signals decay at different rates, and these rates increase with decreasing pH. Probably, the five- and six-coordinated complexes observed in the **19**/Cr(VI) reaction are formed independently and yield the final Cr(III) by two parallel paths.

21

The kinetics of the reduction of Cr(VI) to Cr(III) by D-fructose (**21**) in $HClO_4$ takes place through a complex multistep mechanism that involves the formation of Cr(V) species.[31] At pH > 1, the redox reaction becomes slow and oxo-Cr(V) species remain in solution for a long time, allowing an analysis of the distribution and structure of these intermediates from the EPR spectral features. In the 3–7 pH range and a **21**:Cr(VI) ratio of 19:1, the CW-EPR spectrum taken at 25 °C is composed of two triplets: $g_{iso,1} = 1.9800$ [$^Ha_{iso} = 0.79 \times 10^{-4}\ cm^{-1}$ ($=2.37$ MHz)], $g_{iso,2} = 1.9797$ [$^Ha_{iso} = 0.71 \times 10^{-4}\ cm^{-1}$ ($=2.13$ MHz)], and one doublet at $g_{iso,3} = 1.9792$ [$^Ha_{iso} = 0.58 \times 10^{-4}\ cm^{-1}$ ($=1.74$ MHz)], each with the four weak ^{53}Cr hyperfine peaks at 1.77 mT spacing, typical of five-coordinated oxo-Cr(V) complexes. The different ^{1}H hyperfine coupling constants of the two EPR triplets indicate that these may stem from the two geometric isomers of the Cr(V)-bischelate [CrO(O$^{2(4)}$, O$^{3(5)}$-β-D-fructopyranose)$_2$]$^-$. The doublet at $g_{iso,3}$ indicates that, in this oxo-Cr (V)–**21** species, only one carbinolic proton is coupled to the electron spin, and hence the spectrum can be assigned to the Cr^V-bischelate [CrO(O^{ring},O^2-β-D-fructofuranose)(O$^{2(4)}$,O$^{3(5)}$-β-D-fructopyranose)]$^-$, with one molecule of **21** acting as bidentate ligand through the diol moiety (pyranose form) and the second **21** ligand binds Cr(V) using the diol group and the O ring atom of the furanose form.

3. Complexes with Methyl Glycosides

At pH 5.5 and 7.5, an EPR signal appears when an excess of methyl α-D-galactopyranoside (**22**) or methyl α-D-mannopyranoside (**23**) is added to a Cys:Cr(VI) mixture (1:1 ratio) (Cys = cysteine), but not upon addition of an excess of methyl α-D-glucopyranoside (**24**) or methyl α-D-xylopyranoside (**25**). The potential binding modes in these cyclic glycosides to afford five-coordinated Cr(V) species are shown in Table III.

The observation of a room-temperature CW-EPR signal only for compounds **22** and **23** provides evidence for the higher ability of the *cis*- versus the *trans*-diol to bind Cr(V). This conclusion was taken as the basis for interpreting the EPR spectral features of **23**/Cr(V) and **22**/Cr(V) mixtures in terms of Cr(V)–diol species involving mainly binding through O^2,O^3-**23** and O^3,O^4-**22**. The contribution to the EPR signals from other species was considered to be small or negligible. A low-intensity EPR

TABLE III
Potential Binding Modes of Methyl Glycosides to Cr(V)

Methyl Glycosides	Binding Modes	
22	cis-3,4-Diol	trans-2,3-Diol
23	cis-2,3-Diol	trans-3,4-Diol
24	trans-2,3-Diol	trans-3,4-Diol
25	trans-2,3-Diol	trans-3,4-Diol

Formation of oxo-Cr(V)–diol complexes.

singlet at $g_{iso} = 1.9792$ is observed at the early stages of the ligand-exchange reactions of [CrO(ehba)$_2$]$^-$ [ehba = 2-ethyl-2-hydroxybutanoate^{2-}] with **24** and **25** in a HEPES buffer (pH 7.5). This corresponds to the features expected[80] for [CrO(ehba)(*trans*-O, O-sugar)]$^-$. With time, this singlet is replaced by another weak singlet at $g_{iso} = 1.9799$ due to [CrO(*trans*-O,O-HexOMe)$_2$]$^-$ with HexOMe = methyl glycoside. The relatively poor ability of the *trans*-diol groups to bind the chromium ion causes the disproportionation of [CrO(ehba)$_2$]$^-$ (which is fast at pH 7.5)[81] to occur much faster than the ligand-exchange reaction. As a result, the [CrO(ehba)$_2$]$^-$ EPR signal quickly disappears and only a very weak [CrO(O,O-HexOMe)$_2$]$^-$ signal is observed in the final spectrum. Similarly, the [CrO(O,O-HexOMe)$_2$]$^-$ EPR signal is not observed for the Cr(VI)/Cys reaction in the presence of an excess of **24** or **25** because the Cys/Cr(V) redox reaction is much faster than the Cr(V)–Cys ligand-exchange reactions with these glycosides.[82] From these results, it can be concluded that the kinetic control favors the formation of Cr(V)–diol species of *cis*- over *trans*-diol moieties of six-membered sugar rings.[80]

In excess of α-**23** or α(β)-**22** at pH 5.5 or 7.5, the Cr(VI)/Cys reaction shows two EPR triplets at $g_{iso,1} = 1.9802$ and $g_{iso,2} = 1.9800$ with a ~50:50 $g_{iso,1}/g_{iso,2}$ ratio and $^{Cr}a_{iso} = 16.5(3) \times 10^{-4}$ cm^{-1} (=49.5 MHz) (Fig. 8A). The g_{iso} and $^{Cr}a_{iso}$ values and the ^{1}H hyperfine splitting are consistent with the presence of the two geometrical isomers of [CrO(*cis*-O,O-gly1Me)$_2$]$^-$ (i.e., complexes **I–II** with **23** and complexes **III–IV** with **22**, Scheme 7). The formation of [Cr(O)(*cis*-O,O-gly1Me)$_2$]$^-$ was found to be faster at pH 5.5 than at pH 7.5, and it was interpreted as the consequence of the faster generation of Cr(V) from the redox reaction between Cys and Cr(VI) at the lower pH value.[82]

Figure 8B and C shows the EPR spectra taken at the beginning of the ligand-exchange reaction of the methyl glycoside **23**. At pH 7.5, the rapid ligand-exchange reaction of [CrO(ehba)$_2$]$^-$ with an excess of α-**23** or α(β)-**22** affords the two geometric isomers of [CrO(*cis*-O,O-HexOMe)$_2$]$^-$ as the final Cr(V) species observed in the

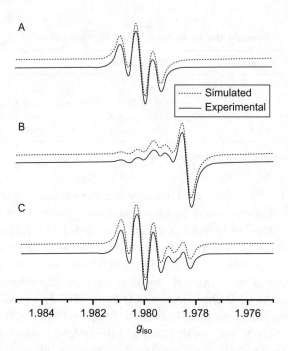

FIG. 8. Experimental and simulated room-temperature (25 °C) X-band CW-EPR spectra of mixtures of (A) Cys:Cr(VI):**23** = 1:1:100, pH 5.5; spectrum recorded after 120 min; (B) [CrO(ehba)$_2$]$^-$:**23** = 1:100, pH 5.5, spectrum recorded after 3.0 min; (C) [CrO(ehba)$_2$]$^-$:**23** = 1:100, pH 7.5, spectrum recorded after 3.0 min. Microwave frequency ≈ 9.67 GHz, mod. ampl. = 0.02 mT. Figure was adapted from ref. 80.

EPR spectra, in approximately 50:50 $g_{iso,1}/g_{iso,2}$ ratio. At pH 5.5, ligand-exchange reactions of [CrO(ehba)$_2$]$^-$ with an excess of α-**23** or α(β)-**22** are not complete, and after 2 h more than 50% of the total Cr(V) is [CrO(ehba)$_2$]$^-$, while oxo-Cr(V)-diolate$_2$ species constitute only approximately 5% of the total Cr(V) amount. This result was attributed, in part, to the lower reactivity of [CrO(ehba)$_2$]$^-$ toward disproportionation and ligand exchange at the lower pH values.

Under more acidic conditions (pH < 1), HexOMe reduces both Cr(VI) and Cr(V). Complexation of Cr(V) takes place prior to the redox steps, but the distribution of Cr(V) species is rather different than at pH 5.5. In the reaction of α-**23** with Cr(VI), at pH < 1 and a 10:1 α-**23**:Cr(VI) ratio, the EPR spectrum is composed[83] of a triplet at $g_{iso,1} = 1.9800$, a doublet at $g_{iso,2} = 1.9798$, and three signals with no resolved ^{1}H hyperfine structure at $g_{iso,3} = 1.9788$, $g_{iso,4} = 1.9746$, and $g_{iso,5} = 1.9716$. The intensity of signals at $g_{iso,4}$ and $g_{iso,5}$ increases relative to those of the $g_{iso,1}$, $g_{iso,2}$, and $g_{iso,3}$

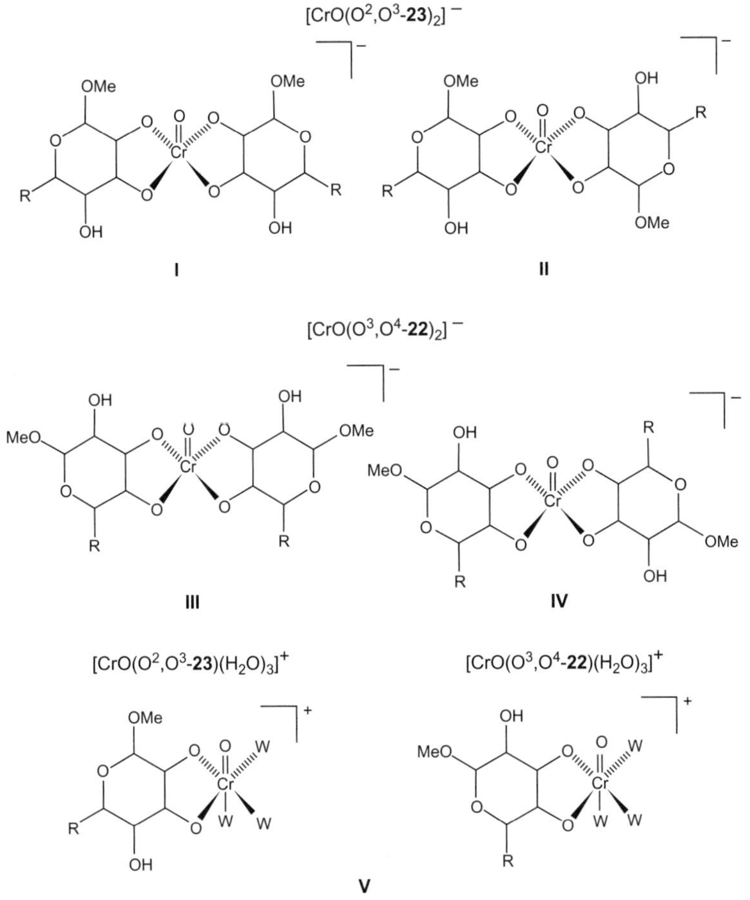

SCHEME 7. Models of oxo-Cr(V)-mono- and bis-chelates formed in the reaction between Cr(VI) and methyl glycosides.

signals with higher [H$^+$]. At [H$^+$] > 0.3 M, the signal at $g_{iso,5}$ is the dominant one and is attributed to six-coordinated monochelates [CrO(O^4,O^6-**23**)(H$_2$O)$_3$]$^+$. Among the six-coordinated complexes, complexes **V** (Scheme 7) are proposed to be the precursors of the subsequent redox steps. The results show that the formation of positively charged Cr(V) monochelates is favored in acidic medium and low glycoside/Cr(VI) ratios, whereas the formation of anionic Cr(V)-bis-chelates is favored at the higher pH and larger glycoside/Cr(VI) ratios.

4. Oxo-Cr(V) Complexes with Other Sugars

26, **27**, **28**

The study of the reaction of 2-deoxyaldoses with Cr(VI) shows that the lack of the 2-OH group favors redox reactions and decreases the capability of the aldose for binding hypervalent Cr.[84,85]

The reaction of 2-deoxy-D-*arabino*-hexose (**26**) and Cr(VI) results in the formation of two oxo-Cr(V) species with $g_{iso} = 1.9781$ and 1.9754 in 2:6 ratio, independently of the pH and concentration of **26**.[85] The two species disappear at comparable rates, meaning that these Cr(V) intermediates are in rapid equilibrium compared to the timescale of their subsequent reduction to Cr(III). The analysis of the EPR parameters indicates that the signals can be assigned to Cr(V)-bis- and monochelates formed with the substrate, namely $[CrO(O^3,O^4\text{-}\mathbf{26})(O^1\text{-}\mathbf{26})(H_2O)]$ and $[CrO(O^1\text{-}\mathbf{26})(H_2O)_3]^{2+}$.[50]

2-Acetamido-2-deoxy-D-glucose (**27**) is oxidized by Cr(VI) to the aldonic acid, and the reaction is proposed to occur by a reaction path involving Cr(V). However, Cr(V) intermediates could not be detected by EPR in the 0.15–1.0 M [H$^+$] range used in the kinetic measurements.[86]

The reaction of Cr(VI) with an excess of 2-deoxy-D-*erythro*-pentose (**28**) in the pH 4–7 range yields only one Cr(V) intermediate species,[79] probably [CrO(**28**)(2-deoxy-D-*erythro*-pentonic acid)]$^-$, with $g_{iso} = 1.9791$ and $^{Cr}a_{iso} = 16.6 \times 10^{-4}$ cm^{-1} (=49.8 MHz). At pH 2, a minor signal with $g_{iso} = 1.9735$ appears, and when [H$^+$] > 0.1 M a third signal at $g_{iso} = 1.9707$ is observed.[79] The g_{iso} values of the two signals at higher field correspond to six-coordinated oxo-Cr(V) complexes. The relative proportions of the three species depend essentially on the acidity. The two (or three) signals decay at comparable rates, and these rates increase, as observed for **26**, with decreasing pH.

29

Using a modulation amplitude of 0.4 mT, two signals could be discerned in the room-temperature CW-EPR spectra of Cr(VI)/3-*O*-methyl-D-glucopyranose (**29**) mixtures in 0.3–0.5 M $HClO_4$: a dominant contribution centered at $g_{iso,1} = 1.9788$ and an additional weak signal at $g_{iso,2} = 1.9712$ with relative intensity < 10% of the main signal.[87] Because of the high modulation amplitude used to observe the oxo-Cr(V) species at this pH, the ^1H hyperfine pattern could not be resolved. Given that the isotropic g values of Cr(V) species are very sensitive to the coordination number and the nature of the donor groups bound to Cr(V),[19,35,50] an estimation of the intermediate Cr(V) species formed in the reaction of Cr(VI) with **29** was made based on g_{iso} values of the EPR signals. The $g_{iso,1}$ signal is typical of five-coordinated oxochromate(V) complexes formed with O-donor ligands, while the $g_{iso,2}$ species is typical of a six-coordinated oxo-Cr(V) complex with two alkoxide donor sites and three water molecules; it can be assigned to a $[CrO(O^1,O^2\text{-}\mathbf{29})(H_2O)_3]^+$ (complex **I**, Scheme 8).[19,35,50] The positive charge of this species is also consistent with its appearance at high $[H^+]$, and it is possibly responsible for the higher rates observed for the Cr(V)–**29** redox reactions at pH < 1.

At pH > 3, oxo-Cr(V) species remain in solution for longer periods of time, and the ^1H hyperfine pattern of the corresponding EPR signals can be resolved at room temperature.[87] Oxo-Cr(V)–**29** complexes formed upon addition of a 100-times molar excess of **29** to Cr(V), generated in either the one-electron reduction of Cr(VI) with glutathione or the ligand-exchange reaction with $[CrO(ehba)_2]^-$, were investigated in the 5.5–7.5 pH range. In this pH range, the CW-EPR spectra taken at the beginning of the reaction were dominated by a triplet at $g_{iso,1} = 1.9792$

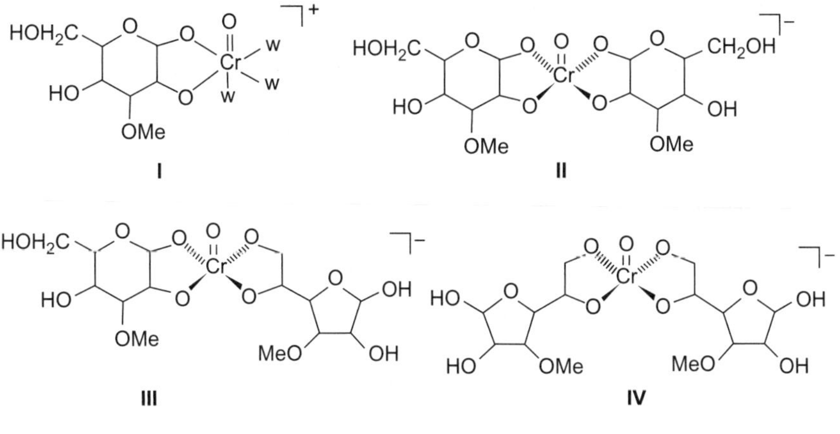

SCHEME 8. Models of oxo-Cr(V) complexes formed with **29**.

[$^H a_{iso,1} = 1.0 \times 10^{-4}$ cm^{-1} ($= 3.0$ MHz)], with a minor component at $g_{iso,2} = 1.9794$ [quintet, $^H a_{iso,2} = 0.9 \times 10^{-4}$ cm^{-1} ($= 2.7$ MHz)]. Fifteen minutes after mixing, the two components at $g_{iso,1}$ and $g_{iso,2}$ were present in 71% and 29% proportions, respectively. The major component of the EPR signal can be attributed to the bis-chelate [CrO(cis-O^1,O^2-29)$_2$]$^-$ (complex **II**, Scheme 8). This result is in line with the reported higher ability of the cis-diol versus the trans-diol for binding to Cr(V).[19,50,80] The minor component at $g_{iso,2} = 1.9794$ can be attributed to the bis-chelate formed with the pyranose and furanose forms of **29**: [CrO(cis-O^1,O^2-**29**)(O^5,O^6-**29**-furanose)]$^-$ (complex **III**, Scheme 8), with four protons coupled to the electronic spin of Cr(V).

With time, the CW-EPR spectral pattern slowly changes; the proportion of the components varies and a third component appears. The CW-EPR spectrum taken 1 h after mixing can be deconvoluted into the earlier-mentioned triplet and quintet, and into a septuplet at $g_{iso,3} = 1.9793$ [$^H a_{iso,3} = 1.0 \times 10^{-4}$ cm^{-1} ($= 3.0$ MHz)], in 6:3:1 ratio. Other minor species also appear, but they were not included in the simulations to avoid over-parameterization.[87] Based on the ^{1}H hyperfine coupling, the three components are suggested to correspond to [CrO(cis-O^1,O^2-**29**)$_2$]$^-$ (complex **II**, Scheme 8), [CrO(cis-O^1,O^2-**29**)(O^5,O^6-**29**-furanose)]$^-$ (complex **III**, Scheme 8), and [CrO(O^5,O^6-**29**-furanose)$_2$]$^-$ (complex **IV**, Scheme 8). At longer reaction times, the proportion of the species at $g_{iso,1}$ decreased and the spectra were dominated by species at $g_{iso,2}$ and $g_{iso,3}$. These five-coordinated Cr(V)-bischelates were still observed 15 h after mixing. Thus, initially, the Cr(V)-bischelate is formed, with Cr(V) bound to the 1,2-cis-diol component of the pyranose form of **29** (kinetic control). However, with time, it transforms into bis-chelates with Cr(V) bound to the 5,6-vic-diol group of the furanose form of the ligand.

5. Complexes with Hydroxy Acids: Alduronic and Glycaric Acids

a. 1-Hydroxy Acids.—Oxo-Cr(V) complexes with such tert-2-hydroxy acids as ehba [2-ethyl-2-hydroxybutanoate^{2-}, **30**] and hmba [2-hydroxy-2-methylbutanoate^{2-}, **31**] are readily isolated and are stable for years in the solid state (low humidity and dark conditions). The relatively stable oxo-Cr(V)–2-hydroxycarboxylato complexes Na[CrO(ehba)$_2$] (**32**)[88] and K[CrO(hmba)$_2$] (**33**)[34] have been isolated, and their structures have been determined by X-ray diffraction. These complexes were the first studied as model compounds for biological activity of Cr (V) complexes.[34,88–91] They have distorted trigonal bipyramidal coordination geometry, with carboxylate and alkoxide donors occupying the axial and equatorials sites, the remaining equatorial site being occupied by an oxo ligand.

The ^{1}H CW-ENDOR results are in agreement with those of X-ray diffraction.[56] These air-stable complexes have half-lives ranking from minutes to days in protic solvents (depending on the pH value and temperature), and about 1 year in such aprotic solvents as dimethyl sulfoxide.[88] In aqueous solutions, they decompose within hours at pH 3–5, but their stability can be enhanced by using ehba or hmba buffers as solvents.[81] EPR and ^{1}H ENDOR studies of these complexes revealed the presence of the *cis* and *trans* isomeric forms of the bischelates complexes, both exhibiting a distorted five-coordinated geometry.[56] These complexes, especially **32**, are useful for ligand-exchange reactions.[87]

Quinic acid [Qa, **34**, (1S,3R,4S,5R)-1,3,4,5-tetrahydroxycyclohexanecarboxylic acid] forms complexes with Cr(V), and the complex K[CrO(Qa)$_2$]·H$_2$O has been isolated and characterized. Analysis of the XAFS data shows a coordination geometry analogous to that of Na[CrO(ehba)$_2$] and K[CrO(hmba)$_2$].[33] Because of the presence of different functional groups, intramolecular competition between 1,2-diol and 2-hydroxy acid for coordination to Cr(V) has been studied as a function of pH in oxo-Cr(V) complexes of **34**. In aqueous solutions at pH values <4.0, K[Cr(O)(Qa)$_2$]·H$_2$O gives two EPR signals [$g_{iso,1}$ = 1.9787, $^{Cr}a_{iso,1}$ = 17.2 × 10^{-4} cm^{-1} (=51.6 MHz); $g_{iso,2}$ = 1.9791, $^{Cr}a_{iso,2}$ = 16.4 × 10^{-4} cm^{-1} (=49.2 MHz)], which are characteristic of 2-hydroxycarboxylato coordination seen with ehba and hmba. These EPR signals are assigned to two geometrical isomers of the [CrO(O^1,O^7-**34**)$_2$]$^-$ linkage isomer (complex **I** in Scheme 9).[63] At pH 5.0–7.5, **34** is able to coordinate Cr(V) by the 2-hydroxy acid and/or the diol moieties. Hence, three additional signals are observed with EPR at g_{iso} 1.9791, 1.9794, and 1.9799, possibly indicating the presence of [CrO(O^1,O^7-**34**)(O^3,O^4-**34**)]$^-$ and [CrO(O^1,O^7-Qa)(O^4,O^5-**34**)]$^-$. Coordination of oxo-Cr(V)

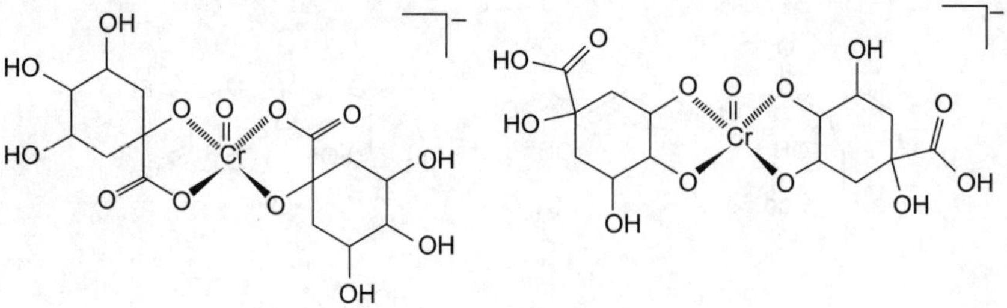

SCHEME 9. Models of the oxo-Cr(V) complexes formed in Cr(VI)/**34**.

by **34** at pH > 7.5 takes place only through the diol groups, as can be deduced from EPR spectra, which show two triplets with $g_{iso} = 1.9800$ and 1.9802. The spectroscopic data for the oxo-Cr(V)–**34** complex seem to suggest the presence of the geometric isomers of the bis-diol oxo-Cr(V)–**34** complex, $[CrO(O^3,O^4\text{-}\mathbf{34})_2]^-$ (complexes **I** and **II** in Scheme 9).

Acid derivatives of saccharides constitute relevant bifunctional ligands, well suited for stabilization of five-membered oxo-Cr(V) chelates, since they provide both 2-hydroxy-1-carboxylate and *vic*-diol sites for potential chelation.[19,50,92]

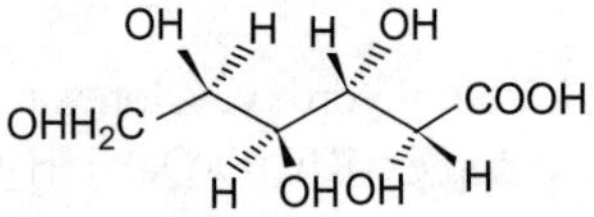

b. Aldonic acids.—The reaction of Cr(VI) with an excess of D-galactonic acid (**35**) affords according to the pH several intermediate $[CrO(\mathbf{35})_2]^-$ linkage isomers.[93] At pH < 2, compound **35** coordinates oxo-Cr(V) through the 2-hydroxy acid moieties, but at higher pH, the oxo-Cr(V)–diolate binding mode is favored over the Cr(V)–2-hydroxy acid coordination mode, such as that observed for the oxo-Cr(V)–**34** system. The spectral parameters are summarized in Table IV.

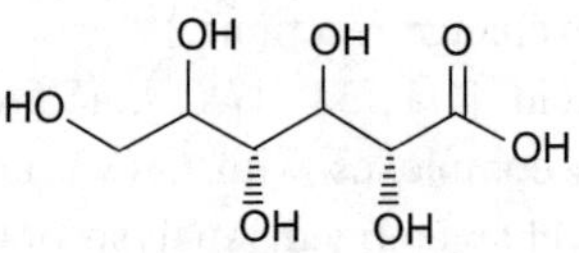

The g_{iso} values obtained for the two signals are those expected for five-coordinated oxochromate(V) complexes. At pH 2–6, the ^{1}H hyperfine splitting of the signal at $g_{iso,1}$ arises from the coupling of two magnetically equivalent protons and may

TABLE IV
Isotropic g and ^{1}H Hyperfine Values of Oxo-Cr(V)/35 Mixtures (m = Multiplicity; t = Triplet; dt = Double Triplet; q = Quintet)

	$g_{iso,1}$ (m)	$^H a_{iso}$	$g_{iso,2}$ (m)	$^H a_{iso}$
pH 2–6	1.9781 (t)	2.10 (0.70)	1.9788 (dt)	2.10, 2.58 (0.70, 0.86)
pH 6–8	1.9795 (q)	2.66 (0.88)	1.978 (dt)	1.96, 2.52 (0.65, 0.84)

The hyperfine values are given in megahertz (10^{-4} cm^{-1}).[94]

be attributed to a complex having two ligand molecules bound to oxo-Cr(V) by their 2-hydroxy acid sites. The ^{1}H hyperfine splitting of the signal at $g_{iso,2}$ may be assumed to arise from two different binding modes of the two ligand molecules, coordinated to Cr(V) via either 1,2-diol or the 2-hydroxycarboxylate donor sites, respectively. In the pH 6–8 range, the CW-EPR spectra are dominated by a signal at g_{iso} = 1.9795 with a splitting pattern typical of a hyperfine coupling to four equivalent protons, which suggests that Cr(V) coordinates with two 35 molecules via the 1,2-diol group. The minor signal ($g_{iso,2}$ = 1.978) can be attributed to the "mixed" isomer having one ligand bound via the 1,2-diol groups and one with the 2-hydroxycarboxylate donor sites.

D-Gluconic acid (36) is another example of a bifunctional acyclic saccharide, and it presents two potential coordination groups: 2-hydroxy-1-carboxylate and *vic*-OH groups. New results indicate that, at pH 3, coordination takes place through the 2-hydroxy-1-carboxylate group to yield three bis-chelate five-coordinated oxochromate(V) complexes. The experimental spectra and their simulations are shown in Fig. 9. Table V summarizes the spectroscopic parameters.

The EPR simulation parameters indicate that the triplets ($g_{iso,1}$ = 1.9783 and $g_{iso,2}$ = 1.9781) can be associated with two bischelated, geometric isomers having two molecules of 36 acting as bidentate ligands through their 2-hydroxy-1-carboxylate component (Scheme 10, complexes **I** and **II**). The quartet with $g_{iso,3}$ = 1.9789 may be related to a mixed coordination species, with one molecule of 36 acting as a bidentate ligand using the 2-hydroxy-1-carboxylate group and the other coordinating via the *vic*-diol sites (Scheme 10, complexes **III–V**). Complementary pulsed EPR/ENDOR studies are being performed to verify these assignments.

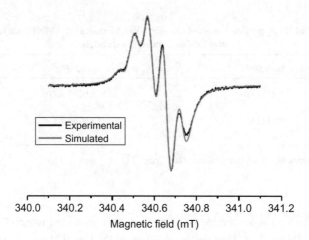

FIG. 9. Experimental and simulated room-temperature X-band CW-EPR spectra of mixtures of oxo-Cr(V)/**36**.

TABLE V
Isotropic g and Hyperfine Values Found for the Oxo-Cr(V)/36 Species Obtained Through Spectral Simulation at pH 3.0 (t = Triplet, q = Quartet)

Multiplicity	g_{iso}	$^H a_{iso}$	$^{Cr} a_{iso}$	LW (mT)	%
t	1.9783	2.13 (0.71)	50.4 (16.8)	0.044	65.8
t	1.9781	1.82 (0.60)		0.036	21.3
q	1.9789	1.91 (0.64)		0.034	12.9

The hyperfine values are given in megahertz (10^{-4} cm^{-1}).[94]

c. Uronic acids.—In a recent study, the oxo-Cr(V) species generated upon reduction of Cr(IV) by GSH in the presence of D-galacturonic acid (**37**) were investigated at acidic pH.[52] A combination of low-temperature CW-EPR, pulsed ENDOR and DFT

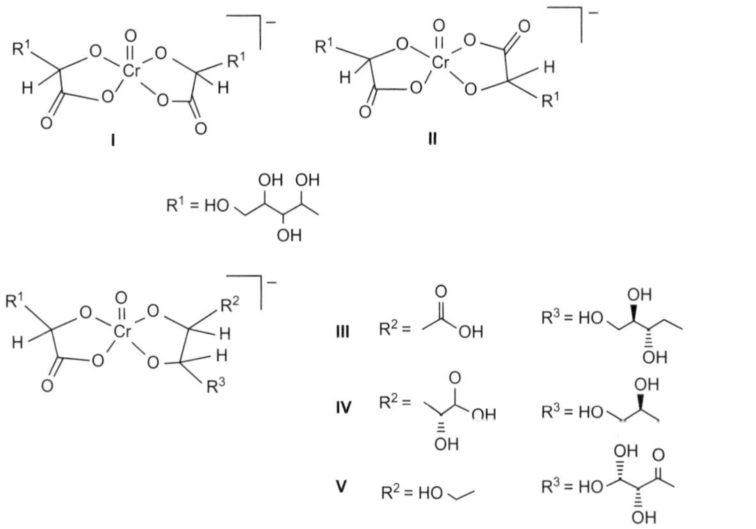

SCHEME 10. Models of oxo-Cr(V) complexes formed with **36**.

was used. Furthermore, experiments were performed in both H_2O and ^2H_2O media, in order to distinguish the contributions of the C—H protons and the exchangeable protons to the spectra. In an earlier study,[37] some of us had interpreted the room-temperature CW-EPR spectrum in terms of three contributions, corresponding to the complexes **I–III** shown in Scheme 11. However, only two contributions could be distinguished at low temperatures (Table VI).[52] Complementary DFT computations confirmed that the experimentally observed set of principal g, ^{53}Cr and ^{1}H hyperfine values matched complexes I and II but argued against the presence of complex **III** in the mixture.

When a large excess of D-glucuronic acid (**38**) over Cr(VI) was used ($\geq$ 2000:1), either in strongly acidic conditions (0.2–0.4 M $HClO_4$) or in the pH 1–4 range, the room-temperature EPR spectra were composed of two triplets at $g_{iso,1} = 1.9783$ and $g_{iso,2} = 1.9782$ and two doublets centered[95] at $g_{iso,3} = 1.9754$ and $g_{iso,4} = 1.9763$. The presence of the weak ^{53}Cr hyperfine peaks with $^{Cr}a_{iso} = 17.2(1) \times 10^{-4}$ cm^{-1} ($=51.5$ MHz) demonstrated that each of these components originated from a chromium-containing complex. At room temperature and pH 1–4, the reaction of Cr(VI) with GSH (1:1 ratio) in the presence of a 2000-times molar excess of **38** leads to oxo-Cr(V) EPR spectra identical to those obtained by direct reaction of Cr(VI). In contrast, under similar conditions in HEPES buffer (pH 7.5), different spectra signals for

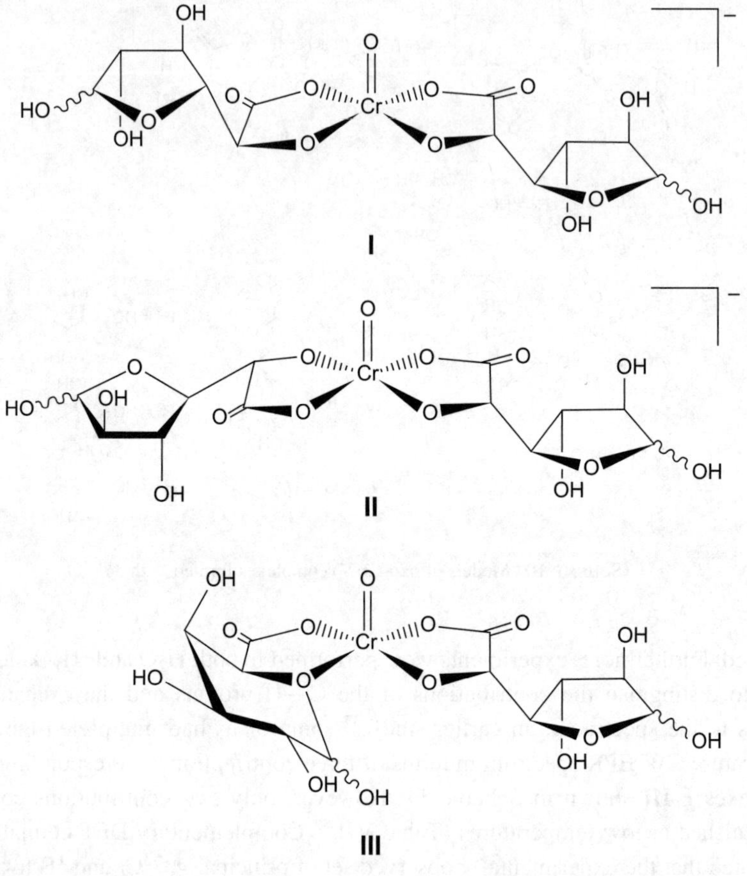

SCHEME 11. Models of oxo-Cr(V) complexes formed with **37** as proposed from room-temperature CW-EPR spectra. It was possible to confirm assignment of only the complexes **I** and **II** with low-temperature CW and pulsed EPR/ENDOR and DFT.[52]

TABLE VI
Experimental Principal g and ^{53}Cr Hyperfine Values Derived from the Low-Temperature EPR Spectra of the Cr(VI)–37 mixture at pH 3

	g_x	g_y	g_z	g_{iso}	$^{Cr}a_x$	$^{Cr}a_y$	$^{Cr}a_z$	$^{Cr}a_{iso}$
Species 1	1.9853	1.9784	1.9726	1.9788	22	23	108	51
Species 2	1.9855	1.9791	1.9718	1.9788	22	23	108	51

The data are taken from ref. 52. The hyperfine values are given in megahertz.

Cr(V)–**38** were observed. This spectrum could be fitted with two triplets at $g_{iso,5} = 1.9798$ and $g_{iso,6} = 1.9796$. In the **38**/Cr(VI) reactions at pH 1–4, the ultimate fate of the chromium was a Cr(III) species, as evidenced by the appearance of the typical broad EPR feature centered at $g \sim 1.98$ during the last stages of the reaction.[95] The ^{1}H hyperfine splitting pattern of the signals having $g_{iso,1}$ and $g_{iso,2}$ is that usually observed when two equivalent carbinolic protons are coupled to the unpaired electron. These couplings seem to suggest that two **38** ligands bind via the 2-hydroxy-1-carboxylate donor sites. A [CrO(O^6,O^5-glucofuranuronate)$_2$]$^-$ complex could match this spectrum. The formation of such an oxo-Cr(V)-bischelate is only possible with **38** bound in the furanose form to the chromium center. However, in the pH range 2–8, **38** exists in aqueous solution as an equilibrium between the pyranose and furanose tautomers, in about 9:1 ratio, with the β anomer being the most stable one.[96,97] Thus, the less-stable furanose form is the one favored in the complexation reaction. The different isotropic ^{1}H hyperfine couplings and g values of the two triplets may correspond to the two geometric isomers of the Cr(V)-bischelate ([CrO(O^5,O^6-glucofuranuronate)$_2$]$^-$) (complexes **I** and **II** in Scheme 12).

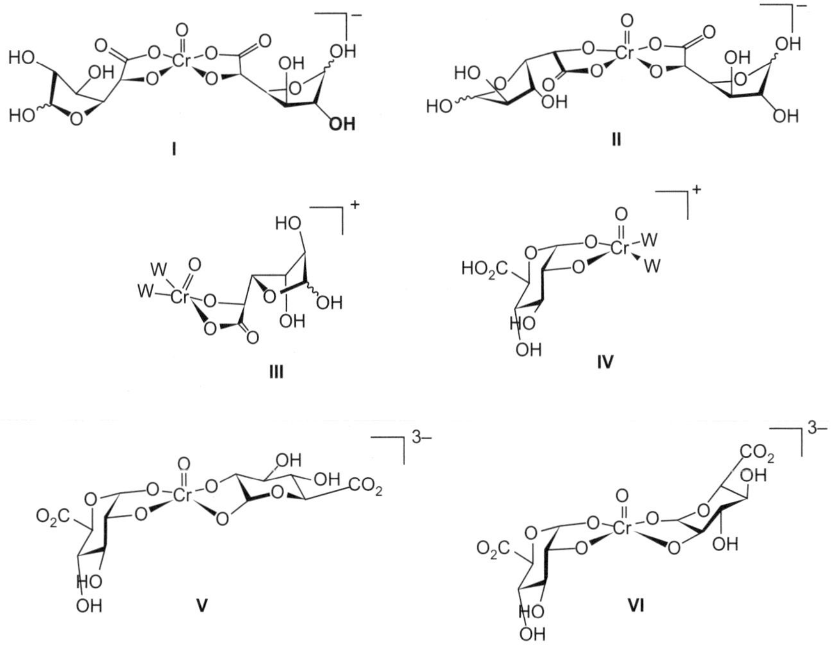

SCHEME 12. Possible models of oxo-Cr(V) complexes formed with **38** in pH 1–4 (**I–IV**) and pH 7.5 (**V** and **VI**).

The doublet signals for the species having $g_{iso,3}=1.9754$ and $g_{iso,4}=1.9763$ stem from the interaction with only one proton with significant s-spin density, and this may point to a monochelate of the form $[CrO(OH_2)_2(O^6,O^5\text{-glucopyranuronate})]^+$ or $[CrO(OH_2)_2(O^1,O^2\text{-glucopyranuronato})]^+$ (Scheme 12, complexes **III** and **IV**, respectively), although it is not clear whether the water protons have a negligible isotropic ^1H hyperfine coupling. Indeed, the ^1H nuclei of the equatorial water molecules in vanadyl pentaaquo complexes are known to have an isotropic proton coupling that depends strongly on the dihedral angle in the O=V—O—H moiety, with isotropic hyperfine couplings up to ~9 MHz.[98,99] Since the oxo-Cr(V) unit is isoelectronic with the oxo-V(IV) unit, a similar behavior may be expected for the water protons in models **III** and **IV** (Scheme 12). It is clear that further analyses involving isotope labeling, pulsed EPR, and even quantum-chemical computations are needed to substantiate these models.

At pH 7.5, the CW-EPR spectra are composed of two triplets with $g_{iso,5}=1.9798$ and $g_{iso,6}=1.9798$. These parameters are indicative of *cis*-1,2-diol binding and may point to the presence of the bischelates **V** and **VI** (Scheme 12). These results thus indicate that **38** can coordinate Cr(V) in different ways depending on the pH: at pH < 4 through the 2-hydroxy-1-carboxylate and at pH 7.5 using *cis*-1,2-diol groups. Several molecules possessing both the *vic*-diol (or cyclic *cis*-diol) and 2-hydroxy-1-carboxylate groups yield exclusively the oxo-Cr(V)-diolate$_2$ complexes at pH 7.5, whereas the oxo-Cr(V)-2-hydroxy-1-carboxylate$_2$ chelates are the major species formed[97] at pH ≤ 5. Similar behavior is also observed with the earlier-mentioned Cr (V)/**34** system, where at pH 7.5, Cr(V) binding to the cyclic-*cis*-diol group is favored.[63]

6. Complexes with Oligosaccharides and Polysaccharides

In addition to polyunsaturated fatty acids, cell walls and membranes also have significant concentrations of glycoproteins, with the glycosylated moieties consisting predominantly of glucose, galactose, and fucose residues. In many cases, the nonreducing ends of the glycoproteins are terminated by sialic acid (*N*-acetylneuraminic, Neu5Ac) groups, which are ketosidically linked to the penultimate sugar residue through the 2-position. Sialoglycoproteins are ubiquitous in animals and humans and occur at the cell surface of the majority of cell classes, where they play crucial roles in cell regulation.[100] Sialic acid has special importance in glycoproteins, since it is a sugar residue that has a negative charge, and it plays an important role in modulating cellular recognition.[100]

Reactions of free sialic acid or sialoglycoproteins of human saliva with Cr(VI) lead to the formation of stable Cr(V) species. The glycerol-like tail at C-6 is exposed to the extracellular environment in cell-bound sialic acid and is the component most likely responsible for the stabilization of Cr(V) by sialoglycoproteins. The Cr(V)–saliva species are formed in the absence of an exogenous reductant and are stable for 2 days at ambient conditions (pH 6–7).[101] Chromium(VI) causes significant DNA damage in human nasal or gastric mucosal cells, which are rich in glycoproteins.[3] Furthermore, human orosomucoid (α_1-acid glycoprotein) is cleaved by a model Cr(V) complex[102] or by Cr(III)/oxidant systems, which are capable of producing Cr(V) intermediates.[103] The epidermal mucus of fish contains glycoproteins that serve for structural and protective purposes and also contains related glycoproteins, such as the protein-bound deaminated sialic acid analogue, 3-deoxy-D-*glycero*-D-*galacto*-2-nonulosonic acid (Kdo).[104] Human skin in its normal and diseased (such as basal cell carcinoma) states also contains significant amounts of sialic acids,[105] which may be of significance with respect to dermal contact with chromates. Further *in vivo* experiments using sialic acid-bound biological matrices are currently being conducted to assess the nature of the Cr(VI)-induced damage to cell-bound sialic acid residues. Since sialic acid residues are involved in the control of fundamental cell processes,[106] any oxidative or alternative damage to these residues might have deleterious consequences on cell behavior. Normal human bronchial epithelial cells generate Cr(V) from both Na_2CrO_4 (soluble) and $ZnCrO_4$ (insoluble). Plants and grazing cattle on Cr-contaminated sites form important links in the process whereby chromium gets into the human food cycle.[13] Evidence has been presented that the observed long-living Cr(V)-EPR signals in mucous cells and those of the respiratory tract are due to complexes stabilized by a combination of extracellular (sialoglycoprotein) and intracellular (D-glucose) carbohydrate species.[20,64] The uptake-reduction model of Cr(VI)-induced genotoxicity was shown to involve Cr(VI) entry into the cells via anionic transport channels and subsequent intracellular reduction.[107]

In the room-temperature CW-EPR spectra of the oxo-Cr(V) species resulting from the interaction of Cr(VI) with the $(1 \to 4)$-α-linked D-galacturonic acid disaccharide **39** at pH 1–3, the isotropic ^{1}H hyperfine coupling could not be resolved.[108] Hence, the spectra were simulated assuming three singlets. At pH 7.0, the EPR signals were better resolved, allowing an evaluation of the different isotropic ^{1}H values and their multiplicity (Table VII).

The $^{Cr}a_{iso}$ value of the oxo-Cr(V)/**39** species was found[108] to be 17.3×10^{-4} cm^{-1} ($=51.9$ MHz), characteristic of oxo-Cr(V) pentacoordinate species.[35] The signals with g_{iso} 1.9780 or 1.9784 may be attributed to bischelates, where **39** binds an oxo-Cr(V) through the carboxylate group and the oxygen atom of the pyranose ring (Scheme 13, complexes **I** and **II**). The possibility of coordination through the

TABLE VII
Isotropic g and ^{1}H Hyperfine Values of Oxo-Cr(V) Intermediates of Cr(VI)/**39** Mixtures
(m = Multiplicity; s = Singlet; t = Triplet)

pH	g_{iso}	m	$^Ha_{iso}$	%
1	1.9788/1.9784/1.9780	s	–	64.3/26.7/9.0
2	1.9788/1.9784/1.9780	s	–	72.2/20.9/6.9
3	1.9788/1.9784/1.9780	s	–	71.0/20.3/8.7
4	1.9800/1.9798/1.9796	t/t/t	2.52/1.53/2.46 (0.84/0.51/0.82)	17.0/22.8/60.2

The relative contribution of each species to the EPR spectrum is also given. The hyperfine values are given in megahertz (10^{-4} cm^{-1}). The data are reproduced from ref. 108.

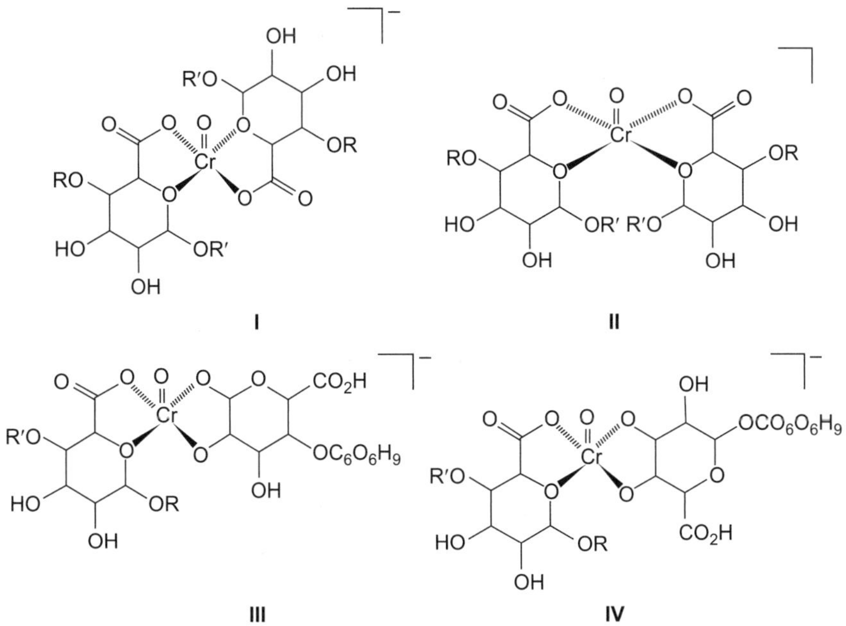

SCHEME 13. Possible models of oxo-Cr(V) complexes formed with **39** at pH 1–3. R=H and R′ = $C_6O_6H_9$ (coordination through the reducing residue) or R = $C_6O_6H_9$ and R′ = H (coordination through the nonreducing residue).

β-hydroxy acid fragment of the pyranose form was rejected because it implies formation of a six-membered chelate ring, which it is not a favorable situation. The signal at g_{iso} = 1.9788 may stem from a bischelate pentacoordinate oxo-Cr(V) species, bound to the carboxylate group and the ring oxygen atom (Scheme 13, complexes **III** and **IV**).

At pH 7.0, the **39**/Cr(VI) reaction mixture leads to detectable oxo-Cr(V) complexes giving room-temperature EPR parameters that may match a coordination of the chromium center to a diol group, leading to the structures suggested in Scheme 14.[108] Note that these assignments should be verified with more detailed EPR and DFT studies.

As expected, oxo-Cr(V) species were also detected in mixtures of Cr(VI) and the D-galacturonic acid trisaccharide **40**.[108] (Table VIII). A clear pH dependence was found.

At pH 2.0 and 3.0, the dominant EPR signal (Table VIII) was assigned to a bischelate of an oxo-Cr(V) pentacoordinate, where the metal is bound by two molecules of **40** through carboxylate groups and a ring oxygen atom [CrO(O^{CO2},

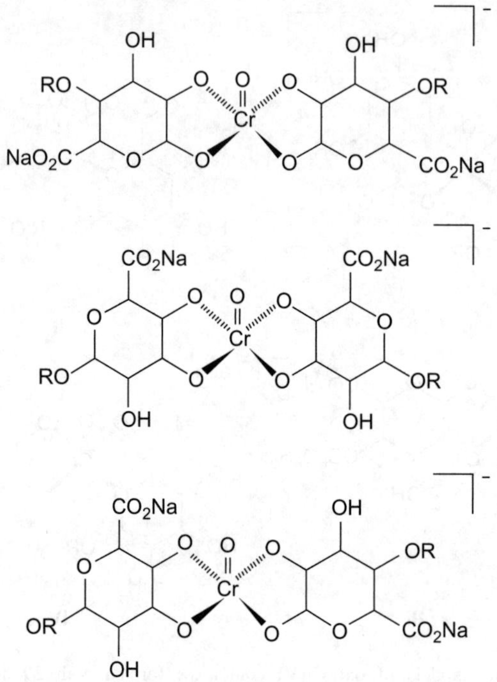

SCHEME 14. Possible models of oxo-Cr(V) complexes formed with **39** at pH 7.

TABLE VIII
Isotropic g and ^1H Hyperfine Values of Oxo-Cr(V) Intermediates of Cr(VI)/**40** Mixtures
(m = Multiplicity; s = Singlet; t = Triplet)

pH	g_{iso}	m	$^H a_{iso}$	%
2	1.9786/1.9790/1.9793	s/s/s	–/–/–	73.8/25.0/1.2
3	1.9788/1.9784/1.9780	s/s/t	–/–/2.34 (–/–/0.78)	68.2/29.0/2.7
4	1.9786/1.9796/1.9800	s/t/t	–/2.2/2.0 (–/0.73/0.67)	61.1/31.6/7.4
7	1.9800/1.9796	t/t	3.1/2.8 (1.02/0.94)	17.0/22.8/60.2

The relative contribution of each species to the EPR spectrum is also given. The hyperfine values are given in megahertz (10^{-4} cm^{-1}). The data are reproduced from ref. 108.

O^{ring}-**40**)$_2$]$^-$. The other minority species observed at pH 2.0 may be linked to the formation of [CrO(O^{CO2},O^{ring}-**40**)(*cis*-O-O-**40**)]$^-$. At pH 3.0, the new triplet signal can match the formation of [CrO(*cis*-O,O-**40**)$_2$]$^-$ species. As the pH increases, the relative contribution of the triplet compounds rises. They were assigned to [CrO(*cis*-O,O-**40**)$_2$]$^-$, with the difference in the EPR parameters (Table VIII) being explained in terms of different geometric conformers, although this proposal remains to be checked by more advanced EPR techniques. Nevertheless, the oxo-Cr(V) species distribution as a function of pH follows the same tendency observed in other uronic acids and polyhydroxocarboxylic compounds.[95]

Room-temperature EPR spectra of mixtures of D-galacturonan ("polyGalA") and Cr(VI) showed more-complex and less-resolved signals than in the former di- and trisaccharide systems.[108] It is clear that the chelating sites and the flexibility of the polymer chain allow the formation of several oxo-Cr(V) species, complicating the EPR analysis at the different pH values. In general, the signals were very wide at every concentration of protons studied, which indicated the contribution of several Cr(V) isomers. However, 20 h after mixing galacturonan with Cr(VI) at pH 2–3, the EPR spectra were composed of only three signals (Fig. 10).

The EPR signals observed at pH 2–3 could be interpreted in terms of oxo-Cr(V) complexes with coordination of two hydroxyl and two carboxylate groups, of a

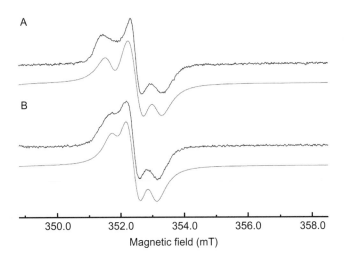

FIG. 10. Experimental (top) and simulated (bottom) room-temperature X-band CW-EPR spectra of mixtures of galacturonan and Cr(VI) taken after 20 h at (a) pH 2 and (b) pH 3. The spectrum is adapted from ref. 108.

$[CrO(O^{CO2},O\text{-polyGal})(H_2O)_2]^+$ species, and of a species where the oxo-Cr(V) group is coordinated with four hydroxyl groups. The contribution of the former species was found to increase strongly with time. The contribution of the latter species amounts to 20% after 20 h of reaction, in contrast to the results for the oligosaccharides in this pH range.

When galacturonan is mixed with Cr(VI) and GSH at pH 7.0, a clear evolution of the EPR spectrum with time can be observed (Fig. 11). From this figure, it becomes clear that Cr(V) produced by a redox reaction between Cr(VI) and GSH is initially coordinated by GSH ($g_{iso} = 1.9857$). In a second, slower step, the GSH ligand is exchanged by the acidic polymer ($g_{iso} = 1.9796$). The latter isotropic g value suggests a coordination of the chromium ion in the latter complex by the hydroxyl groups of the ligand.

IV. Conclusions

All of the afore-cited studies in which Cr(V) is shown to form complexes with different biologically relevant ligands, such as monosaccharides, glycosides, other carbohydrates, carbohydrate derivatives, nitrogen- and/or sulfur-containing molecules, and sialoglycoproteins, demonstrate the necessity of obtaining a clear identification of the structures and chemical properties of these systems. EPR can play an essential role in this context, as has become clear from the abundant examples given in this article. Although room-temperature CW-EPR has been used extensively as a fast technique for "fingerprinting," the advantages of the more advanced CW and

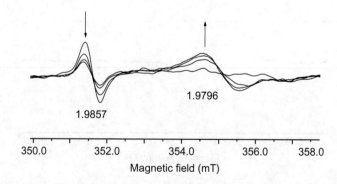

FIG. 11. Evolution of the experimental room-temperature X-band CW-EPR spectra of mixtures of galacturonan, Cr(VI), and GSH taken at pH 7. The spectrum is adapted from ref. 108.

pulsed EPR/ENDOR techniques are currently largely underexploited in this field. Combinations of these methods with state-of-the-art quantum-chemical computations should further increase these possibilities. We are confident that such an approach will give a new impetus to this research field.

ACKNOWLEDGMENTS

L. F. S. acknowledges the Nacional Research Council of Argentina (CONICET), and the Nacional University at Rosario (UNR) for financial support of the research on oxo-chromium (V)–sugar complexes, and the Santa Fe Province Programs of Promotion to the Scientific and Technological Activities for biosorption studies of waste water containing hypervalent chromium. L. F. S. and S. V. D. acknowledge the bilateral agreement MINCYT (Argentina)–FWO (Belgium) program for scientific research. S. V. D. and J. C. G. acknowledge the Erasmus Mundus external cooperation program.

REFERENCES

1. D. Egelman, Corporate corruption of science—The case of chromium(VI), *Int. J. Occup. Environ. Health*, 12 (2006) 169–176.
2. H.-W. Kuo, J.-S. Lai, and T.-L. Lin, Nasal septum lesions and lung function in workers exposed to chromic acid in electroplating factories, *Int. Arch. Occup. Environ. Health*, 70 (1997) 272–276.
3. J. Blasiak, A. Trzeciak, E. Malecka-Panas, Drzewoski J-, T. Iwanienko, I. Szumiel, and M. Wojewodzka, DNA damage and repair in human lymphocytes and gastric mucosa cells exposed to chromium and curcumin, *Teratog. Carcinog. Mutagen.*, 19 (1999) 19–31.
4. B. S. Khangarot and D. M. Tripathi, The stereoscan observations of the skin of catfish, *Saccobranchus fossilis*, following chromium exposure, *J. Environ. Sci. Health*, A227 (1992) 1141–1148.
5. R. H. Liebross and K. E. Wetterhan, *In vivo* formation of chromium(V) in chick embryo liver and red blood cells, *Carcinogenesis*, 13 (1992) 2113–2120.
6. D.-S. Guo, X.-Y. Juan, and J.-B. Wu, Influence of Cr(III) and Cr(VI) on the interaction between sparfloxacin and calf thymus DNA, *J. Inorg. Biochem.*, 101 (2007) 644–648.
7. G. L. Eichhorn and Y. A. Shin, Interaction of metal ions with polynucleotides and related compounds. XII. The relative effect of various metal ions on DNA helicity, *J. Am. Chem. Soc.*, 90 (1968) 7323–7328.
8. H. Arakawa, R. Ahmard, M. Naomi, and H. A. Tajmir-Riahi, A comparative study of calf thymus DNA binding to Cr(III) and Cr(VI) ions, *J. Biol. Chem.*, 274 (2000) 10150–10153.
9. K. Wetterhahn Jennette, Microsomal reduction of the carcinogen chromate produces chromium(V), *J. Am. Chem. Soc.*, 104 (1982) 874–875.
10. A. Levina, R. Codd, and P. A. Lay, Chromium in cancer and dietary supplements, in G. Hanson and L. Berliner, (Eds.) *High Resolution EPR. Applications to Metalloenzymes and Metals in Medicine, Biological Magnetic Resonance*, Vol. 28, Springer, New York, 2009, pp. 551–579.
11. G. R. Borthry, W. E. Antholine, J. M. Myers, and C. R. Myers, Reductive activation of hexavalent chromium by human lung epithelial cells: Generation of Cr(V) and Cr(V)-thiol species, *J. Inorg. Biochem.*, 102 (2008) 1449–1462.

12. S. Boyko and D. M. L. Goodgame, The interaction of soil fulvic and chromium(VI) produces relatively long lived, water soluble chromium(V) species, *Inorg. Chim. Acta*, 123 (1984) 189–191.
13. D. M. Goodgame and A. M. Joy, Formation of chromium(V) during the slow reduction of carcinogenic chromium(VI) by milk and some of its constituents, *Inorg. Chim. Acta*, 135 (1987) L5–L7.
14. G. Micera and A. Dessi, Chromium adsorption by plant roots and formation of long-lived species: An ecological hazard?*J. Inorg. Biochem.*, 34 (1988) 157–166.
15. J. Liu, J. Jiang, X. Shi, H. Gabrys, T. Walezak, and H. M. Swartz, Low frequency EPR study of chromium(V) formation from chromium(VI) in living plants, *Biochem. Biophys. Res. Comm.*, 206 (1995) 829–834.
16. R. Codd, P. A. Lay, N. Ya Tsibakhashvili, T. L. Kalabegishvili, I. G. Murusidze, and H.-Y. N. Holman, Chromium(V) complexes generated in *Arthrobacter oxydans* by simulation analysis of EPR spectra, *J. Inorg. Biochem.*, 100 (2006) 1827–1833.
17. X. Shi, N. S. Dalal, and V. Vallyathan, One-electron reduction of carcinogen chromate by microsomes, mitochondria and *Escherichia coli*: Identification of Cr(V) and •OH radical, *Arch. Biochem. Biophys.*, 290 (1991) 381–386.
18. S. Bellú, S. García, J. C. González, A. M. Atria, L. F. Sala, and S. Signorella, Removal of chromium (VI) and chromium(III) from aqueous solution by grainless stalk of corn, *Sep. Sci. Technol.*, 43 (2008) 3200–3220.
19. R. Codd, C. T. Dillon, A. Levina, and P. A. Lay, Studies on the genotoxicity of chromium from the test tube to the cell, *Coord. Chem. Rev.*, 216/217 (2001) 533–577.
20. R. Codd, J. A. Irwin, and P. A. Lay, Sialoglycoprotein and carbohydrate complexes in chromium toxicity, *Curr. Opin. Chem. Biol.*, 7 (2003) 213–219.
21. M. Mitewa and P. R. Bontchev, Chromium(V) coordination chemistry, *Coord. Chem. Rev.*, 61 (1985) 241–272.
22. E. G. Samuel, K. Srinivasan, and J. K. Kochi, Mechanism of the chromium-catalyzed epoxidation of olefins. Role of oxochromium(V) cations, *J. Am. Chem. Soc.*, 107 (1985) 7606–7617.
23. R. S. Czernuszewicz, V. Mody, A. Czader, M. Galezowski, and D. T. Gryko, Why the chromyl bond is stronger than the perchromy bond in high-valent oxochromium(IV, V) complexes of tris(pentafluorophenyl)corrole, *J. Am. Chem. Soc.*, 131 (2009) 14214–14215.
24. E. Szajna-Fuller, Y. L. Huang, J. L. Rapp, G. Chaka, V. S. Y. Lin, M. Pruski, and A. Bakac, Kinetics of oxidation of an organic amine with a Cr(V) salen complex in homogeneous aqueous solution and on the surface of mesoporous silica, *Dalton Trans.*, 17 (2009) 3237–3246.
25. D. M. L. Goodgame, P. B. Hayman, and D. E. Hathway, Carcinogenic chromium(VI) forms chromium (V) with ribonucleotides but not with deoxy ribonucleotides, *Polyhedron*, 1 (1982) 497–499.
26. H. A. Headlam and P. A. Lay, EPR spectroscopic studies of the reduction of chromium(VI) by methanol in the presence of peptides. Formation of long-lived chromium(V) peptide complexes, *Inorg. Chem.*, 40 (2001) 78–86.
27. A. Levina, L. Zhang, and P. A. Lay, Formation and reactivity of chromium (V). Thiolato complexes: A model for the intracellular reactions of carcinogenic chromium(VI) with biological thiols, *J. Am. Chem. Soc.*, 132 (2010) 8720–8731.
28. S. Gez, R. Luxenhofer, A. Levina, and P. A. Lay, Chromium(V) complexes of hydroxamic acids: Formation, structures and reactivities, *Inorg. Chem.*, 44 (2005) 2934–2943.
29. R. Kapre, K. Ray, I. Sylvestre, T. Weyhermüller, S. DeBeer George, F. Neese, and K. Wieghardt, Molecular and electronic structures of oxo-bis(benzene-1,2-dithiolato)chromate(V) monoanions. A combined experimental and density functional study, *Inorg. Chem.*, 45 (2006) 3499–3509.
30. A. Levina and P. A. Lay, Solution structures of chromium(V) complexes with glutathione and model thiols, *Inorg. Chem.*, 43 (2004) 324–335.

31. S. García, L. Ciullo, M. S. Olivera, J. C. González, S. Bellú, A. Rockenbauer, L. Korecz, and L. F. Sala, Kinetics and mechanism of the reduction of chromium(VI) by D-fructose, *Polyhedron*, 25 (2006) 1483–1490.
32. C. Milsmann, A. Levina, H. H. Harris, G. J. Foran, P. Turner, and P. A. Lay, Charge distribution in chromium and vanadium catecholato complexes: X-ray absorption spectroscopic and computational studies, *Inorg. Chem.*, 45 (2006) 4743–4754.
33. R. Codd, A. Levina, L. Zhang, T. W. Hambley, and P. A. Lay, Characterization and X-ray absorption spectroscopic studies of bis(quinato(2-)]oxochromate(V), *Inorg. Chem.*, 39 (2000) 990–997.
34. M. Krumpolc, B. G. DeBoer, and J. Roček, A stable chromium(V) compound. Synthesis, properties and crystal structure of potassium bis(2-hydroxy-2-methylbutyrato)-oxochromate(V) monohydrate, *J. Am. Chem. Soc.*, 100 (1978) 145–153.
35. G. Barr-David, M. Charara, R. Codd, R. P. Farrell, J. A. Irwin, P. A. Lay, R. Brumley, S. Brumby, J. Y. Ji, and G. R. Hanson, EPR characterization of the Cr(V) intermediates in the Cr(VI/V) oxidations of organic substrates and of relevance to Cr-induced cancers, *J. Chem. Soc. Faraday Trans.*, 91 (1995) 1207–1216.
36. P. O'Brian, J. Pratt, F. J. Swanson, P. Thornton, and G. Wang, Chromium(V) containing complex from the reaction of glutathione with chromate, *Inorg. Chim. Acta*, 169 (1990) 265–269.
37. J. C. González, V. Daier, S. I. García, B. A. Goodman, A. M. Atria, L. F. Sala, and S. Signorella, Redox and complexation chemistry of the Cr(VI)/Cr(V)-D-galacturonic acid system, *J. Chem. Soc. Dalton Trans.* (2004) 2288–2296.
38. B. M. Hoffman, ENDOR of metalloenzymes, *Acc. Chem. Res.*, 36 (2003) 522–529.
39. G. Jeschke, Electron paramagnetic resonance: Recent developments and trends, *Curr. Opin. Solid State Mater. Sci.*, 7 (2003) 181–188.
40. P. Kuppusamy and J. L. Zweier, Cardiac applications of EPR imaging, *NMR Biomed.*, 17 (2004) 226–239.
41. D. Goldfarb, High-field ENDOR as a characterization tool for functional sites in microporous materials, *Phys. Chem. Chem. Phys.*, 8 (2006) 2325–2343.
42. S. Van Doorslaer and E. Vinck, The strength of EPR and ENDOR techniques in revealing structure-function relationships in metalloproteins, *Phys. Chem. Chem. Phys.*, 9 (2007) 4620–4638.
43. L. H. Sutcliffe, The design of spin probes for electron magnetic resonance spectroscopy and imaging, *Phys. Med. Biol.*, 43 (1998) 1987–1993.
44. J. P. Klare and H.-J. Steinhoff, Spin labeling EPR, *Photosynth. Res.*, 102 (2009) 377–390.
45. J. A. Weil, J. R. Bolton, and J. E. Wertz, Electron Paramagnetic Resonance, (1994) Wiley, New York.
46. M. Brustolon, and E. Giamello, (Eds.) Electron paramagnetic resonance. A practitioner's toolkit, 1st ed. Wiley, New Jersey, 2009.
47. F. Neese, Prediction of molecular properties and molecular spectroscopy with density functional theory: From fundamental theory to exchange-coupling, *Coord. Chem. Rev.*, 253 (2009) 526–563.
48. K. Möbius, A. Savitsky, A. Schnegg, M. Plato, and M. Fuchs, High-field EPR spectroscopy applied to biological systems: Characterization of molecular switches for electron and ion transfer, *Phys. Chem. Chem. Phys.*, 7 (2005) 19–42.
49. A. Schweiger and G. Jeschke, Principles of Pulse Electron Paramagnetic Resonance, (2001) Oxford University Press, Oxford.
50. S. Signorella, J. C. González, and L. F. Sala, EPR Spectroscopic characterization of Cr(V)-saccharide complexes, *J. Arg. Chem. Soc.*, 90 (2002) 1–19.
51. P. Höfer, A. Grupp, H. Nebenführ, and M. Mehring, Hyperfine sublevel correlation (HYSCORE) spectroscopy—A 2D electron spin resonance investigation of the squaric acid radical, *Chem. Phys. Lett.*, 132 (1986) 279–282.

52. M. F. Mangiameli, J. C. González, S. I. García, M. I. Frascaroli, S. Van Doorslaer, P. Salas, and L. F. Sala, New insights on the mechanism of oxidation of D-galacturonic acid by hypervalent chromium, *Dalton Trans.*, 40 (2011) 7033–7045.
53. H. Fujii, T. Yoshimura, and H. Kamada, ESR studies of oxochromium(V) porphyrin complexes: Electronic structure of the $Cr^V=O$ moiety, *Inorg. Chem.*, 36 (1997) 1122–1127.
54. J. R. Dethlefsen, A. Døssing, and E. D. Hedegård, Electron paramagnetic resonance studies of nitrosyl and thionitrosyl and density functional theory studies of nitride, nitrosyl, thionitrosyl, and selenonitrosyl complexes of chromium, *Inorg. Chem.*, 49 (2010) 8769–8778.
55. M. Branca, G. Micera, U. Segre, and A. Dessi, Structural information on chromium(V) complexes of 1,2-diols in solution, as determined by isotropic and anisotropic 1H ENDOR spectroscopy, *Inorg. Chem.*, 31 (1992) 2404–2408.
56. M. Branca, A. Dessí, G. Micera, and D. Sanna, EPR and ^1H ENDOR of the solution of equilibria of bis(2-ethyl-2-hydroxybutanoato(2-))oxochromate(V) and bis(2-hydroxy-2-methylbutanoato(2-))oxochromate(V), *Inorg. Chem.*, 32 (1993) 578–581.
57. G. R. Borthiry, W. E. Antholine, J. M. Myers, and C. R. Myers, Addition of DNA to CrVI and cytochrome b_5 containing proteoliposomes leads to generation of DNA strand breaks and Cr^{III} complexes, *Chem. Biodivers.*, 5 (2008) 1545–1557.
58. D. A. Cummings, J. McMaster, A. L. Rieger, and P. H. Rieger, EPR spectroscopic and theoretical study of chromium(I) carbonyl phosphine and phosphonite complexes, *Organometallics*, 16 (1997) 4362–4368.
59. J. Telser, L. A. Pardi, J. Krzystek, and L.-C. Brunel, EPR spectra from 'EPR-silent' species: High-field EPR spectroscopy of aqueous chromium(II), *Inorg. Chem.*, 37 (1998) 5769–5775.
60. M. Branca, A. Dessi, H. Kozlowski, G. Micera, and J. Swiatek, Reduction of chromate ions by glutathione tripeptide in the presence of sugar ligands, *J. Inorg. Biochem.*, 39 (1990) 217–226.
61. M. Branca and G. Micera, Reduction of chromium(VI) by D-galacturonic acid and formation of stable chromium(V) intermediates, *Inorg. Chim. Acta*, 153 (1988) 61–65.
62. R. Brambley, J. Y. Li, and P. A. Lay, Ligand-exchange reactions of chromium(V): Characterization of the ligand-exchange equilibria of bis(2-ethyl-2-structure of hydroxybutanoato(2-))oxochromate(V) in aqueous 1,2-ethanediol and solution bis(1,2-ethanediolato(2-))oxochromate(V), *Inorg. Chem.*, 30 (1991) 1557–1564.
63. R. Codd and P. A. Lay, Competition between 1,2-diol and 2-hydroxy acid coordination in Cr(V)-quinic acid complexes: Implications for stabilization of Cr(V) intermediates of relevance to Cr(VI)-induced carcinogenesis, *J. Am. Chem. Soc.*, 121 (1999) 7864–7876.
64. V. Roldán, V. Daier, B. Goodman, M. Santoro, J. C. González, N. Calisto, S. Signorella, and S. Sala, Kinetics and mechanism of the reduction of chromium(VI) and chromium(V) by D-glucitol and D-mannitol, *Helv. Chim. Acta*, 83 (2000) 3211–3228.
65. M. Santoro, E. Caffaratti, J. M. Salas-Peregrin, L. Korecz, A. Rockenbauer, L. Sala, and S. Signorella, Kinetics and mechanism of the chromic oxidation of *myo*-inositol, *Polyhedron*, 26 (2007) 169–177.
66. S. Signorella, V. Daier, M. Santoro, S. García, C. Palopoli, J. C. Gonzalez, L. Korecz, A. Rockenbauer, and L. Sala, The EPR pattern of [CrO(cis-1,2-cyclopentanediolato)$_2$]$^-$ and [CrO(trans-1,2-cyclopentanediolato)$_2$]$^-$, *Eur. J. Inorg. Chem.*, 7 (2001) 1829–1833.
67. R. P. Farrell and P. A. Lay, New Insights into the structure and reactions of chromium (V) complexes. Implications for chromium(VI) and chromium(V) oxidations of organic substrates and the mechanism of chromium-induced cancers, *Comments Inorg. Chem.*, 13 (1992) 133–175.
68. S. Signorella, V. Daier, S. García, R. Cargnello, J. C. González, M. Rizzotto, and L. F. Sala, The relative ability of aldoses and deoxyaldoses to reduce Cr(VI) and Cr(V). A comparative kinetic and mechanistic study, *Carbohydr. Res.*, 316 (1999) 14–25.
69. L. F. Sala, J. C. González, S. I. García, M. I. Frascaroli, and M. F. Mangiameli, Hypervalent chromium oxidation of carbohydrates of biological importance, *Global J. Inorg. Chem.*, 2 (2011) 18–38.

70. V. Roldán, J. C. González, M. Santoro, S. García, N. Casado, S. Olivera, J. C. Boggio, J. M. Salas Peregrin, S. Signorella, and L. F. Sala, Kinetics and mechanism of the oxidation of disaccharides by Cr (VI), *Can. J. Chem.*, 80 (2002) 1676–1686.
71. S. Signorella, C. Palopoli, M. Santoro, S. García, V. Daier, J. C. González, V. Roldán, M. I. Frascaroli, M. Rizzotto, and L. F. Sala, Kinetics and mechanistic studies on the chromic oxidation of carbohydrates. Implications on environmental contamination by chromium(VI) in soils (review), *Res. Trends*, 7 (2001) 197–207.
72. S. Signorella, S. García, M. Rizzotto, A. Levina, P. A. Lay, and L. F. Sala, The EPR pattern of Cr(V) complexes of D-ribose derivatives, *Polyhedron*, 24 (2005) 1079–1085.
73. S. Signorella, R. Lafarga, V. Daier, and L. F. Sala, The reduction of Cr^{VI} to Cr^{III} by the α- and β-anomer of D-glucose in dimethylsulphoxide. A comparative study, *Carbohydr. Res.*, 324 (2000) 12–135.
74. S. Signorella, S. García, and L. Sala, An easy experiment to compare factors affecting the reaction rate of structurally related compounds, *J. Chem. Ed.*, 76 (1999) 405–408.
75. L. Sala, S. Signorella, M. Rizzotto, M. Frascaroli, and F. Gandolfo, Oxidation of D-mannose and L-rhamnose by Cr(VI) in perchloric acid. A comparative study, *Can. J. Chem.*, 70 (1992) 2046–2052.
76. M. Branca, A. Dessi, H. Kozlowski, G. Micera, and M. V. Serra, In vitro interaction of mutagenic chromium(VI) with red blood cells, *FEBS. Lett.*, 257 (1989) 52–54.
77. M. Rizzotto, M. Frascaroli, S. Signorella, and L. Sala, Oxidation of D-mannose and L-rhamnose by Cr (VI) in aqueous acetic acid medium, *Polyhedron*, 15 (1996) 1517–1523.
78. M. Rizzotto, V. Moreno, S. Signorella, V. Daier, and L. F. Sala, Electron spin resonance and potentiometric studies of the interaction of Cr(VI) and Cr(V) with D-ribose-5-phosphate and nucleotides, *Polyhedron*, 19 (2000) 417–423.
79. V. Daier, S. Signorella, M. Rizzotto, M. Frascaroli, C. Palopoli, C. Brondino, J. Salas Peregrin, and L. Sala, Kinetics and mechanism of the reduction of Cr^{VI} to Cr^{III} by D-ribose and 2-deoxy-D-ribose, *Can. J. Chem.*, 77 (1999) 57–64.
80. M. Rizzotto, A. Levina, M. Santoro, S. García, M. Frascaroli, S. Signorella, L. F. Sala, and P. A. Lay, Redox and ligand-exchange chemistry of chromium(VI/V)-methyl glycoside systems, *J. Chem. Soc. Dalton Trans.* (2002) 3206–3213.
81. A. Levina, P. A. Lay, and N. E. Dixon, Disproportionation and nuclease activity of bis[2-ethyl-2-hydroxybutanoato(2-)]-oxochromate(V) in neutral aqueous solutions, *Inorg. Chem.*, 39 (2000) 385–395.
82. P. A. Lay and A. Levina, Kinetics and mechanism of Cr(VI) reduction to Cr(III) by L-cysteine in neutral aqueous solution, *Inorg. Chem.*, 25 (1996) 7709–7717.
83. S. Signorella, M. I. Frascaroli, S. García, M. Santoro, J. C. González, C. Palopoli, V. Daier, N. Casado, and L. F. Sala, Kinetics and mechanism of the chromium(VI) oxidation of methyl-α-glucopyranoside and methyl-α-D-mannopyranoside, *J. Chem. Soc. Dalton Trans.* (2000) 1617–1623.
84. M. Rizzotto, S. Signorella, M. Frascaroli, V. Daier, and L. Sala, Chromic oxidation of 2-deoxy-D-glucose by Cr(VI). Comparative study, *J. Carbohydr. Chem.*, 14 (1995) 45–51.
85. S. Signorella, M. Rizzotto, M. I. Frascaroli, C. Palopoli, D. Martino, A. Bousecksou, and L. Sala, Comparative study of oxidation by Cr(VI) and Cr(V), *J. Chem. Soc. Dalton Trans.* (1996) 1607–1611.
86. L. Sala, C. Palopoli, and S. Signorella, Oxidation of 2-acetamido-2-deoxy-D-glucose by Cr(VI) in perchloric acid, *Polyhedron*, 14 (1995) 1725–1730.
87. M. I. Frascaroli, J. M. Salas-Peregrin, L. F. Sala, and S. Signorella, Kinetics and mechanism of the chromic oxidation of 3-O-methyl-D-glucopyranose, *Polyhedron*, 28 (2009) 1049–1056.
88. R. J. Judd, T. W. Hambley, and P. A. Lay, Electrochemistry of the quasi-reversible bis[2-ethyl-2-hydroxybutanoato(2-)]oxo-chromate(V) and (IV) and bis[2-hydroxy- methylbutanoato(2-)]oxochromate-(V) and (IV) redox couples and the crystal molecular structure of sodium bis[2-ethyl-2-hydroxybutanoato(2-)]oxochromate (V) sesquihydrate, *J. Chem. Soc. Dalton Trans.* (1989) 2205–2210.

89. K. D. Sugden and K. E. Wetterhahn, Reaction of chromium(V) with the EPR spin traps 5,5-dimethylpyrroline N-oxide and phenyl-*N-tert*-butylnitrone resulting in direct oxidation, *Chem. Res. Toxicol.*, 10 (1997) 1397–1405.
90. A. Levina, G. Barr-David, R. Codd, P. A. Lay, N. E. Dixon, A. Hammershoi, and P. Hendry, *In vitro* plasmid DNA cleavage by chromium(V) and –(IV) 2-hydroxycarboxylato complexes, *Chem. Res. Toxicol.*, 12 (1999) 371–381.
91. M. Krumpolc and J. Roček, Synthesis of stable chromium(V) complexes of tertiary hydroxyl acids, *J. Am. Chem. Soc.*, 101 (1979) 3206–3209.
92. B. Gyurcsik and L. Nagy, Metal complexes of carbohydrates and their derivatives: Coordination equilibrium and structure, *Coord. Chem. Rev.*, 203 (2000) 81–149.
93. S. Signorella, M. Santoro, C. Palopoli, M. Quiroz, J. Salas-Peregrin, and L. Sala, Kinetics and mechanism of the oxidation of D-galactono-1,4-lactone by Cr(VI) and Cr(V), *Polyhedron*, 17 (1998) 2739–2749.
94. J. C. González, Luis F. Sala, Silvia I. García, María I. Frascaroli, and Sabine Van Doorslaer, unpublished results.
95. J. C. González, S. García, S. Bellú, J. M. Salas-Peregrín, A. M. Atria, L. Sala, and S. Signorella, Redox and complexation chemistry of the $Cr^{VI}/Cr^V/Cr^{IV}$-D-glucuronic acid system, *Dalton Trans.*, 39 (2010) 1–14.
96. M. L. D. Ramos, M. M. Caldeira, and V. M. S. Gil, NMR study of uronic acids and their complexation of with molybdenum(VI) and tungsten(VI) oxoions, *Carbohydr. Res.*, 286 (1996) 1–15.
97. M. L. Ramos, M. M. Caldeira, and V. M. S. Gil, Peroxo complexes of sugar acids with oxoions of Mo (VI) and W(VI) as studied by NMR spectroscopy, *J. Chem. Soc. Dalton Trans.* (2000) 2099–2103.
98. N. M. Atherton and J. F. Shackleton, Proton ENDOR of $VO(H_2O)_5^{2+}$ in $Mg(NH_4)_2(SO_4)_2 \cdot 6H_2O$, *Mol. Phys.*, 39 (1980) 1471–1485.
99. S. C. Larsen, DFT calculations of proton hyperfine coupling constants for $[VO(H_2O)_5]^{2+}$: Comparison with proton ENDOR data, *J. Phys. Chem. A*, 105 (2001) 8333–8338.
100. R. Schauer, S. Kelm, G. Reuter, P. Roggentin, and L. Shaw, Biochemistry and role of sialic acids, in A. Rosenberg, (Ed.), *Biology of the Sialic Acids,* Plenum Press, New York, 1995, pp. 7–67.
101. R. Codd and P. A. Lay, Chromium (V)-sialic (neuraminic) acid species are formed from mixtures of chromium(VI) and saliva, *J. Am. Chem. Soc.*, 123 (2001) 11799–11800.
102. H. Y. Shrivastava and B. U. Nair, Protein degradation by peroxide catalyzed by chromium(III): Role of coordinated ligand, *Biochem. Biophys. Res. Commun.*, 270 (2000) 749–754.
103. H. Y. Shrivastava and B. U. Nair, Chromium (III)-mediated structural modification of glycoprotein: Impact of the ligand and the oxidants, *Biochem. Biophys. Res. Commun.*, 285 (2001) 915–920.
104. M. Kimura, Y. Hama, T. Sumi, M. Asakawa, R. B. N. Narasinga, A. P. Horne, S.-C. Li, Y.-T. Li, and H. Nakagawa, Characterization of a deaminated neuraminic acid-containing glycoprotein from the skin mucus of the loach, *Misgurnus anguillicaudatus, J. Biol. Chem.*, 269 (1994) 32138–32143.
105. C. Fahr and R. Schauer, Detection of sialic acids and gangliosides with special reference to 9-*O*-acetylated species in basilomas and normal human skin, *J. Invest. Dermatol.*, 116 (2001) 254–260.
106. S. Ueno, N. Susa, Y. Furukawa, and M. Sugiyama, Formation of paramagnetic chromium in liver of mice treated with dichromate(VI), *Toxicol. Appl. Pharmacol.*, 135 (1995) 165–171.
107. A. Levina and P. A. Lay, Mechanistic studies of relevant to the biological activities of chromium, *Coord. Chem. Rev.*, 249 (2005) 281–298.
108. J. C. González, S. I. García, S. Bellú, A. Atria, J. M. Salas-Peregrin, A. Rockenbauer, L. Korecz, S. Signorella, and L. F. Sala, Oligo and polyuronic acid interactions with hypervalent chromium, *Polyhedron*, 28 (2009) 2719–2729.

SYNTHESIS AND PROPERTIES OF SEPTANOSE CARBOHYDRATES

Jaideep Saha and Mark W. Peczuh

Department of Chemistry, University of Connecticut, Storrs, Connecticut, USA

I. Introduction	122
II. Synthesis of Septanose-Containing Mono-, Di-, and Oligo-Saccharides	127
1. Cyclization via C—O Bond Formation	127
2. Preparation and Reactions of Carbohydrate-Based Oxepines	138
3. Miscellaneous Septanose Syntheses	149
4. Transannular Reactions of Septanose Sugars	152
III. Conformational Analysis of Septanose Monosaccharides	154
1. Observed Low-Energy Conformations	156
2. Determination of Septanose Conformations	161
IV. Biochemical and Biological Investigations Using Septanose Carbohydrates	163
1. Olgonucleotides Containing Septanosyl Nucleotides	164
2. Protein–Septanose Interactions	172
3. Additional Reports	176
V. Outlook	177
Acknowledgments	178
References	178

Abbreviations

Ac, acetyl; AONs, antisense oligonucleotides; B, boat; Bn, benzyl; Bz, benzoyl; C, chair; CD, circular dichroism; CO, carbon monoxide; ConA, concanavalin A; DAST, diethylaminosulfur trifluoride; DFT, density functional theory; DMDO, dimethyldioxirane; DMT, dimethoxytriphenylmethyl; DNA, deoxyribonucleic acid; dsDNA, double-stranded DNA; E, envelope; Fmoc, fluorenylmethyloxycarbonyl; GlcNAc, N-acetylglucosamine; ITC, isothermal titration calorimetry; k_{cat}, catalytic rate constant; K_a, association constant; K_i, inhibition constant; K_M, Michaelis constant; LiSPh, lithium thiophenolate; LPS, lipopolysaccharide; μM, micromolar; MMT, monomethoxytriphenylmethyl; n-BuLi, n-butyllithium; NIS, N-iodosuccinimide; NMR, nuclear magnetic resonance; ONs, oligonucleotides; TsOH, p-toluenesulfonic acid; QM/MM, quantum-mechanics/molecular mechanics; RCM, ring-closing metathesis; RNA, ribonucleic acid; RNaseH, ribonuclease H; S_N1, unimolecular nucleophilic substitution; S_N2, bimolecular nucleophilic substitution; ssRNA, single-stranded RNA; STD, saturation transfer difference; T, twist; TB, twist boat; TBDPS, *tert*-butyldiphenylsilyl; TBS, *tert*-butyldimethylsilyl; TC, twist chair; TfOH, trifluoromethanesulfonic acid; T_m, thermal melting temperature; TMS, trimethylsilyl; trityl, triphenylmethyl; TMSOTf, trimethylsilyl trifluomethanesulfonate; UV, ultraviolet; $W(CO)_6$, tungsten hexacarbonyl

I. INTRODUCTION

The molecules that form the foundation of living systems are often organized into four categories. They are the primary metabolites: nucleic acids, proteins, carbohydrates, and lipids. The categories can be grouped together in different ways, based on features that they have in common. For example, nucleic acids, proteins, and polysaccharides are polymeric. Nucleic acids and proteins are further related because they are templated polymers. Other classification systems are also possible.[1] Interest in the development of size-expanded versions of biomolecules has grown over the past 10–20 years. The most developed examples are of oligonucleotides (ONs) and polypeptides that incorporate unnatural, homologated monomers.[2–4] For example, new DNA helices have been observed when the purine and pyrimidine bases were enlarged by inserting an additional benzene ring into their structures.[5–13] The new ONs showed useful properties, such as more efficient stacking of base pairs due to increased surface area, and fluorescent properties that make them useful for detection of natural ONs. Peptides derived from expanded (β) amino acids have also been prepared. The β-peptide oligomers, or oligopeptides that contain mixtures of α- and β-amino acids, manifest unique structures that are similar to natural α-helices. The helix mimics have been utilized in the disruption of key protein–protein interactions that involve the docking of an α-helix from one protein to a cognate binding domain on the other protein.[14–17] For both expanded nucleic acids and expanded peptides, the new unnatural oligomers adopt structures that are reminiscent of the natural motifs. Furthermore, these new classes of molecules are also able to interact with natural biopolymers in a rational way. Conceptually, extension of this theme to oligo- and poly-saccharides is straightforward. It would entail expansion of the monosaccharide components and their incorporation into oligosaccharides, either with other expanded monomers or mixed with natural monosaccharides. In practice, a number of issues, ranging from synthesis to conformational analysis to molecular recognition, must be investigated before septanose carbohydrates may be deployed as tools for glycobiology or as potential therapeutics. This *Advances in Carbohydrate Chemistry and Biochemistry* article assesses efforts that have been made toward the synthesis, conformational analysis, and molecular-recognition events of septanose sugars. We update recent reviews of septanoses[18,19] and organize the material into themes that may stimulate other researchers to entertain related questions.

An "expanded monomer approach" applied to carbohydrates leads to the septanose sugars. The majority of carbohydrates found in eukaryotic organisms, namely glycolipids, structural polysaccharides, glycoproteins, and other glycoconjugates, have the sugar components mainly in the pyranose ring form. Adding another carbon atom to the pyranose ring homologates it to a septanose, having seven atoms in its ring. It is important to stress the difference between heptose and septanose, both of which are accepted IUPAC terms. Heptoses are simply seven-carbon sugars. An example is L-*glycero*-D-*manno*-heptose (**1**) (also shown in its hemiacetal form, **2**), a component of the lipopolysaccharide (LPS) core structure of Gram negative bacteria. Neuraminic acid (**3**)[20,21] is a nonose, or nine-carbon sugar. While **2** and **3** are examples of higher-order monosaccharides, the ring structure they adopt is still a pyranose. In contrast, those cyclic forms of monosaccharides having a seven-membered ring are termed septanoses.[22] Depending on which hydroxyl group participates in the cyclization of an aldohexose such as D-glucose (**4**), for example, there will result the oxetose **5** (3-hydroxyl), the furanose **6** (4-hydroxyl), the pyranose **7** (5-hydroxyl), or the septanose **8** (6-hydroxyl) (Fig. 1). Aldoheptoses normally adopt the pyranose ring form on account of the stability of the six-membered ring. Estimates of relative energies place furanoses and septanoses approximately 3–5 kcal/mol higher in energy than pyranoses.[19] Septanoses as the free hemiacetal forms have been observed in situations where the 5-position of a hexose or heptose is either deoxygenated or protected. Population of the seven-membered ring structure may be variable and is probably influenced by solvent dependence.[23,24] The septanose ring is maintained, however, when the anomeric carbon atom is locked in a glycosidic linkage (full acetal), which prevents equilibration between ring forms. Upon hydrolytic cleavage of the glycosidic bond, septanosides undergo reequilibration back to the pyranose reducing sugar.[25] Incorporation of septanose monosaccharides as glycosides into a large variety of glycoconjugates is therefore readily feasible.

A distinction must be made between septanosides that lack (6-unsubstituted) or contain (6-substituted) an exocyclic hydroxymethyl group at C-6. In the same way that D-glucose can form a septanose ring by connecting the 6-hydroxyl group to the anomeric center as in Fig. 1, *all* of the common aldohexoses can adopt a septanose

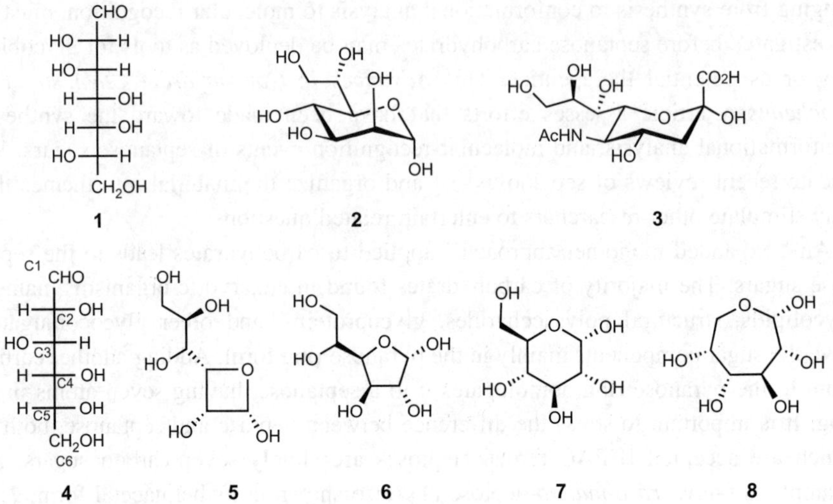

FIG. 1. Higher-order sugars: L-*glycero*-D-*manno*-Heptose (**1**, open-chain and **2**, α-pyranose) and *N*-acetylneuraminic acid (**3**, α-pyranose form). Structures **4–8** depict D-glucose (**4**) as its open-chain Fischer projection, and as the α-oxetose (**5**), α-furanose (**6**), α-pyranose (**7**), and α-septanose (**8**) ring forms.

ring form. Such septanosides thus lack a 5-hydroxymethyl group that is present in a hexopyranoside. Those septanosides that lack an exocyclic hydroxymethyl group may be designated as "6-unsubstituted." The 6-substituted systems, on the other hand, possess a hydroxymethyl group exocyclic to the ring. Considering, for example, the cyclization of D-glucose (**1**) through its 6-hydroxyl group, such substitution at C-6 would appear to be a minor structural difference, but it is important on several grounds, as is presented throughout this article. Perhaps most important of these factors is the potential interactions between septanoside-containing carbohydrates and such natural proteins as lectins and enzymes that act on carbohydrates. An analysis of natural protein–carbohydrate interactions shows that interaction between the exocyclic hydroxymethyl group of glycopyranosides and their protein counterparts is an essential component of the carbohydrate–protein association. If septanosides are to serve as mimics of pyranosides, this exocyclic group may prove to be an important structural feature. On the other hand, if septanosides were to be naturally present in biological systems, they would most probably be of the C-6-unsubstituted variety. This argument is based on the prevalence of the principal aldohexoses in biological systems, namely D-glucose, D-galactose, and D-mannose. Although septanosides in natural biological systems have not yet been detected, their presence is not outside the realm of possibility. One interesting question that has not thus far been addressed concerns the potential immunogenicity of septanosides.

Two research areas that are formally peripheral to septanose carbohydrates but involve themes that overlap with this field merit mention here. First, a number of routes have been described for the synthesis of polyhydroxylated azepanes (such as **9–13**), as shown in Fig. 2. Those reported have relied on strategies ranging from chemoenzymatic syntheses[26] to cyclization via amine alkylation,[27] or C—C bond-forming reactions.[28] These azepanes (**9–13**) were assayed for their glycosidase- and protease-inhibitory activity, a factor that in fact motivated their synthesis. The azepanes incorporate structural features of the natural products nojirimycin (**14**) and fagomine (**15**). These two compounds are imino sugars, a class of potent glycosidase inhibitors; they are surveyed in detail in this volume by Stütz and Wrodnigg.[28a] The ring-nitrogen atom in these molecules, upon protonation, approximates the positive charge of the oxacarbenium ion that is recognized as an intermediate structure in the hydrolysis of a glycoside. The other major design feature of **9–12** is the seven-membered ring, which was considered to be sufficiently flexible to adopt conformations that could imitate the putative oxacarbenium intermediate at the glycosidase active site. A later investigation invoked the role of the exocyclic hydroxymethyl group that is part of the glycopyranose and is part of the interaction motif with the protein.[29,30] Observed K_i values in the micromolar range for the polyhydroxylated azepanes provided support for the validity of the design principles.

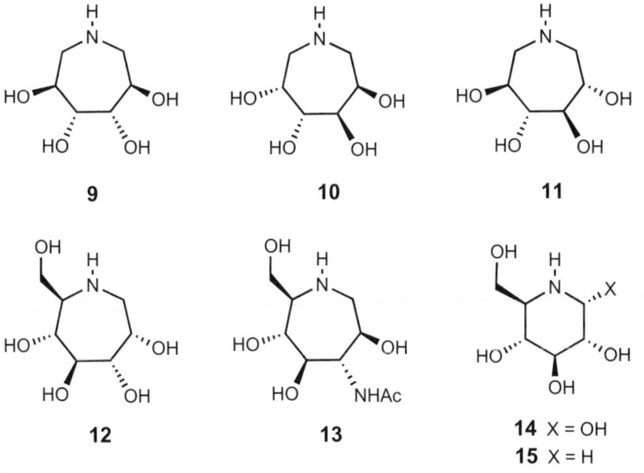

FIG. 2. Polyhydroxylated azepanes **9–13** that have been assayed as glycosidase inhibitors. These structures were inspired by the natural-product glycosidase inhibitors nojirimycin (**14**) and fagomine (**15**).

A recent report on the *N*-acetylglucosaminidase-inhibitory activity of azepane **13** provides a valuable case study. Azepane **13** mimics the functionality and stereochemistry of the the natural GlcNAc substrate. The hydroxyl group at C-2 of **13** mimics the anomeric group that would be present in GlcNAc, the C-3 group the *N*-acetyl group at C-2 on GlcNAc, and so on. The measured K_i value of **13** for the protein *Bt*GH84, a model of the human *N*-acetylglucosaminidase, was 89 μM.[30] The report also included a crystal structure of the complex of **13** bound to *Bt*GH84. The inhibitor **13** adopted a conformation in the complex that was different than its favored conformation when unbound. The authors suggested that the enzyme did not select the bound conformation from a Boltzmann distribution of unbound conformers. Rather, the ligand binds initially and then changes its conformation to optimize interactions with the protein. The argument suggests that the flexibility for azepanes that was presented as a design feature of the polyhydroxyazepanes does, in fact, operate in this manner.

The second area ancillary to septanose carbohydrates deals with the synthesis and biological activity of fused polycyclic marine toxin ethers, such as brevetoxin and ciguatoxin. A correlation between low-energy conformations of oxepanes, specifically chair (*C*) and twist-chair (*T*) conformations, and the overall molecular topology of brevetoxin fragments has been developed.[31] Conformations of the large rings in the fused polyether systems can change the shape of the molecule from linear to "*C*" to "*T*" architectures. The sensitivity of the oxepane conformation to such factors as the size (six-, seven-, or higher-membered) of adjacent rings, and the stereochemistry of their attachment was also explored. Whereas the context of fused-ring systems does not accord with the glycosidic linkages of septanose carbohydrates, the observation of low-energy conformers that depend on local effects is analogous to the conformational analysis of septanose carbohydrates that is to be presented here.

A long list of questions related to septanose carbohydrates remains to be explored. Fundamental questions regarding the chemistry and biological activity of septanosides should provide deeper insights into these aspects. Additionally, questions about how best they might be applied in drug development, tissue engineering, and materials research, and the like, remain wide open. This article aims to orient the reader toward the main ideas in the field of septanose carbohydrates. Presented first are several strategies for the synthesis of septanosides. They are organized according to how the seven-membered ring of the septanose is formed and, where applicable, approaches are then presented within each section for the formation of septanosyl glycosides. Next, efforts to determine low-energy conformations of septanoses are described. Conformational data from X-ray crystallography and solution experiments are discussed. Finally, examples of septanose sugars in biomolecular assays are cataloged. Included in this section are nucleic acids that incorporate septanose nucleosides as well as the binding of various septanosides to lectins and enzymes that act on carbohydrates.

II. SYNTHESIS OF SEPTANOSE-CONTAINING MONO-, DI-, AND OLIGO-SACCHARIDES

Fundamental to investigations of septanose carbohydrates is the ability to prepare them dependably with sufficient efficiency and selectivity. Cataloged here are general strategies employed by various groups for synthesizing stable, isolable septanosides. Septanosides have been synthesized by three major strategies, with subtle variations as demanded by the specific substrates. The first and most widely used strategy is the cyclization of a suitable acyclic precursor via C—O bond formation. Often, under appropriate reaction conditions, the hydroxyl group at C-6 and the aldehydic C-1 of an aldohexose can effect cyclization and form a septanose. The C-6-unsubstituted septanosides are formed most readily by this route when starting from natural aldohexoses. A second method entails the ring opening of cyclopropanated glycopyranoses, which themselves are prepared from glycals. The ring expansion is usually acid or base catalyzed in the presence of an appropriate nucleophile. Oxepines are central to the third strategy. Carbohydrate-based oxepines are seven-membered oxacycles derived from either pyranoses or furanoses, and which can be subsequently functionalized to septanosides via epoxidation and nucleophilic ring opening under mild conditions. The reactivity pattern parallels that of glycals themselves. Each of these strategies is detailed here with illustrative examples.

1. Cyclization via C—O Bond Formation

a. Monosaccharide Septanoses.—Cyclization via the 6-hydroxyl group of aldohexoses has been the most commonly employed strategy for synthesizing septanoses. The route usually involves selective protection of the 6-hydroxyl group of the pyranose sugar, protecting the remaining hydroxyl groups, and then opening the ring to the acyclic form and protecting the aldehyde as an acetal. Deprotection of the 6-hydroxyl group then allows cyclization through nucleophilic attack on the aldehydic carbon atom, giving rise to the septanose. Micheel and Sückfull reported the first synthesis of a galactose-based septanoside by this strategy, as depicted in Scheme 1.[32] 6-Deoxy-6-iodo-1,2;3,4-di-O-isopropylidene-α-galactopyranose (**16**) was first deacetonated, and then the aldehyde group was protected as the diethyl dithioacetal and the remaining hydroxyl groups as acetates to give compound **17** (66%, three steps). Deprotection of the dithioacetal and hydrolytic replacement of the 6-iodo group, using cadmium carbonate and mercuric chloride, gave the aldehydo derivative, which upon acetylation yielded the galactoseptanose **18** as a mixture of anomers in 15% yield over two steps.

Micheel then slightly modified his synthesis, starting from D-galactose diethyl dithioacetal (**19**) (Scheme 2).[33] Selective tritylation of the 6-hydroxyl group followed

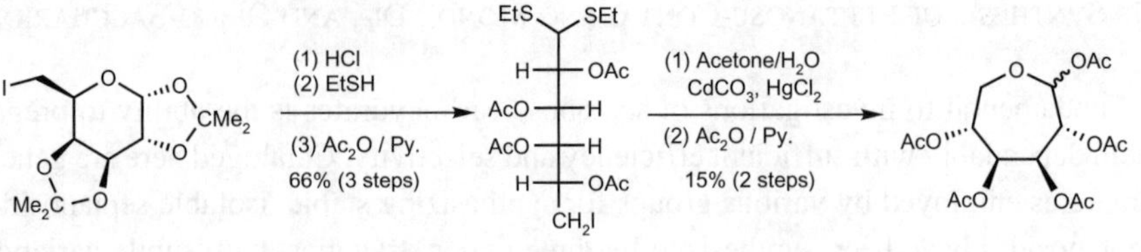

SCHEME 1. Micheel's septanose synthesis.

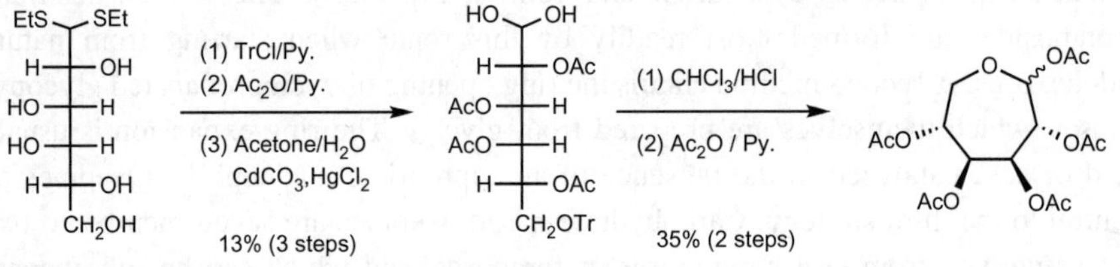

SCHEME 2. Micheel's modified septanose synthesis.

by acetylation of the remaining secondary hydroxyl groups proceeded nearly quantitatively. Demercaptalation afforded the aldehydrol **20** (13%). Removal of the 6-substituent (39%) provided the same intermediate aldehydrol as in the earlier synthesis. Acetylation of this product then gave the septanose **18** as an α:β mixture. The strategy is similar to original one (Scheme 1), except for the difference in the protection mode at C-6. It is noteworthy that product **18** in both syntheses depicted in Schemes 1 and 2 arose by trapping a septanose hemiacetal by acetylation. This suggested that the hemiacetal form is significantly populated in the equilibrium with its acyclic hydroxyaldehyde tautomer.

While investigating the degradation of 2,3,4,5-tetra-*O*-methyl-D-glucose (**23**) (Scheme 3), Anet observed the adventitious formation of the glucoseptanose by NMR spectroscopy; this observation was parlayed into the first stereoselective synthesis of a glucose-derived septanoside.[24] Compound **23** was itself prepared from 6-*O*-benzyl-D-glucose (**21**), which was first converted into the diethyl dithioacetal with subsequent peracetylation, followed by conversion of the dithioacetal into the dimethyl acetal with concomitant deprotection of the acetates to give tetrol **22**. The hydroxyl groups in **22** were methylated to give the tetramethyl ether. Hydrogenolysis of the 6-*O*-benzyl group and hydrolysis of the dimethyl acetal yielded aldehydo

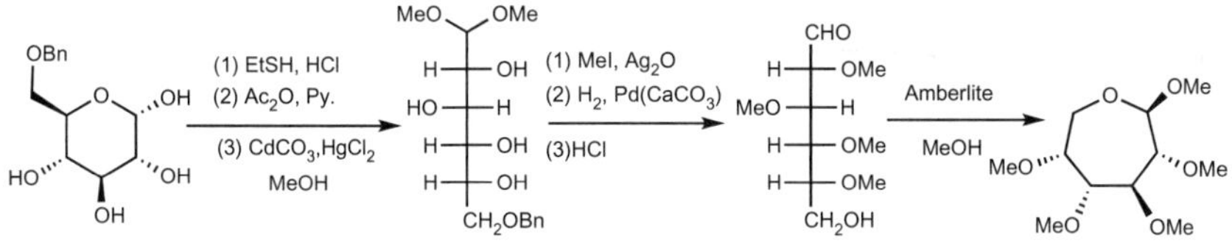

SCHEME 3. Anet's septanoside synthesis.

derivative **23**. The β-septanose (hemiacetal) form of **23** was noted by ^{1}H NMR in both chloroform-d and deuterium oxide solutions (see later). Upon treatment of **23** with methanol in the presence of an acidic ion-exchange resin, the dimethyl acetal was reformed, but not isolated, and this underwent cyclization to give methyl 2,3,4,5-tetra-O-methyl-β-D-glucoseptanoside (**24**). Yields for the synthetic steps in the conversion of **21–24** were not reported in the manuscript. Nonetheless, the synthesis clearly illustrates the C—O bond-forming strategy and it is expected to be relatively efficient, based on the chemoselectivity afforded by the protecting groups used.

Stevens and coworkers have made significant and sustained contributions to the field of septanose synthesis and structure elucidation.[34–41] Their investigations involving septanose sugars span nearly 30 years from the mid-1970s to the 2000s. Syntheses of the septanose form for five of the eight isomeric aldohexoses (all C-6 unsubstituted) were detailed by the group. When considered as an integrated body of work, the efforts by Stevens have established a sound foundation for the entire field in terms of synthesis and conformational analysis of septanoses. Even as Stevens has himself continued to explore seven-membered ring sugars, his work has inspired the development of new methodologies toward the synthesis of septanosides.

The synthetic routes reported by Stevens have largely utilized D-glucose (**7**) as the starting material. Treatment of **7** with acidified acetone, for example, formed 1,2;3,4-di-O-isopropylidene-α-D-glucoseptanose (**26**, 1.9%) and 2,3;4,5-di-O-isopropylidene-D-glucoseptanose (**27**, 1.1%) as minor products isolated by chromatography, along with the major product, the well-known 1,2;5,6-di-O-isopropylidene-α-D-glucofuranose (**25**, > 60%) (Scheme 4).[41,42] As detailed later, selective refunctionalization of **26** and **27** led to L-*ido*-, L-*altro*-, D-*gulo*-, and D-*galacto*-septanoside derivatives.

Even though the yields of the septanose derivatives in these reactions are decidedly modest, the products were isolated in sufficient quantities to permit analysis of their structures and conformations. Importantly, these routes provided enough of the respective septanoses to allow the preparation of alkylated and acylated glycoseptanosyl isomers.

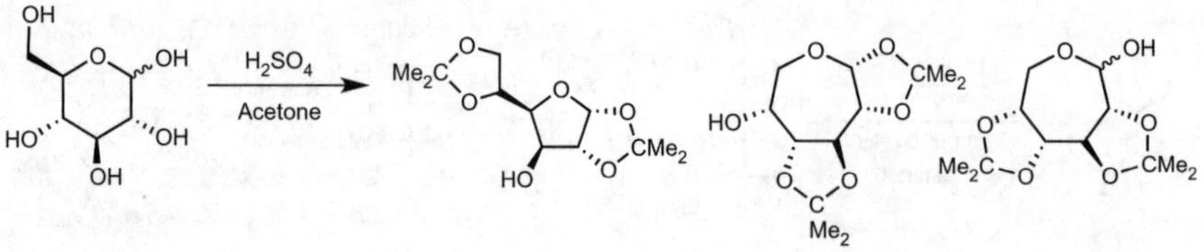

SCHEME 4. Septanose minor products in the acetonation of D-glucose.

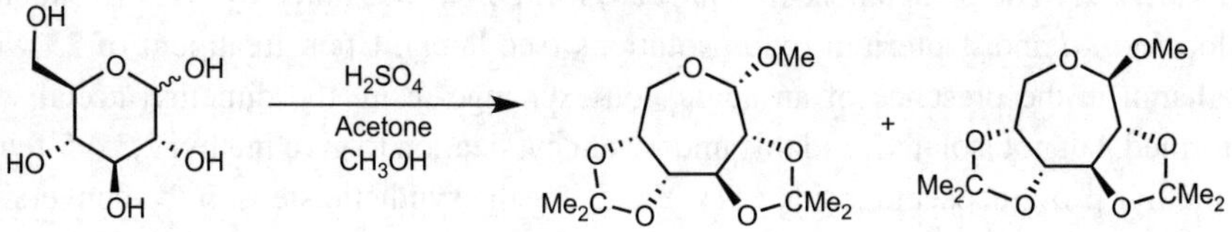

SCHEME 5. Acetonated methyl D-glucoseptanosides from D-glucose.

A similar treatment of D-glucose (**7**) with a mixture of acetone and methanol in the presence of acid led to a mixture of products, among them methyl 2,3:4,5-di-*O*-isopropylidene-α-D-glucoseptanoside (**28**) (4.2%) and its β anomer **29** (2.8%).[41] An alternative route where formation of the dithioacetal, selective C-6 functionalization and cyclization was utilized gave **28** and **29** in 17% and 9% yields (Scheme 5).[35,36,38,43]

Treatment of methyl 2,3:4,5-di-*O*-isopropylidene-β-D-glucoseptanoside (**29**) with dilute HCl gave a mixture of methyl 2,3-*O*-isopropylidene-β-D-glucoseptanoside (**30**) (46%), methyl 3,4-*O*-isopropylidene-β-D-glucoseptanoside (**32**) (12%), methyl 4,5-*O*-isopropylidene-β-D-glucoseptanoside (**33**) (24%), and methyl β-D-glucoseptanoside (**31**) (6.6%) along with unreacted starting material (54%). The yields for the new products were calculated based on recoved starting material. Longer reaction times resulted in a higher yield of the deacetonated product **31**. Hydrolysis of methyl 2,3:4,5-di-*O*-isopropylidene-α-D-glucoseptanoside (**28**) gave a similar distribution of products as did **29**, but in this case, the major product isolated was methyl 4,5-*O*-isopropylidene-α-D-glucoseptanoside (**34**) (45%), the anomer of **33**, which was isolated as the minor product under similar conditions, starting with β anomer **29** (Scheme 6, inset).[35] Products from these syntheses were then used to prepare the corresponding L-*ido*, L-*altro*, D-*gulo*, and D-galactoseptanosides.

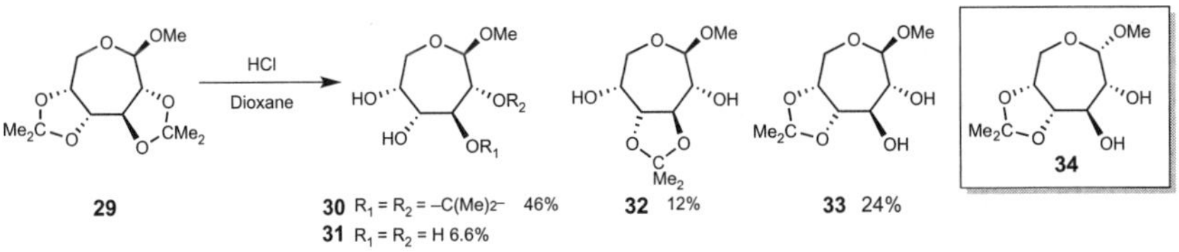

SCHEME 6. Hydrolysis of acetonated glucoseptanosides.

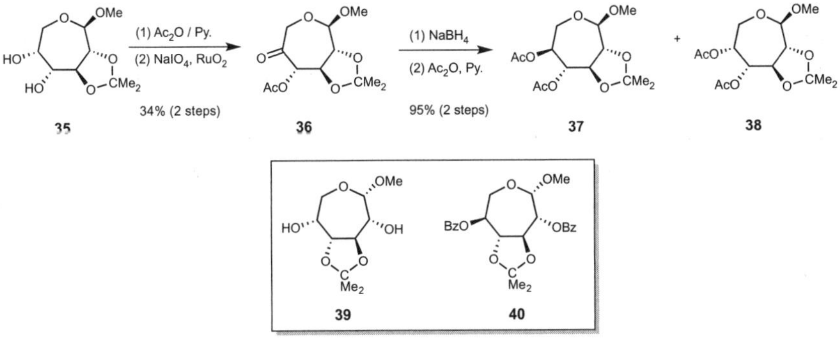

SCHEME 7. Inversion at C-5 for D-*gluco* to L-*ido* conversion.

Epimerization of the C-5 hydroxyl group of D-glucose converts it into L-idose. Methyl 2,3-*O*-isopropylidene-β-D-glucoseptanoside (**35**, Scheme 7), on treatment with one equivalent of acetic anhydride in pyridine selectively gave methyl 4-*O*-acetyl-2,3-*O*-isopropylidene-β-D-glucoseptanoside as the sole acetylated product in 40% yield; the 5-hydroxyl group was left intact and could therefore be functionalized. Oxidation with RuO$_4$ gave ketone **36** (86%), which was then reduced with NaBH$_4$ and acetylated to give the 5-epimers methyl 4,5-di-*O*-acetyl-2,3-*O*-isopropylidene-α-L-idoseptanoside (**37**) and methyl 4,5-di-*O*-acetyl-2,3-*O*-isopropylidene-β-D-glucoseptanoside (**38**) in almost 2:3 ratio (95% overall yield).[36] Methyl 3,4-*O*-isopropylidene-α-D-glucoseptanoside (**39**) was similarly transformed to afford methyl 2,5-di-*O*-benzoyl-3,4-*O*-isopropylidene-β-L-idoseptanoside (**40**). Benzoylation of **39** gave the 3-monobenzoate as the predominant (72%) product, allowing further transformation at the free 5-hydroxyl group. *p*-Toluenesulfonylation on the 5-hydroxy group afforded the corresponding tosylate in nearly quantitative yield, and this was subjected to reaction with lithium benzoate. Displacement of the 5-sulfonyloxy group yielded the 5-inverted L-*ido* septanoside **40** (78% over two steps). Inversion of configuration at C-5 starting from **39** was also performed through a sequential oxidation of the 5-hydroxyl group and a reduction–acetylation step (similar to **35**), as described in the paper.[35]

5-*O*-Benzyl-1,2-*O*-isopropylidene-4-*O*-tosyl-α-D-galactoseptanose (**42**, Scheme 8) proved to be a key intermediate in the synthesis of D-galactoseptanose and D-guloseptanose analogues from D-glucose. Relative to D-glucose, D-galactose is epimeric at C-4, while D-gulose is epimeric at both C-3 and C-4. Synthesis of the galactose derivative commenced with benzyl protection of 1,2;3,4-di-*O*-isopropylidene-α-glucoseptanose (**26**) followed by selective hydrolysis of the 3,4-isopropylidene group, which gave 5-*O*-benzyl-1,2-*O*-isopropylidene-α-D-glucoseptanose (**41**) in 43% yield.[36] Tosylation of **41** gave primarily the 4-*O*-tosyl derivative **42** (49%), along with the C-3 regioisomer and the ditosylate, each in approximately 15% yield. Treatment of **42** with sodium benzoate gave a mixture of 3-*O*-benzoyl and 4-*O*-benzoyl derivatives of 5-*O*-benzyl-1,2-*O*-isopropylidene-α-D-galactoseptanose. Saponification of this mixture gave 5-*O*-benzyl-1,2-*O*-isopropylidene-α-D-galactoseptanose (**43**) in 72% yield over the two steps. 3,4-Anhydro-5-*O*-benzyl-1,2-*O*-isopropylidene-α-D-galactoseptanose (**44**, Scheme 9) arose as a minor product (22%) during the benzoate substitution on **42**. Addition of methoxide under refluxing conditions gave a mixture of 5-*O*-benzyl-1,2-*O*-isopropylidene-4-*O*-methyl-α-D-galactoseptanose (**45**, identified as its 3-*O*-acetyl derivative) (23%) and 5-*O*-benzyl-1,2-*O*-isopropylidene-3-*O*-methyl-α-D-guloseptanose (**46**) (33%).

Using the C-5 epimeriztion strategy used to convert D-glucose into L-idose, but starting with a D-galactoseptanose derivative (such as **47**) delivered the L-altrose isomer **50** (Scheme 10). Treatment of 5-*O*-benzyl-1,2-*O*-isopropylidene-α-D-galactoseptanose

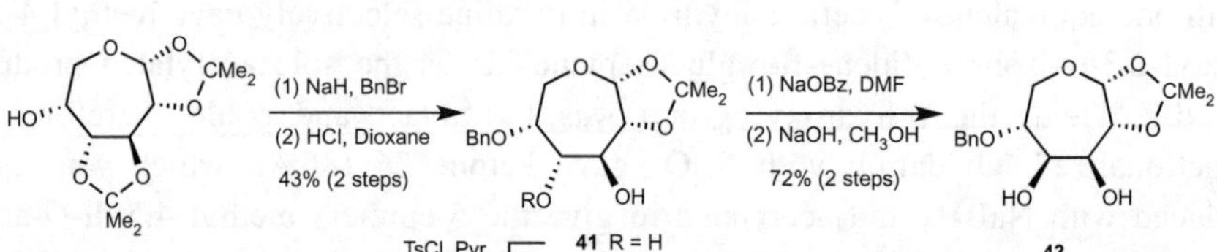

SCHEME 8. D-Glucoseptanose to D-galactoseptanose conversion.

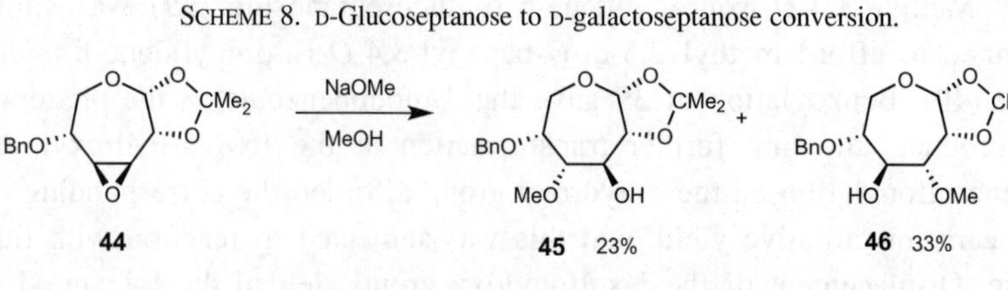

SCHEME 9. D-Guloseptanose derivative.

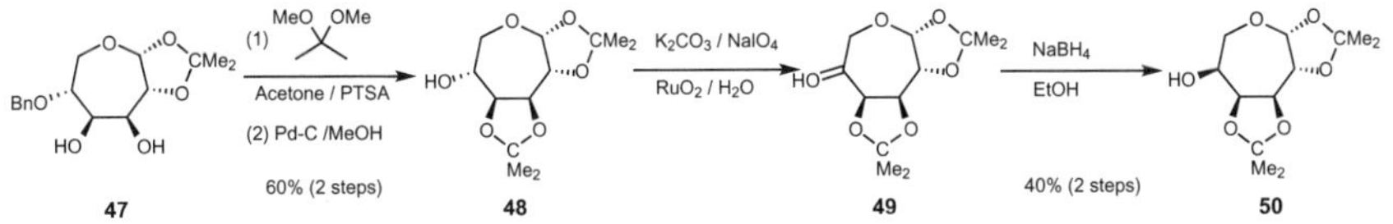

SCHEME 10. L-Altroseptanose derivative.

(**47**) with 2,2-dimethoxypropane and acidified acetone yielded the 1,2:3,4-di-O-isopropylidene derivative. Hydrogenolytic removal of the 5-benzyl group yielded **48** (60% over two steps). Subsequent oxidation of the alcohol provided ketone **49**, which was subjected to sodium borohydride reduction to give a separable mixture of the starting material D-galactoseptanose **48** (29%) and the L-altroseptanose isomer **50** (40%).

b. Strategies Used to Prepare Septanose-Containing Disaccharides.—A route to a variety of novel septanosides reported by McAuliffe and Hindsgaul[44] drew on the preparation of hexofuranosides via cyclization of O,S-acetals that had previously reported by Wolfrom et al.[45] The synthesis of hexofuranosides began with protected hexose dithioacetals, such as the peracetylated D-galactose diethyl dithioacetal **51** (Scheme 11), whose reaction with acetyl chloride and $POCl_3$ transformed it into the 2,3,4,5,6-penta-O-acetyl-1-chloro-1,1-dideoxy-1-ethylthio-D-galactose aldehydrol (**52**). This was subsequently converted in 93% yield over two steps into the O,S-acetal **53** by glycosylation with ethanol using silver carbonate as activator. Deacetylation of **53** followed by treatment with mercuric oxide and mercuric chloride facilitated regio- and stereoselective cyclization to yield ethyl β-D-galactofuranoside (**54**) (87%). The same strategy was used by Wolfrom et al. to synthesize methyl β-D-glucofuranoside.[46,47] Hindsgaul expanded the scope of the method by synthesizing glycopyranosyl furanosides via the same sequence. Glycosylation of a pyranosyl acceptor with 2,3,4,5,6-penta-O-acetyl-1-chloro-1,1-dideoxy-1-ethylthio aldehydrol derivatives of D-glucose, D-galactose, and D-mannose, respectively, was followed by susbsequent cyclization.[4] Selectivity for the furanoside was attributed to the kinetic preference for forming five- rather than six-membered rings. Also, the authors argued that the regioselectivity was due to the highly favorable positioning of the C-4 hydroxyl group with respect to the anomeric carbon in the acyclic precursor.

Having established the novel glycosylation–cyclization sequence, the opportunity to synthesize septanose glycosides through appropriate protecting-group manipulations presented itself. As in the Micheel and Anet examples, selective protection of the 6-hydroxyl group of the precursor to cyclization was required. To test the feasibility of the approach, 2,3,4,5,6-penta-O-acetyl-D-galactose diethyl dithioacetal **51** was converted first into the chloro-ethylthio acetal and subsequently to the

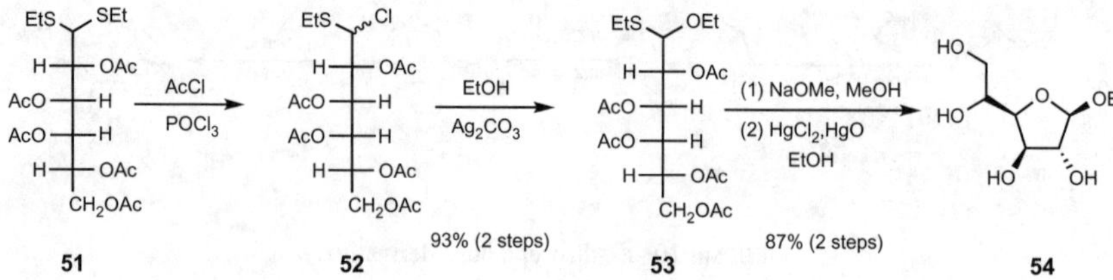

SCHEME 11. D-Glucofuranoside synthesis.

O,S-acetal **55** in 56% yield (Scheme 12). Removal of the acetates set up the selective protection of the 6-hydroxyl group as its TBDPS ether, followed by acetylation of the remaining hydroxyl groups to provide compound **57** in 86% yield over the three steps. The silyl group at C-6 was then removed to unveil the alcohol **58** (70%), setting the stage for the cyclization reaction. Cyclization of **58** was effected by treatment with NIS and TfOH to yield methyl β-D-galactoseptanoside (**59**) as the sole product (85%).[48,49] Diastereoselectivity in the reaction arose from anchimeric assistance by the neighboring acetate group at C-2 and was set in the glycosylation reaction that formed **55**.

The generality of the glycosylation–cyclization strategy was demonstrated by the preparation of a family of glycoseptanosyl glycopyranose derivatives.[48] Steps to the disaccharides mirrored the synthesis of methyl 2,3,4,5-tetra-O-acetyl-β-D-galactoseptanoside (**59**) in Scheme 12. Chloro-ethylthio donors were prepared from D-galactose, D-glucose, and D-mannose diethyl dithioacetals, respectively. Glycosylation of 1,2:3,4-di-O-isopropylidene-α-D-galactose, methyl 2,3,6-tri-O-benzyl-α-D-glucoside, and methyl 2-O-benzyl-4,6-O-benzylidene-α-D-glucoside with these donors gave the protected glycoseptanosyl glycopyranose derivatives **60–62** (Fig. 3). The stereoselectivity was again controlled through anchimeric assistance by the C-2 acetate group at the glycosylation step. Products corresponding to only one anomer were isolated for **60** and **61**, whereas **62** was produced as a 4:1 (α:β) mixture. This was the first preparation of glycoseptanosyl glycopyranose derivatives reported in the literature.

Our group recently reported the synthesis of 2-acetamido-2-deoxyglycoseptanosides from protected D-glucose, D-mannose, D-galactose, and D-xylose pyranosyl starting materials.[50] The route was designed to utilize a 6-hydroxyaldehyde that would be accessed by a one-carbon homologation of an aldopyranose precursor with subsequent transformation into glycosides via a tandem cyclization–glycosylation in the presence of various alcohols. The strategy would employ a late-stage C—O bond formation from acyclic hydroxyaldehydes. Incorporation of the amino functionality at the beginning of the synthesis would provide a new class of ring-expanded

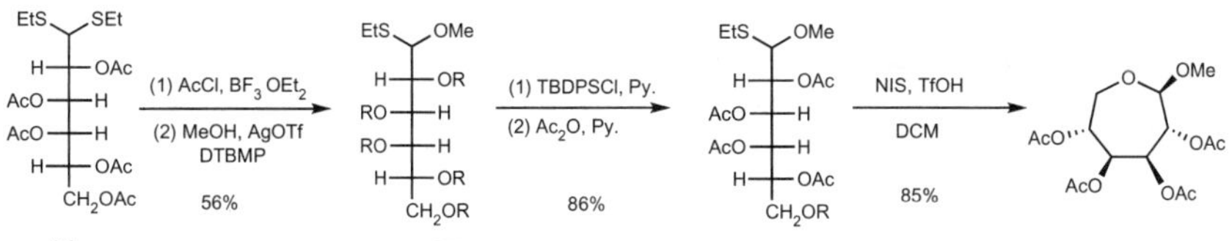

SCHEME 12. Methyl β-D-galactoseptanoside tetraacetate (**59**).

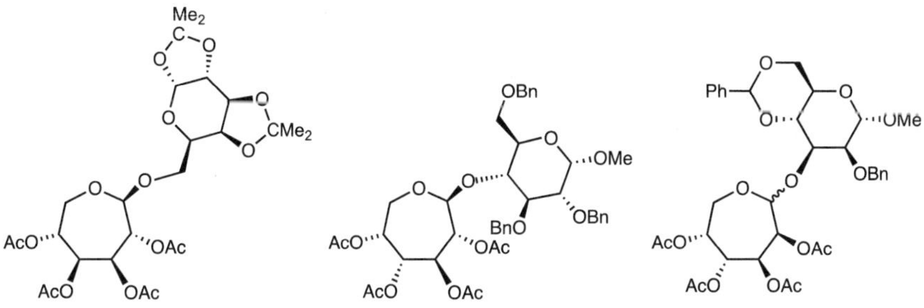

FIG. 3. Septanosyl glycopyranose derivatives **60–62** prepared by the Hindsgaul glycosylation–cyclization route.

2-amino-2-deoxyglycosides. The preparation of methyl 2-acetamido-3,4,5,7-tetra-*O*-acetyl-2-deoxy-α-D-*glycero*-D-idoseptanoside (**68**) is outlined in Scheme 13. The synthesis started with conversion of D-glucose 2,3,4,6-tetrabenzyl ether (**63**) into the corresponding *N*-benzylglycosylamine **64** (74%). Diastereoselective addition of vinylmagnesium bromide followed by Fmoc protection of the allylic amine gave compound **65** (58%). Ozonolysis of the alkene then yielded acyclic aldehyde **66** (91%). This 6-hydroxyaldehyde then underwent cylization in acidic methanol, and subsequent removal of the Fmoc group gave methyl 3,4,5,7-tetra-*O*-benzyl-2-benzylamino-2-deoxy-α-D-*glycero*-D-*ido*-septanoside (**67**) in 68% yield. A two-step hydrogenolysis of the *O*- and *N*-benzyl protecting groups followed by peracetylation (68%) provided the septanoside **68**.

This route differs significantly from the other methods described (which also relied on C—O cyclization) in terms of stereoselectivity, substrate selection, and ease of preparation of the intermediates. For example, most of the 1,6 C—O cyclization strategies for the preparation of seven-membered systems reported earlier utilized a

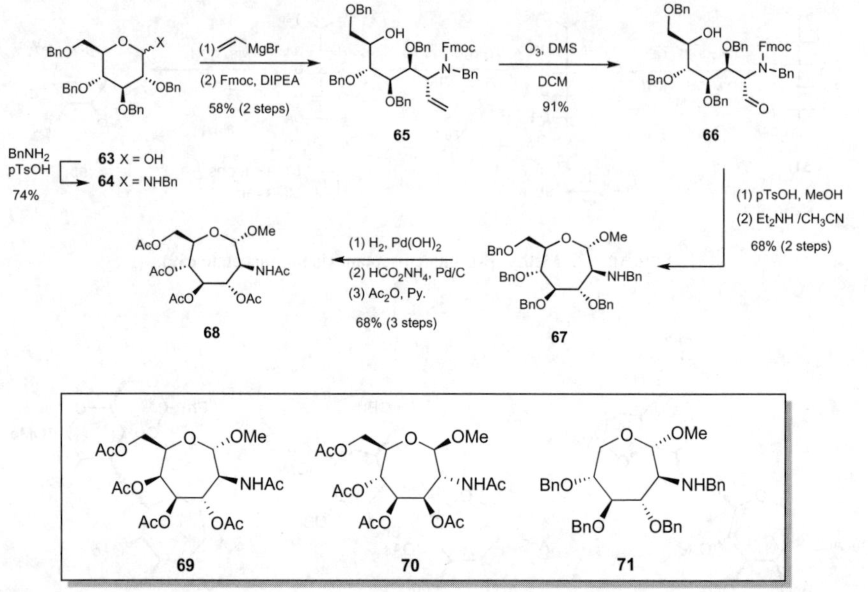

SCHEME 13. Preparation of methyl 2-acetamido-3,4,5,7-tetra-*O*-acetyl-2-deoxy-α-D-*glycero*-D-*ido*-septanoside (**68**).

primary 6-hydroxyl group, and the resultant septanoses thus lacked an exocyclic hydroxymethyl group. Since the latter group is important in hexopyranose systems for interactions between proteins and carbohydrates,[51] we developed an efficient method (involving either a primary *or* a secondary 6-hydroxyl group to afford the corresponding septanosides) to prepare such compounds as **68–71**. Significantly, the route showed high selectivity in the two steps involved in the generation of a new stereocenter, namely the Grignard addition to the imine/glycosylamine, and in the formation of the glycosidic bond. Further evaluation of the scope of the sequence with other glycopyranosyl starting materials, such as D-galactose, D-mannose, and D-xylose, provided septanosides with similar efficiency and selectivity.

Although the aforementioned acid-mediated tandem cyclization–glycosylation method was successfully utilized for the preparation of different amino glycoseptanosides when simple alcohols were used in excess, attempted glycosylation with more-complex alcohols, such as sugar-derived alcohols, to generate septanose-containing di- and oligosaccharides was not fruitful. Also, the use of such strong acids as *p*-toluenesulfonic acid restricted the choice of protecting groups in both the donors and the acceptors subjected to glycosylation. Mindful of the shortcomings of the method, we undertook a detailed investigation to devise an alternative route for preparing more elaborate amino sugar analogues, for study in broader biological contexts. At this point, we looked to donor types and coupling protocols that were conventional in the pyranose field, and this necessitated finding a suitable septanosyl donor.

Considering the equilibrium between **66** and its cyclic hemiacetal form as observed by NMR, we anticipated that it would be essential to find conditions that could favor reaction of only the cyclic hemiacetal form, and essentially "cap" the hemiacetal into an active donor. We began the reaction screening with a "toolbox" approach, considering methods commonly used for preparing pyranosyl donors. Initially we chose acid-mediated transformations, and subsequently the preparation of septanosyl thioglycosides that could be accessed under acidic conditions. Reaction of the aldehyde **66** with ethanethiol or thiophenol under various conditions resulted in the formation of acyclic dithioacetals rather than the desired cyclic thioseptanosides. Treatment of **66** with trichloroacetonitrile led only to unreacted starting material or removal of the Fmoc group. These outcomes collectively indicated a difference in reactivity of the septanose hemiacetals from analogous aldopyranoses; use of either acid or base permitted only the acyclic form (namely **66**) in the equilibrium to undergo subsequent transformations (see later).

Functionalization of the septanose was successful using neutral reaction conditions.[52] In fact, when diethylaminosulfur trifluoride (DAST) was used as the reagent for cyclization of **66**, the cyclic septanosyl fluoride **72** was isolated in 48% yield, and optimization of the conditions raised this to 85% (Scheme 14). Septanosyl fluorides derived from D-galactose (**74**) and D-mannose (**75**) were prepared in the same way. The glucose-derived septanosyl fluoride **72** was next evaluated as donor in a glycosylation reaction with 1,2:3,4-di-*O*-isopropylidene-α-D-galactose as a model acceptor. After screening different activators and reaction conditions, AgClO$_4$–SnCl$_2$ was found to be the combination that gave the best yields. Glycosylation provided α-septanosides exclusively, which demonstrated high selectivity of this transformation. The selectivity for glycosylation was attributed to interference by the bulky

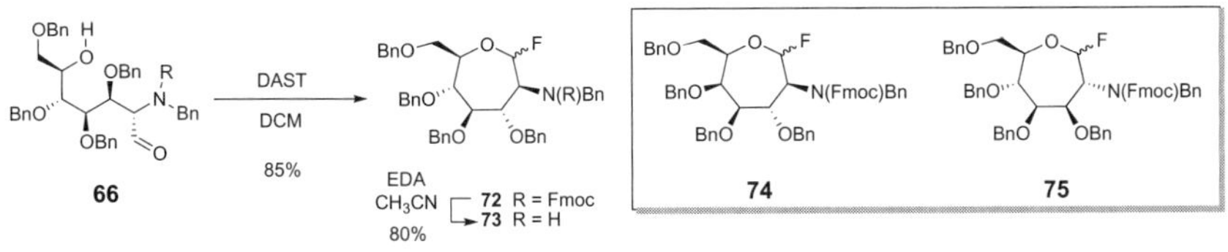

SCHEME 14. Protected D-glucoseptanosyl fluoride as donor for coupling reactions.

2-amino functionality, which drove formation of the glycosyl bond to the *anti* orientation. A collection of glycosides obtained through this protocol (**76–86**) is depicted in Fig. 4. This methodology delivered a wide array of seven-membered 2-amino-2-deoxy glycosides derived from both D-glucose and D-galactose coupled to a variety of aglycons.

2. Preparation and Reactions of Carbohydrate-Based Oxepines

a. Oxepines via Cyclopropanation and Ring Expansion.—Cyclizations in the previous section all involved C—O bond formation; in those examples, intramolecular attack by oxygen at an electrophilic carbon site formed the ring. In this section, seven-membered ring heterocycles loosely referred to as oxepines are central to the synthesis of septanosides. Formally, oxepine refers to oxacycloheptatriene (**87**), which can equilibrate with the corresponding arene oxide (**88**). Recent literature has

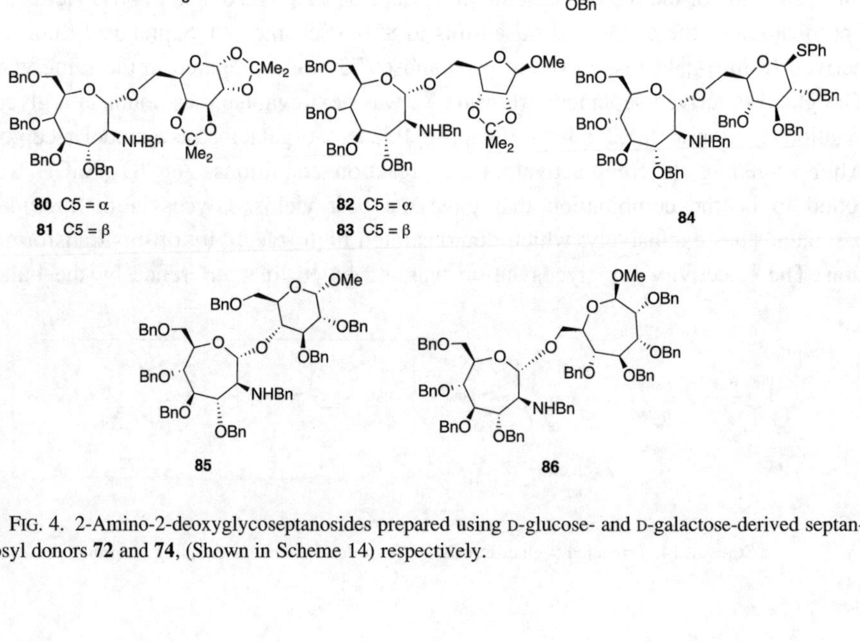

FIG. 4. 2-Amino-2-deoxyglycoseptanosides prepared using D-glucose- and D-galactose-derived septanosyl donors **72** and **74**, (Shown in Scheme 14) respectively.

used the term oxepine in reference to such substituted dihydro- and tetrahydro-oxepines as **89** and **90** (Fig. 5).

One major strategy for the synthesis of oxepines starts with cyclopropanation of a six-membered ring, followed by ring expansion. The remaining double bonds in oxepines can be functionalized to give septanose sugars. Septanosides described here are of the C-6-substituted variety. Hoberg and Bozell reported one of the first syntheses of a carbohydrate-type oxepine via cyclopropanation of a glycal.[53] Scheme 15 illustrates several aspects of the strategy. A modified Simmons–Smith reaction on 6-protected D-glucal (**91**) and 6-deoxy-D-glucal (**92**) gave, after acetylation, the cyclopropanated products **93** and **94** in 85–87% yield (Scheme 15). It was anticipated that cyclopropane ring opening to form oxacarbenium ion **95** in the presence of a cyanide nucleophile would form such species as the glyco-3-enoseptanosyl cyanide **96**. In fact, for **94**, where the C-6 position was deoxygenated, the reaction did provide the cyanide **96**, in 49% yield. When C-6 was oxygenated, however, as with **91**, an intramolecular cyclization to form the 1,7-anhydro derivative **97** occurred (78%). In this case, intramolecular trapping of the oxacarbenium ion by the C-7 OTBS group was preferential to intermolecular attack by a nucleophile.

Later work by Hoberg showed that switching from the original protecting groups of a TBS ether at C-7 and acetate at C-5 to a 4,6-*O*-dibutylsilylidene protection mode prevented the intramolecular reaction.[54,55] Cyclopropanation of

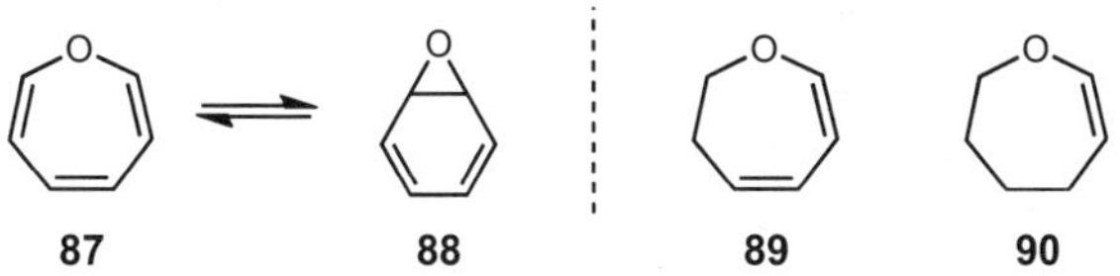

FIG. 5. Left: oxepine (**87**) in equilibrium with arene oxide (**88**). Right: general structures (**89** and **90**) that have been referred to as oxepines.

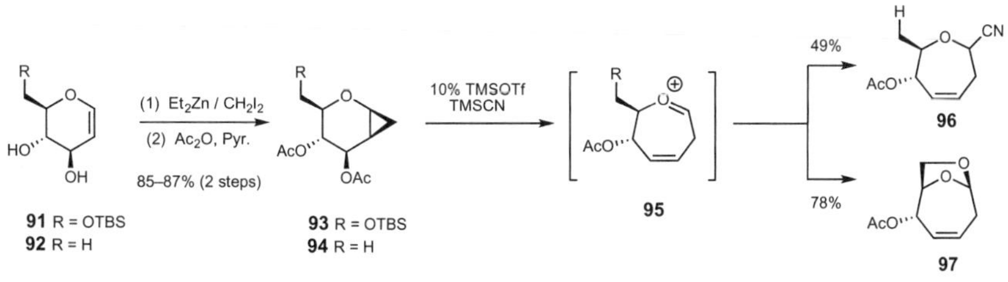

SCHEME 15. Carbohydrate-type oxepine via cyclopropanation of a glycal.

4,6-O-dibutylsilylidene-protected glycals afforded such products as **98** and **103** (Scheme 16). Ring opening of **98** using a variety of nucleophiles in the presence of TMS triflate delivered the corresponding oxepines (**99–102** are representative) as 2:1–1:1 anomeric mixtures in 82–93% yields. The planar nature of the oxocarbenium ion was invoked to explain the low selectivity in nucleophilic addition in this system. Most recently, synthesis of **104** from TMS triflate-mediated addition onto **103** allowed for the functionalization of the double bond of the oxepine. Shown in Scheme 16 is the *cis*-dihydroxylation of alkene **104** under conventional conditions to prepare the diol **105** efficiently. Additional reactions performed on the double bond of **104** included halogenation, epoxidation, cyanohydrin formation, and hydroboration oxidation, all in yields from 40% to 90%. The new compounds are a unique class of *C*-glycosyl heptoseptanoside analogues.

A related approach to oxepine synthesis through ring expansion of cyclopropanated sugars has been reported by Nagarajan *et al.*[56] Their route differs from Hoberg's because the cyclopropanation utilizes a dihalocarbene instead of the unfunctionalized carbenoid used in the Simmons–Smith reaction. As a result of this change, the cyclopropanes were able to be open under basic rather than acidic conditions. Cyclopropanation of D-glucal tribenzyl ether (**106**) with dibromocarbene was stereoselective, forming the new ring *anti* to the bulky C-3 substituent and giving **107** in 84% yield (Scheme 17). Ring opening using methanolic base was concomitant with solvolysis of dibromide **107** and resulted in the ring-expanded methyl oxepine **108** in 67% yield as an inseparable anomeric mixture. The reaction proceeded through nucleophilic capture of the allylic cation formed by loss of the *endo*-halogen atom. While the Nagarajan report did not carry such oxepines as **108** forward to completely saturated septanosides, it offered the possibility of such a strategy.

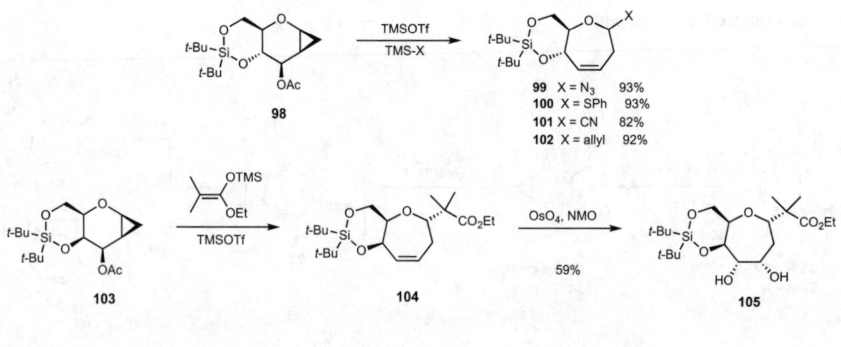

SCHEME 16. *C*-Glycosyl heptoseptanoside analogues.

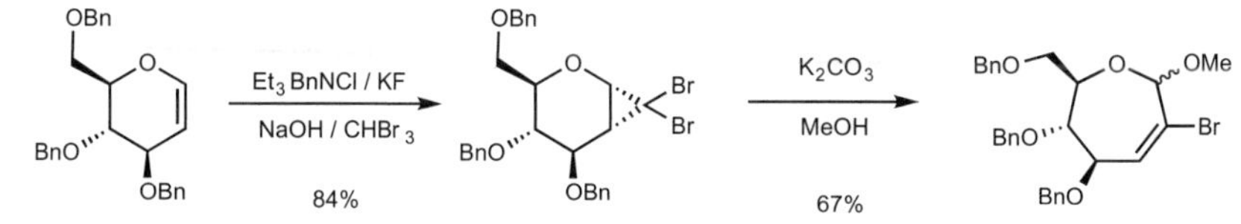

SCHEME 17. Cyclopropanation utilizing a dihalocarbene.

Oxepines can serve as key intermediates in the synthesis of aldoseptanoses. For example, dihydroxylation of the double bond of compound **104** in Scheme 16 affords the 2-deoxy-*C*-septanosyl product **105**. Jayaraman and coworkers have effectively combined the cyclopropanation route to oxepines together with subsequent functionalization to provide a powerful approach for preparing genuine septanoses.[57] Such protected 2-hydroxyglycals as the tetrabenzyl ether **109** serve as the starting materials in their syntheses (Scheme 18). The 2-hydroxyglycals can be prepared from the corresponding glycosyl bromides via straightforward transformations.[58,59] Dihalocyclopropanation of **109** proceeded to give **110** as the sole product (81%), whose basic methanolysis resulted in the diastereoselective formation of oxepine **111** in 94% yield. The stereoselective formation of the α-glycoside is noteworthy because basic methanolysis of **107**, which lacked the 2-benzyloxy group, gave a mixture of anomers. Lithium–halogen exchange on **111** with *n*-BuLi and subsequent protonation gave the septanoside **112**. Epoxidation of alkene **112** with dimethyl dioxirane (DMDO) gave rise to the α-hydroxy ketone **113** (67% over two steps). Stereoselective reduction of the ketone with NaBH$_4$, followed by hydrogenolytic removal of the benzyl protecting groups, delivered methyl α-D-*glycero*-D-*talo*-septanoside (**114**) in 66% yield (two steps). When starting from the D-*galacto* configured glycal, methyl-α-D-*glycero*-L-*altro*-septanoside (**115**) was similarly prepared.

b. **Septanose-Containing Di- and Trisaccharides by the Ring-Expansion Approach.**—Jayaraman's group has extended their method by varying[60] the nucleophile in the ring-opening step of the reaction (Scheme 19). Reaction of **116** in the presence of furanoside, pyranoside, phenol, and azide nucleophiles delivered oxepines **117** in 58–97% yield. Diastereoselectivities in the transformation of **116** to **117** were variable, based on the nature of the nucleophile. The furanoside and pyranoside nucleophiles gave the α anomer exclusively (as was the case when methoxide was the nucleophile), whereas phenol and azide nucleophiles gave α/β mixtures (Fig. 6). The authors suggested that the π-orbitals of phenol and azide favor addition by an S$_N$1-type mechanism onto an oxocarbenium ion, whereas the alcoholic nucleophiles

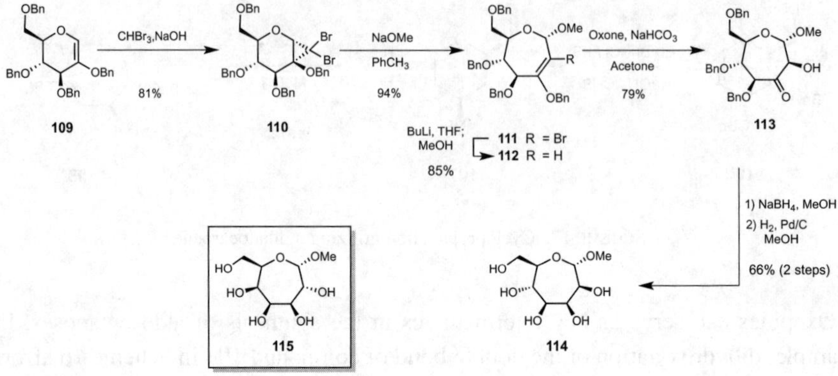

SCHEME 18. 2-Hydroxyglycal-based septanoside syntheses.

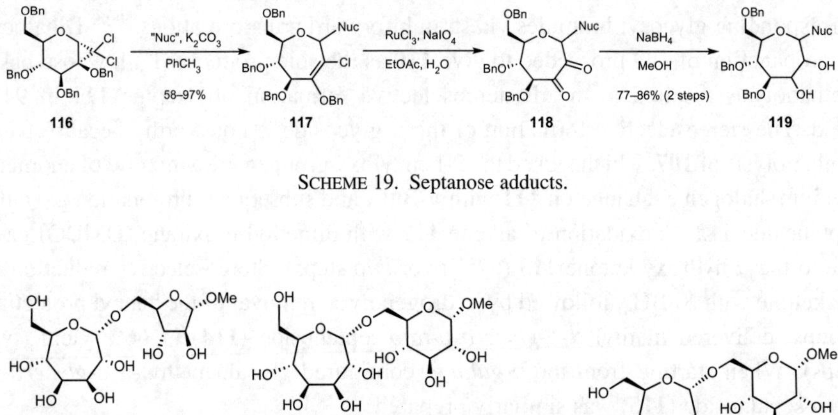

SCHEME 19. Septanose adducts.

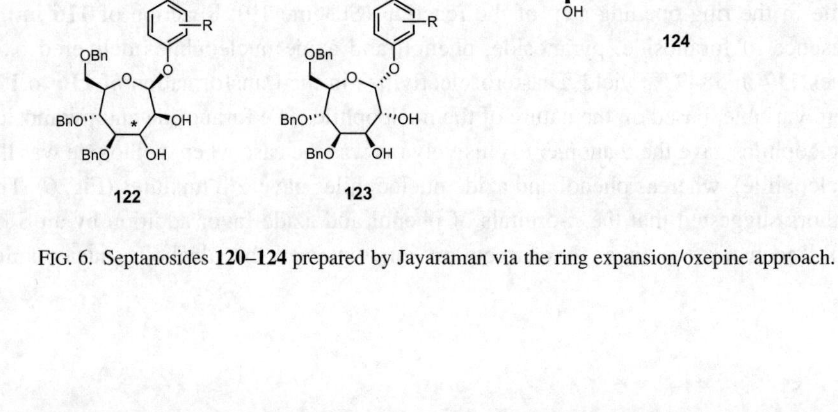

FIG. 6. Septanosides **120–124** prepared by Jayaraman via the ring expansion/oxepine approach.

operate via a kinetic anomeric effect. Nonetheless, oxepines **117** were oxidized to the corresponding 2,3-diketones **118** using a perruthanate reagent generated *in situ*. Reduction by sodium borohydride provided the protected septanosides **119** (77–86%, two steps). The diastereoselectivity of the diketo reductions was dependent on the anomeric configuration; α-septanosides gave only one diastereomer, while the β-septanosides were epimeric at C-3 (see **120–123**, Fig. 6). Selected examples were also reported where benzyl protecting groups were removed to yield completely deprotected septanosides. Synthesis of trisaccharide **124** and some related septanosides, using this same strategy, has also been described.[61] Overall, the method reported provides an effective route to prepare septanose-containing disaccharides, since ring expansion and formation of the glycosidic linkage occur simultaneously.

c. Oxepines by Ring-Closing Metathesis and Cycloisomerization Reactions.

In the previous section, the cyclopropanation of a glycal to form a bicyclic intermediate was followed by a ring-expansion reaction en route to each oxepine. Presented here are routes that afford oxepines either by ring-closing metathesis (RCM) reactions or by cycloisomerization of terminal alkynes.

Sturino and coworkers[28] prepared polyoxygenated oxepanes, which are 1,6-anhydroaldoses but which might be considered 1-deoxyseptanoses, by an RCM approach. An example starting from 6-*O*-benzyl-2,3-*O*-isopropylidene-D-ribose (**125**, Scheme 20) is illustrative; several examples of the sequence were shown. Wittig methylenation of **125** gave alcohol **126** in 70% yield. As a result of alkenation, the 4-hydroxyl group (pentose numbering) no longer formed a hemiacetal and was therefore available for further functionalization. Allylation of the exposed hydroxyl group gave **127** (77%), which was a precursor for RCM. Metathesis using Grubbs' first-generation ruthenium catalyst gave oxepine **128** in 97% yield. Dihydroxylation of **128** and isopropylidene protection of the subsequent diol provided the oxepane derivative **129** in 52% yield over two steps. An earlier report from van Boom *et al*.[62] utilized essentially the same RCM-based approach for the synthesis of oxepines (Scheme 21). First, 2,3,5-tri-*O*-benzyl-D-arabinofuranose (**130**) was converted into the alkene **131**

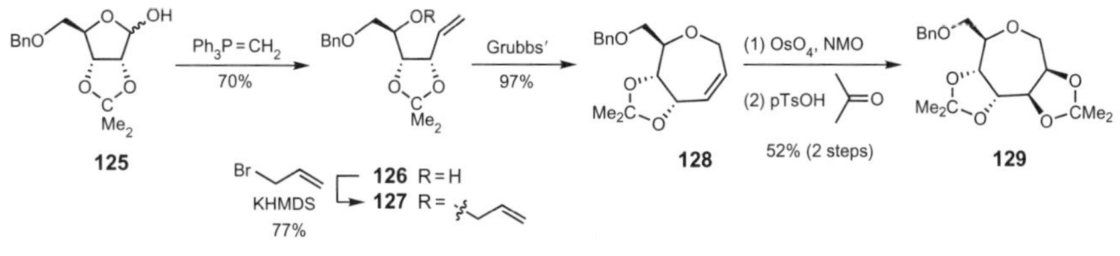

SCHEME 20. Polyoxygenated oxepanes.

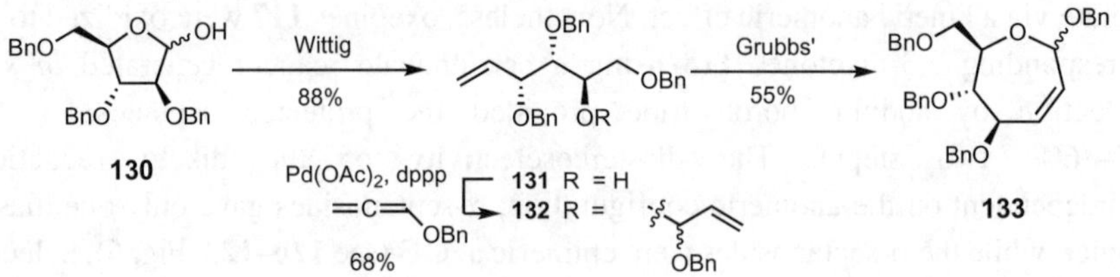

SCHEME 21. Oxepine **133** by RCM.

(88%), which on treatment with benzyloxy-1,2-propadiene in the presence of catalytic palladium yielded the allylic acetal **132** (68%) as a mixture of diastereomers. The authors also allylated at this step (not shown). Acetal formation is shown because it is essentially a glycoside in the developing oxepine/septanoside. RCM of **132** gave oxepine **133** in 55% yield. The versatility of the method was demonstrated by synthesizing oxepines akin to **133**, starting from other protected furanoses such as 2,3;5,6-di-*O*-isopropylidene-D-mannofuranose and and 6-*O*-trityl-2,3-*O*-isopropylidene-β-D-ribose. The authors did not continue the septanoside synthesis by functionalization of the oxepine via dihydroxylation; the likelihood for success, however, seems high based on the precedent shown in Scheme 20.

Oxepines presented thus far (such as **108**, **112**, **128**, or **133**) share one uniting characteristic; the unsaturated alkene functionality in their rings was distal to the oxygen atom, as with **134** of **135** in Fig. 7. Oxepines where the unsaturated functionality is adjacent to the ring-oxygen atom afford structures akin to **136**. The similarity between this type of oxepine and glycals (for example **137**) rests primarily on their both being cyclic enol ethers. One early conjecture was that oxepines of this type should have reactivity similar to that of glycals. To test this idea, our group came up with methods to synthesize carbohydrate-based oxepines with an eye to their further derivatization.[63]

The RCM strategy adopted is illustrated in Scheme 22. A pyranose sugar was used as a starting material in this approach instead of the furanose was used in previous syntheses. Wittig methylenation of 2,3,4,5-tetra-*O*-benzyl-D-glucose (**63**) gave the alkene **138** (74%). Subsequent formation of a vinyl ether at the alcohol site generated diene **139** (80%). RCM of this diene required use of the Schrock catalyst and generated oxepine **140** 92% yield. This more highly reactive metathesis catalyst was necessitated by the presence of the vinyl ether in the diene; reaction with either of the Grubbs catalysts was considered to form an unreactive Fisher carbene that did not continue through the cyclization. The route was sufficiently general to afford a variety of carbohydrate-based oxepines **140**–**146** (Fig. 8). A method for preparing

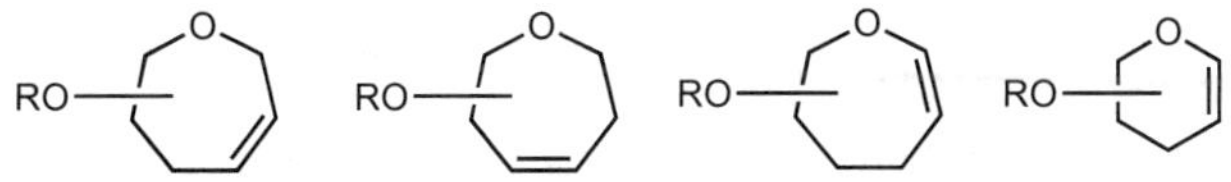

FIG. 7. "Oxepine" isomers **134–136** that vary as to which ring C—C bond is unsaturated; the substituted dihydropyran **137** possesses the enol ether functionality of a glycal, akin to oxepine **136**.

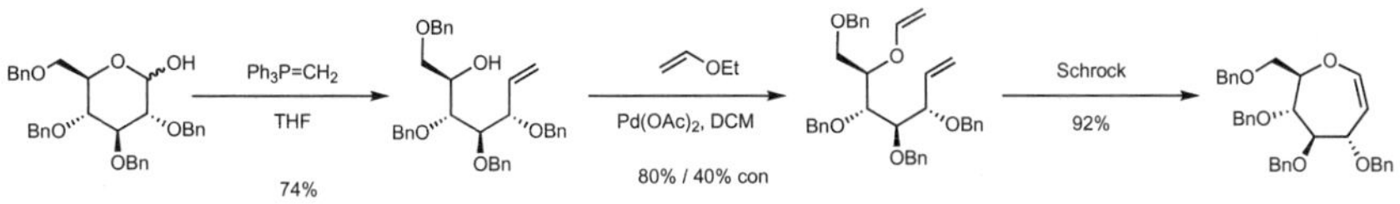

SCHEME 22. Oxepine by the RCM strategy.

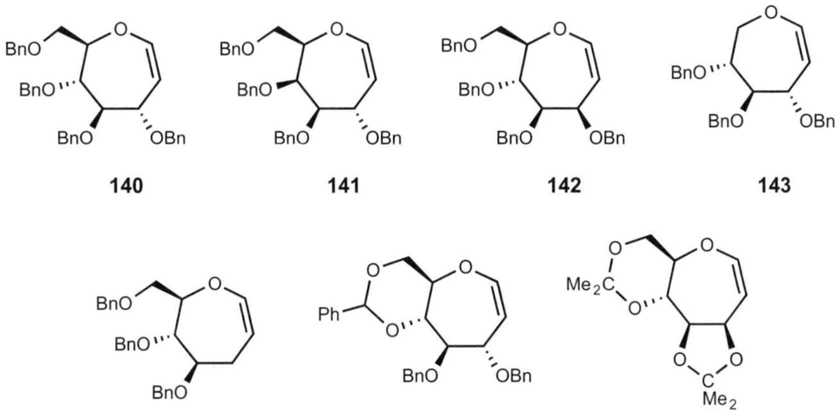

FIG. 8. Oxepines **140–146** prepared using the RCM strategy illustrated in Scheme 22.

oxepines via the cyclization and subsequent elimination of protected hydroxyacetals[64,65] was also used to prepare **140** and **144** but was not as general.[66]

A complementary route to carbohydrate-based oxepines was developed by the McDonald group.[67] It is based on the *endo*-selective cycloisomerization of alkynyl alcohols in the presence of molybdenum or tungsten catalysts to give the cyclic enol ethers.[68] The initial report of this strategy as applied to the synthesis of carbohydrate-based oxepines used such alkynyl diols as **148**, for example, as the starting material for cycloisomerization (Scheme 23). Alkynyl diol **148** was prepared from 2,3-*O*-isopropylidene-D-ribose (**147**) by Wittig chloromethylation followed by dehydrohalogenation (31% over two steps). Photoinitiated dissociation of one of the CO ligands from the $W(CO)_6$ catalyst in the presence of triethylamine formed the tungsten complex $Et_3NW(CO)_5$ (Scheme 23, box), which is the active catalyst for the cycloisomerization; the same catalytically active complex may also be generated from a stable Fisher carbene under thermal conditions.[69] In the presence of this catalyst, compound **148** was converted into **149** and then **150** after acetate protection of the final hydroxyl group (56% over two steps). Cyclization occurred with excellent regioselectivity through the primary hydroxyl group to deliver the *endo* product. According to the authors, the rigidifying isopropylidene group was necessary for successful oxepine formation. This report of the photocycloisomeriztion of alkynols to deliver oxepines showed several examples of the strategy with yields for the isomerization ranging from 61% to 83%.

In the hopes of accessing similar molecules that would contain C-6 functionalization, our group explored the cycloisomerization of such alkyne hemiketals as **152**.[70] Compound **152** was prepared by addition of ethynyltrimethylsilane to 5-*O-tert*-butyldiphenylsilyl-2,3-*O*-isopropylidene-D-ribonolactone (**151**), followed by desilylation (25% over two steps). Triethylamine-mediated cycloisomerization[71] provided an oxepinone, compound **153**, in 41% yield. 1,2-Reduction of the enone functionality followed by acetylation under standard conditions provided **154** in 56% yield over two steps as a 3:1 ratio of diastereomers (the favored diastereomer is shown in Scheme 24). A small group of oxepines were prepared by this method. Variability in the yield of the cyclization step, which was moderate at best, has prevented this route from being applied more generally for the preparation of oxepines.

Of the three types of oxepines shown in Fig. 7, those of the enol ether type (**136**) seem to be the most versatile in terms of their ability to form septanose glycosides.

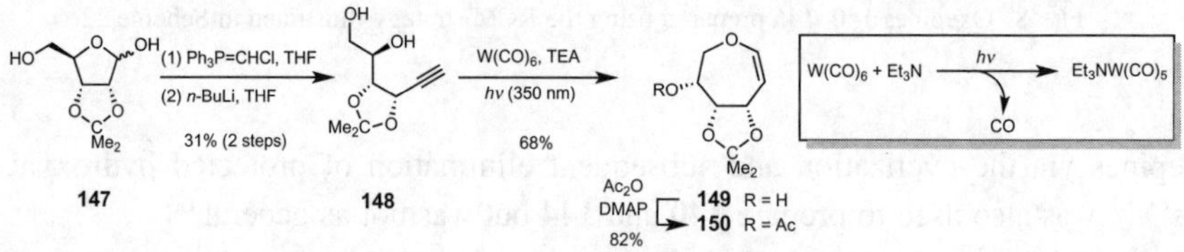

SCHEME 23. *endo*-Selective cycloisomerization of alkynyl alcohols in the presence of molybdenum or tungsten catalysts.

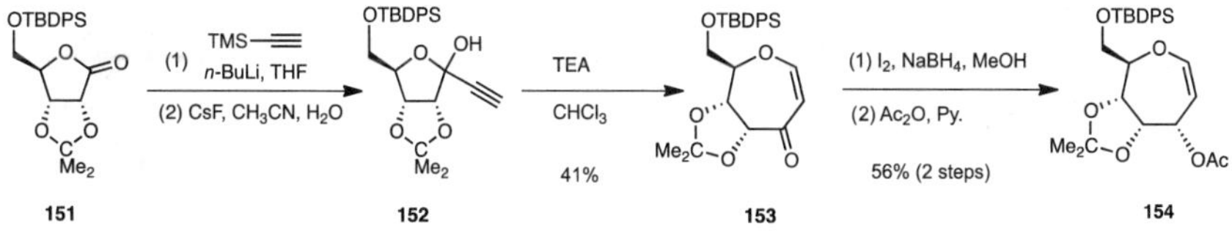

SCHEME 24. Oxepines by cycloisomerization of alkyne hemiketals.

Therefore, when considering efficiency and scope, the cycloisomerization route and the RCM routes seem to be the most useful procedures to form those oxepines. Specifically, the cycloisomerization route is best for C-6-unsubstituted oxepines, while the RCM route is most amenable to C-6 substitution. The enol ether functionality is central to their subsequent reactivity and is reminscient of the common six-membered ring glycals that are well established as useful starting materials for the synthesis of pyranose glycosides. The next section draws examples from the literature showing glycal-like reactivity of carbohydrate-based oxepines.

d. Glycosylations Enabled by Oxepines.—Oxepines have served indirectly as glycosyl donors for the synthesis of septanosides in two ways. First, epoxidation of oxepines with dimethyldioxirane generates 1,2-anhydroseptanoses, which subsequently serve as glycosyl donors in the presence of nucleophilic acceptors. Ring opening of 1,2-anhydroseptanoses has been conducted under neutral, anionic, and acidic conditions to provide a variety of glycosides. Second, oxepines, via 1,2-anhydroseptanoses, have been converted into thioglycosides. Addition of a sulfur nucleophile, in the form of LiSPh, delivers a thioglycoside which can then itself be used as a donor in subsequent glycosylations. Examples of both glycosylation strategies are provided in this section.

Following the work of others,[72–74] we investigated the diastereoselectivy of epoxidation by DMDO on oxepines **140–143**.[75] The model emerged from a combination of reaction-product analysis and DFT calculations. To explain the selectivities observed, it invoked the orientation of electronegative substituents (such as benzyloxy groups) relative to the molecular plane, the lone pairs on the ring-oxygen atom of the oxepine, and the conformation of the ring. These factors are interrelated and electronic in nature. It was argued that the steric bulk of the benzyl protecting groups makes only a minor contribution to the selectivity. Figure 9 summarizes the conclusions of the article. It was demonstrated that oxepines **141** and **142**, derived from D-galactose and D-mannose, were highly selective for formation of the respective α epoxides **155** and **156** (10:1 and >25:1 α:β). The glucose-derived oxepine **140**, on the other hand, showed only a low (3:1) selectivity for the α epoxide **157**. The D-xylose-based oxepine **143** gave β epoxide **158** with high (>25:1) selectivity. Wei's majority-rules model

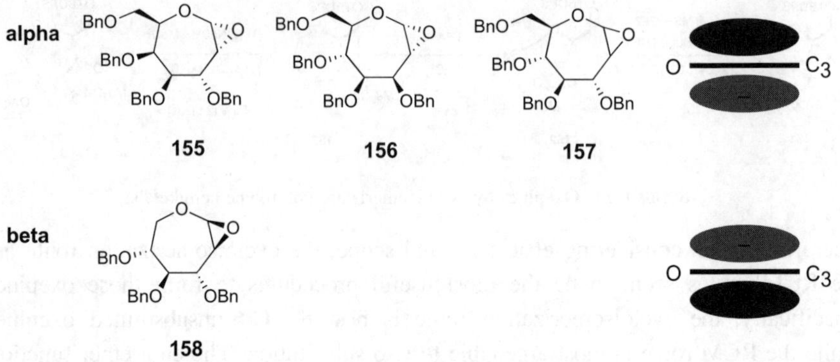

FIG. 9. A model for diastereoselectivity in epoxidation of carbohydrate-based oxepines. (For color version of this figure, the reader is referred to the Web version of this chapter.)

says that the π-cloud of the reacting alkene will distort so that the greater electron density is opposite to the majority of electronegative groups (such as benzyloxy groups). The cartoons on the right in Fig. 9 graphically represent the π-cloud distortion. This model implicitly includes the contributions of the lone pair on the ring-oxygen atom. The DMDO reagent approaches the oxepine in a way that the electrophilic oxygen that is to become the epoxide oxygen is antiperiplanar to the lone pair of electrons on the ring-oxygen atom. This analysis of the origins of diastereoselectivity of epoxidation has provided a framework that should allow the prediction of reaction products for new oxepines.

Such epoxides as **155–158** are correctly termed 1,2-anhydro sugars. They are reactive species and can permit glycoside formation under appropriate reaction conditions. Epoxide **158**, prepared from oxepine **143** (Scheme 25),[76,77] was ring opened by using sodium methoxide and gave methyl α-D-idoseptanoside (**159**) in 89% yield. Other anionic and neutral nucleophiles, such as 2-propanol, 1,2:3,4-di-O-isopropylidene-α-D-galactose, ethanethiol, sodium azide, and allylmagnesium bromide, were used to open the epoxide ring of **158** with yields that ranged from 33% to 89%. Epoxide opening with lithium thiophenoxide delivered the phenyl thioseptanoside **160** (70%).[78] Protection of the 2-hydroxyl group as either the benzyl ether **161** (87%) or the acetate **162** (89%) furnished the second class of glycosyl donors, which were utilized in a subsequent investigation.

The phenyl thioseptanoside donors **161** and **162** were used to glycosylate a series of alphatic alcohols, phenols, and furanosyl derivatives, and also pyranosyl acceptors to give predominantly α-septanosyl glycosides in good yields (44–72%). In fact, thioseptanoside **162** was used as a donor to prepare the first reported diseptanoside,

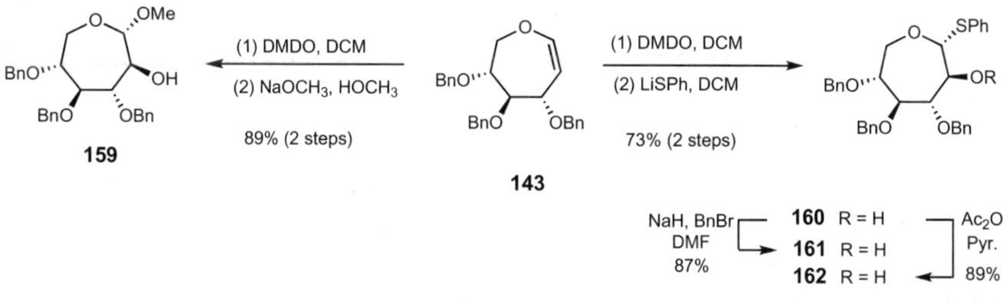

SCHEME 25. Glycosylation via 1,2-anhydroseptanose derivatives.

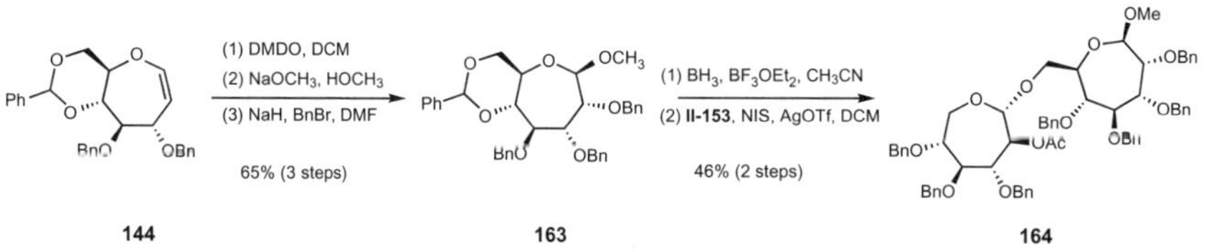

SCHEME 26. Septanoside used as a donor to prepare a diseptanoside, compound **164**.

compound **164** (Scheme 26). The acceptor was prepared from benzylidene-protected oxepine **144**. Epoxidation of **144**, followed by methanolysis, gave a methyl septanoside; benzyl protection of the 2-hydroxyl group then gave **163** in 65% yield over three steps. Selective opening of the benzylidene ring gave methyl 2,3,4,5-tetra-*O*-benzyl-β-D-*glycero*-D-*gulo*-septanoside (not shown), which was carried on to the glycosylation reaction. Under the conditions established with the other acceptors, the 1 → 7 α-linked diseptanoside **164** was prepared. Selective benzylidene opening and glycosylation took place in 46% yield over the two steps.[79]

Recently McDonald *et al.* reported the synthesis of (1 → 5)-linked D-mannoseptanosyl di- and trisaccharides.[80] Once again they utilized differentially protected phenyl 1-thio-α-D-mannoseptanoside as donor and methyl α-D-mannoseptanoside as acceptor to synthesize the di- and tri-α-septanosides exclusively. The strategy used for their synthesis mirrored that depicted in Scheme 26. The septanoside structures and their properties in a glycosidase assay are presented in Section IV.

3. Miscellaneous Septanose Syntheses

In addition to the major strategies previously outlined, other methods have also been utilized to synthesize septanosides. Among these, a bis-aldol addition of nitroalkanes onto pyranose-derived dialdehydes has been utilized to synthesize 3-nitro- and 4-nitro-septanosides.[81–84] Such septanose derivatives are useful as they serve as precursors to amino sugars, which have potential biological interest. Furthermore, a nitro sugar, evernitrose,[85] is present in the antibiotic natural product everninomicin.[86]

The bis-Henry aldol approach is best illustrated by recent work by Osborn and Turkson.[81] Shown in Scheme 27 is the sequence starting from methyl 4,6-*O*-benzylidene-α-D-glucopyranoside (**165**) and methyl 2,6-di-*O*-benzyl-α-D-glucopyranoside (**168**). Sodium periodate-mediated oxidation of the vicinal diols in **165** and **168** gave the corresponding dialdehydes **166** and **169**. Addition of nitromethane to each aldehyde, first in an intermolecular fashion and then intramolecularly to form the seven-membered ring, furnished 5,7-*O*-benzylidene-3-deoxy-3-nitro-α-D-*glycero*-D-*altro*-septanoside (**167**) and 2,7-di-*O*-benzyl-4-deoxy-4-nitro-α-D-*glycero*-D-*talo*-septanoside (**170**) in 43% and 35% yields, respectively, over the two steps. Good regioselectivity in the bis-Henry reation was observed; that is, the 2,3-diol (**165**) gave the 3-nitro septanoside, whereas the 3,4-diol (**168**) gave the 4-nitro product. A detailed rationalization of the regioselectivity was not provided. Addition reactions using bulkier nitroalkanes led to increasing amounts of the corresponding dioxepans (such as **171**). This has been attributed to the relief of steric strain and greater thermodynamic stability of the dioxepanes, as well as the propensity of bulky nitroalkanes to condense directly with one of the aldehyde groups.[83,84]

Another method utilized to synthesize septanoses is the ring expansion of cyclic ketones by treatment with diazomethane.[19,87–89] The examples in Scheme 28 are illustrative. Thus, reaction of pyranosid-4-ulose **172** with an excess of diazomethane delivered the septanosid-4-ulose **173** in 60% yield. Although it was not reported, reduction of **173** would have provided the corresponding 5-deoxyseptanoside. The second example in Scheme 28 illustrates the reduction of the ring-expansion product. Ring expansion of methyl 4,6-*O*-benzylidene-2-deoxy-α-D-*threo*-hexopyranosid-3-

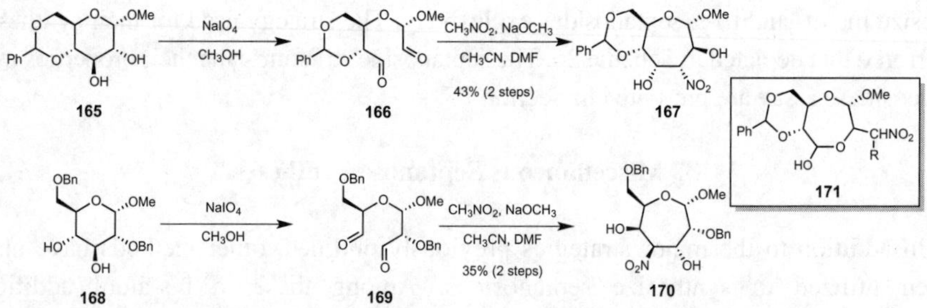

SCHEME 27. Nitroseptanosides by nitromethane condensation with dialdehydo sugars.

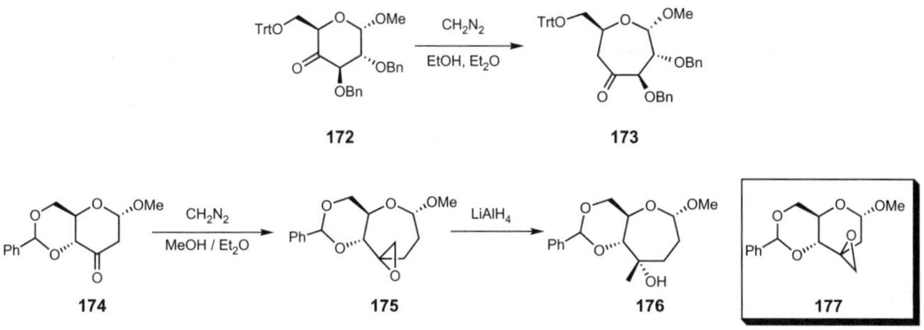

SCHEME 28. Septanosides by ring expansion of pyranosiduloses.

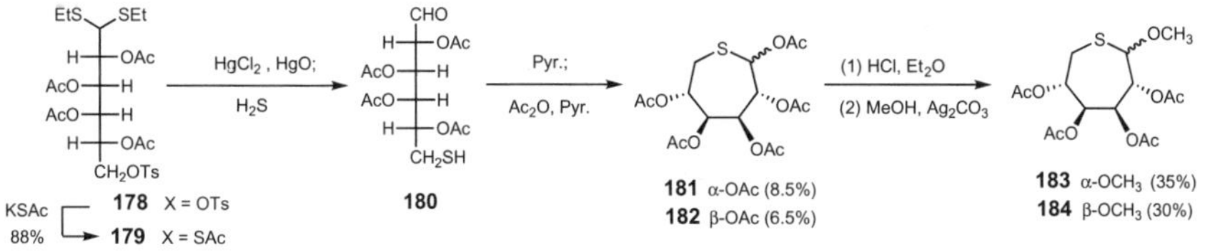

SCHEME 29. 6-Thioseptanosides.

ulose (**174**) resulted in a seven-membered ring ketone analogous to **173**. However, this reacts further *in situ* with the excess of diazomethane to give epoxide **175** (30%). Reduction of the epoxide then affords the methyl septanoside **176** (66%). The epoxide **177** was also generated during the conversion of **174** into **175**; it arises via direct epoxidation on the keto group of the starting material. The preparation of septanosides via pyranosidulose ring expansion suffers from a number of shortcomings.[19] Primary among them is the inability to control or predict product distribution of the diazomethane reaction products; furthermore, unwanted products such as those from epoxidation of the pyranosidulose and septanosidulose diminish the yield of the septanosidulose product.

Even though they are in a class of their own, thio sugars are chemically very similar to normal oxygen-based sugars. Whistler and Campbell[90] synthesized D-galactose-based 6-thioseptanosides **181–184** (Scheme 29). Treatment of 2,3,4,5-tetra-*O*-acetyl-6-*O*-tosyl-D-galactose diethyl dithioacetal (**178**) with potassium thioacetate gave the 6-*S*-acetyl-6-thio derivative **179** in 88% yield. This was treated with mercuric chloride and then hydrogen sulfide to yield 2,3,4,5-tetra-*O*-acetyl-6-thio-*aldehydo*-D-galactose **180**. Establishment of mutarotational equilibrium in pyridine and subsequent acetylation gave the α and β septanoses **181** and **182** in 8.5% and 6.5% yields, respectively (over two steps). Glycosyl chloride formation followed by methanolysis on the α/β mixture gave the corresponding methyl galactoseptanosides **183** and **184** in 35% and 30% yields, respectively.

4. Transannular Reactions of Septanose Sugars

There are several instances where reactive intermediates present during the synthesis of septanosides have reacted intramolecularly to give ring-contracted or bicyclic products.[91] Similar intramolecular reactions (such as the formation of 1,6-anhydro sugars) have precedents in the pyranose literature. Such cases of intramolecular reactivity described in the literature notably involve novel ethers or thioethers as the nucleophilic species in the ring contractions.

The first example was a deliberate attempt to effect ring contraction of a polyoxygenated thiepane via generation of an episulfonium ion.[92,93] The thiepane **185** (Scheme 30) underwent different pathways depending on the reaction conditions. When triphenylphosphine and CBr_4 were used, the episulfonium ion **186** gave rise to the bromotetrahydrothiopyran **187** via a double inversion. The stereospecificity in the formation of **187** supported the episulfonium pathway rather than a sulfur-carbenium species that would have given rise to diastereomers via an S_N1 process. Under Mitsonobu conditions, the thiepane **185** gave rise to bicyclic derivative **188** and a tetrahydrothiofuran derivative **189**. Compound **188** arose from intramolecular reaction of the hydroxyl group on an alkoxyphosphonium intermediate (en route to the

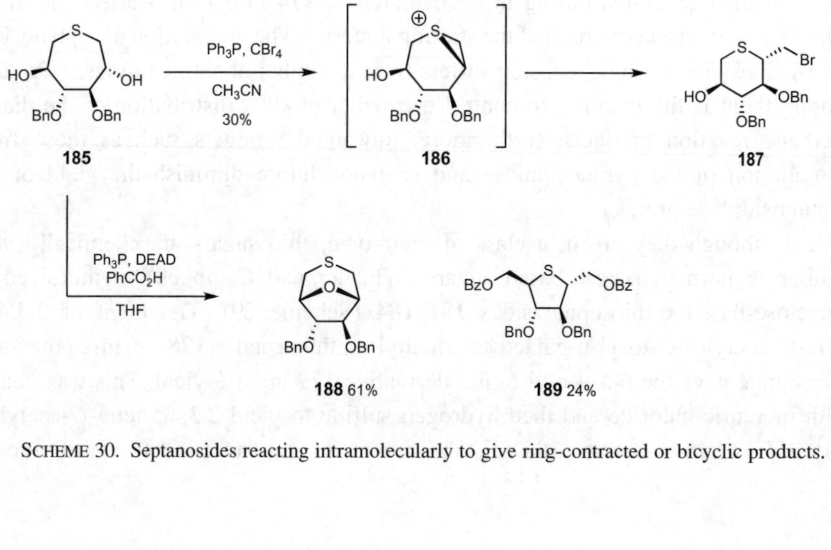

SCHEME 30. Septanosides reacting intramolecularly to give ring-contracted or bicyclic products.

episulfonium ion). The tetrahydrothiofuran derivative **189**, on the other hand, came about by iterative formation of episulfonium ions. The first arose from **185** and the second from **187**; this took place because of the excess of reagents used.

We have observed intramolecular reactions of septanose species under a variety of reaction conditions using a number of starting materials. Figure 10 shows the putative reactive intermediates **190–193** that were present during iodoglycosidation of the D-xylose-derived oxepine **143**, epoxidation of the oxepine **142**, and TMS-triflate activation of septanosyl fluorides **72** and **75**, respectively. Products **194–197** were obtained in a range of yields depending on whether or not other nucleophiles were present in the reaction. All of the intermediates are electrophilic and therefore

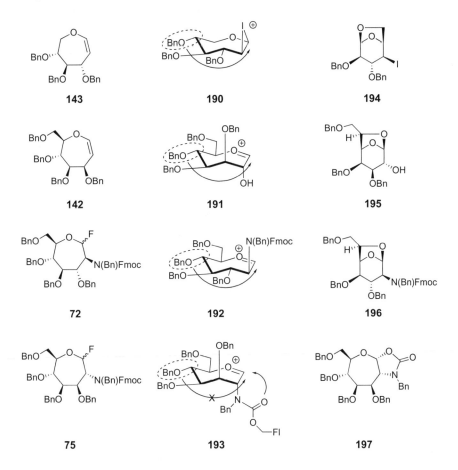

FIG. 10. Precursors, electrophilic intermediates, and products of transannular reactions of septanoses.

poised for attack by intramolecular nucleophiles that may be present. We have argued that the primary nucleophile in each case is the 5-benzyloxy group (circled in **190–192**). The only case where the product structure did not reflect the inherent nucleophilicity of the group at C-5 was when the manno-configured septanosyl fluoride **75** was subjected to standard glycosylation conditions. The fused oxazolidinone **197** was isolated, irrespective of the choice of activators and in reactions where a nucleophile was either present or absent. Trapping by the carbamate oxygen atom in the case of **75** occurred because the disubstituted amine was on the same side as the nucleophile and prevented its attack. Moreover, the carbamate was the nucleophile only because the C-5 group was blocked; this is because the glucose-configured system (**72** converted into **196**) delivered the bicyclic product that arose via attack of the 5-benzyloxy group. The fact that products **194–197** were obtained provided strong evidence for formation of the cyclic oxocarbenium ions **190–193**.

III. Conformational Analysis of Septanose Monosaccharides

Conformational analysis of six-membered ring systems is a key component of general organic chemistry, including the conformational analysis of pyranose sugar systems. Whereas the two chair conformations of cyclohexane are equi-energetic and the difference in energy of simple monosubstituted cyclohexanes may be $\geq$ 1–4 kcal/mol, the estimated energy difference between the two chair conformations of β-glucose (all-equatorial versus all-axial) is up to 9.6 kcal/mol.[94] Greater substitution on the ring thus raises the energy difference between the chair conformers. Implicit in this observation is expectation that transition states between the conformers will have correspondingly higher energies. By limiting the number of available nearby-in-energy conformations, the flexibility of the ring will also be diminished. The same general pattern holds for seven-membered ring systems. So although conformational analyses of simple model seven-membered rings such as cycloheptane,[95–99] oxepane,[96,100,101] and 1,3-/1,4-dioxepanes,[102,103] among others,[97] have shown that these rings can adopt multiple low-energy conformations, this conclusion does not necessarily hold true for septanose sugars, which are more highly substituted. As this section demonstrates, collected solution- and solid-state data show that a given septanose primarily adopts a single low-energy conformation. Furthermore, that stable conformation will be largely dependent on the stereochemistry of the ring substituents.

Conformations taken up by seven-membered rings are termed twist (T), chair (C), twist-boat (TB), and boat (B) conformers, akin to the related conformers for six-membered rings (Fig. 11). Most often a twist form is the lowest-energy conformation in a given system. For cycloheptane, which is highly symmetrical, all of the T conformers are identical and therefore isoenergetic. Septanoses have lower symmetry because of the multiple hydroxyl subsituents in a defined stereochemistry around the seven-membered ring. The consequence of this lower symmetry is that septanoses have multiple, distinct T conformations. The different T conformers should therefore have different energies. Fourteen unique T conformations may be defined for a given septanose. The naming of individual T conformers begins by identifying three coplanar atoms. These three atoms by default define the molecular plane; the remaining four atoms lie either above the molecular plane or below it. Examples of two specific T conformers are given in Fig. 11 for septanoside **198**. The molecular plane for **198** is defined by three atoms, the ring oxygen atom, C-1, and C-2. Above the plane are C-3 and C-4, and C-5 and C-6 are below it. Based on the naming system described by Stoddart,[104] the conformer of **198** shown is therefore defined as $^{3,4}T_{5,6}$. It should be noted that TC was used by Stoddart to connote twist–chair conformers, here denoted as T conformers. We use the T nomenclature to be consistent with IUPAC. The other conformer shown for **198** is $^{5,6}T_{3,4}$. The ring oxygen atom, C-1, and C-2 again define the molecular plane. Starting from the $^{3,4}T_{5,6}$ conformer, partial bond rotations of the C-3—C-4, C-4—C-5, and C-5—C-6 bonds will result in the $^{5,6}T_{3,4}$ conformer. The C-1 atom sits on an axis of pseudosymmetry; substituents attached to C-1 are isoclinal and are affected similarly by steric effects from the top or bottom face of the ring.[105] The nomenclature for chair conformers follows that of the T forms, but for a C conformer, four atoms define the molecular plane. As a result, one atom is either

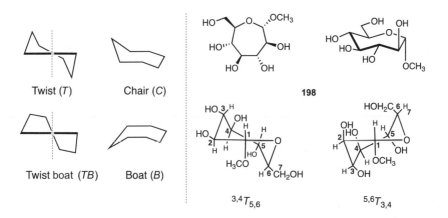

FIG. 11. Left: Low-energy conformations of seven-membered rings. Right: Different views of methyl α-D-*glycero*-D-*ido*-septanoside (**198**). Shown are wedge-and-dash, chair conformation, and two-labeled T conformers, illustrating how the T nomenclature works.

above or below the plane (superscript = above or subscript = below) and the other two atoms are on the opposite side.

Interconversion between the T conformers of seven-membered rings is a complex process. Five-membered furanoses interconvert between low-energy skew (S) conformations via envelope (E) conformations through pseudorotation.[106] A pseudorotational itinerary in the form of a two-dimensional wheel can describe these interconversions. The axes of the wheel are defined by two parameters, the phase angle (P) and the puckering amplitude τ_m; specific S and E conformers are at energetic minima at specific values of P and τ_m. An additional angular variable is necessary to describe the interconversion of conformers for pyranoses. The parameters that are used to describe a three-dimensional sphere are the puckering amplitude (Q) and two angular variables, θ and ϕ.[107] In their description of pyranose systems, Cremer and Pople indicated that a "surface of a hypersphere in N-3 dimensions" describes the conformations in higher rings, namely seven-membered rings. Bocian and Strauss used angular variables (ρ, θ, ϕ_1, ϕ_2) to describe a toroidal surface which is populated with seven-membered ring conformers.[96,99] A cross-section of the torus delivers the relative energies of twist (T) and chair (C) conformers as a function of the pseudorotational phase angle. The TB and B conformers may also interconvert in a similar fashion, but this energy surface is higher in energy than the T–C surface. We have ignored the shortcomings of the pseudorotational wheel as a tool for contemplating the interconversion of T conformations for septanoses. In a practical sense, the wheel simply provides a sense of which conformers are geometrically nearby the conformer of interest; there are no reports that attempt to chart the pathways of the interconversion between conformers. Our use of the pseudorotational wheel is rationalized by the desire to visualize which T and C conformers may be near each other in conformational space, not necessarily how those conformers interconvert with each other. The wheel in Fig. 12 collects the 14 T conformations and 14 C conformations of septanose carbohydrates in the same way.

1. Observed Low-Energy Conformations

Pakulski has tabulated structural and conformational information for a large number of septanosides in his two recent reviews of seven-membered ring sugars.[18,19] Conformations were organized based on data collected from the solid state (via X-ray crystallography) or in solution (via analysis of NMR coupling constants), or both. In general, the solid-state and solution data have consistently arrived at the same low-energy conformation. Figures 13 and 14 compile septanosides whose conformations have been reported, list the observed conformers, and track these relative to the

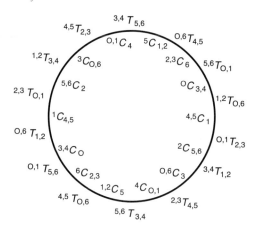

FIG. 12. Pseudorotational wheel depicting 14 twist (*T*) and 14 chair (*C*) conormers of seven-membered rings.

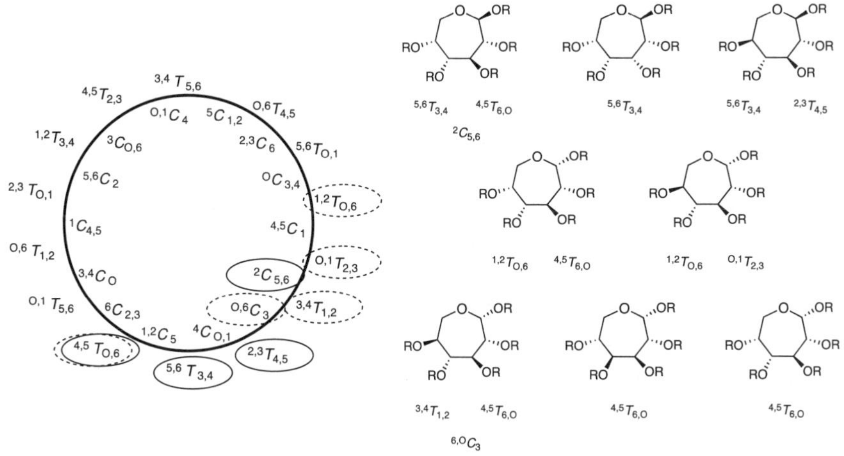

FIG. 13. Conformational map of C-6-unsubstituted septanosides. Left: A pseudorotational wheel of septanose conformations. Solid ovals mark conformations of β-septanosides, dashed ovals α-septanosides. Right: Septanosides and the conformers that have been observed for them.

pseudorotational wheel introduced in Fig. 12. Structures in the figures are organized based on whether or not they contain substitution at C-6, their stereochemistry, and their anomeric configuration.

A number of observations are in order upon consideration of Figs. 13 and 14. First, all the molecules in the figures adopt one or two (at most three) conformations. This is

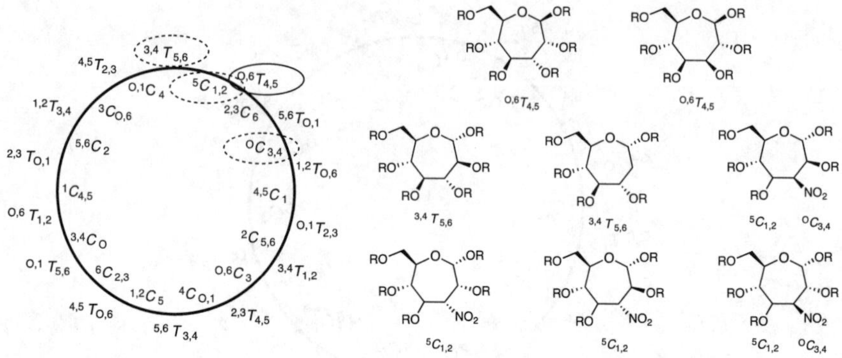

FIG. 14. Conformational map of C-6-substituted septanosides. Left: A pseudorotational wheel of septanose conformations. Solid ovals mark conformations of β-septanosides, dashed ovals α-septanosides. Right: Septanosides and the conformers that have been observed for them.

in stark contrast to the suggestion that septanoses may be flexible in the way that simpler seven-membered ring systems are. Rather, the data show that septanose monosaccharides primarily adopt one low-energy conformation. Second, the low-energy conformations adopted are largely T conformers. Third, related structures are clustered at similar locales on the pseudorotational wheel. For example, the C-6-unsubstituted rings in Fig. 13 adopt conformations that are relatively close to each other on the pseudorotational wheel. Again the implication is that, despite the particulars of interconversion, conformations near to each other on the pseudorotational wheel are therefore conformationally similar to each other—or at least relative to conformers that are distant from them on the wheel. Furthermore, the C-6-unsubstituted rings occupy a region of conformational space that is distinct from that of the C-6-substitued rings. It should be noted that some of the examples in both figures have benzylidene or isopropylidene protecting groups as part of their structures; that is, the R groups of the structures in Figs. 13 and 14 include fused cyclic moieties. The conformational preferences of the smaller rings could potentially govern the conformation of the septanose in these cases. However, in cases where direct comparisons can be made, similar conformations are adopted for a given septanoside in the presence or the absence of these rigidifying groups. Fourth, the stereochemistry of ring carbon atoms in septanoses determines the favored conformations. Such a relationship between stereochemistry and conformation is perhaps obvious, but noteworthy. It explains, to some degree, the clustering of structures to specific areas on the pseudorotational wheel and provides a rationale for predicting the favored conformations of structures that have not yet been determined experimentally.

In addition to the influence of fused rings and the stereochemistry of the ring carbon atoms, a number of other factors contribute to the determination of low-energy conformations of septanoses, in particular the anomeric effect.[108–114] Estimates for the magnitude of the anomeric effect in pyranosides range from ~0.4 to 3.2 kcal/mol, depending on the electronic nature of the anomeric substituent and the dielectric constant of the solvent.[115–118] Work by Desilets and St. Jacques on a methoxybenzoxepine puts the value in a similar range (0.65–0.7 kcal/mol) for seven-membered rings.[113] This work thus provides reasonable precedent for similar influence of the anomeric effect for septanosides. The anomeric effect requires only that sufficient orbital overlap be possible. Based on the ring-puckering mode and hence the placement of the ring oxygen atom, the consequences of where the aglycone resides may be different, but still be governed by the same effect. Additionally, because septanoses have one more (substituted) carbon atom in the ring, a summation of other forces may supersede the influence of the anomeric effect in energetic terms.

Nonetheless, septanoses can readily adopt appropriate T (or C) conformations that allow for the alignment of ring-oxygen lone pairs with the C—O bond at the anomeric position. In fact, for a given septanose, this preference can direct the distribution of conformers that is populated. The lack of steric bias of substitutents at the isoclinal position of a T conformer can work in concert with the anomeric effect to further restrict the accessible conformational space. The related exo-anomeric effect, which restricts rotation of the C—O anomeric bond is also operative.[119] Computational analysis of septanose conformations has also provided evidence of intramolecular hydrogen bonding. For example, the minimized conformer of septanoside **198** shows intramolecular hydrogen bonds between hydroxyl groups on C-2, C-3, and C-4 in a geared fashion (Figs. 15 and 17). However, the overall contribution of intramolecular hydrogen bonding between ring hydroxyl groups may be dubious for conformations in solution, because of the ability of these groups to hydrogen bond with protic (aqueous) solvent molecules.[120] In addition, in molecules where it is applicable, the exocyclic hydroxymethyl group can hydrogen-bond to the ring oxygen atom. An interaction such as this, together with gauche interactions, can contribute to the favored rotamer conformations about the C-6—C-7 bond. Overall, those factors that are operative in pyranose conformational analysis seem to carry over to the conformational analysis of septanoses.

The crystal structure of methyl α-septanoside **210** illustrates many of these features (Fig. 15). The overall conformation of the molecule as detailed in Fig. 15 is $^{3,4}T_{5,6}$. The —OCH$_3$ aglycone is positioned quasi-axially with a dihedral angle [O(lp)—O—C-1—O(C-1)] of − 158.3°. Alignment of the ring-oxygen lone pair with the σ* of the glycosidic (C—O) bond suggests that the anomeric effect influences the conformation of the molecule. Atoms that define the molecular plane for the $^{3,4}T_{5,6}$

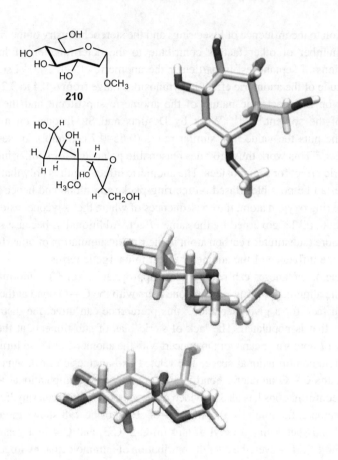

FIG. 15. Different perspectives of the methyl 2-deoxy-α-D-*gluco*-septanoside (**210**) structure from X-ray data. (See Color Insert.)

conformation are the ring oxygen atom, C-1, and C-2. Based on this, C-1, the anomeric center, is in the isoclinal position. An intramolecular hydrogen bond between the C-3 and C-4 hydroxyl groups is evident, while the other hydroxyl groups form intermolecular hydrogen bonds that assist in crystal packing. The intramolecular hydrogen bond evidences the gearing observed in computational models. The exocyclic hydroxymethyl group is oriented as a *gg* rotamer that is reminiscent of pyranose and furanose examples.[51,121–123] This alignment is further evidence of electronic effects such as this gauche effect. These same factors also influence the structures of septanoses in solution, as outlined in the preceding section.

2. Determination of Septanose Conformations

Presented here is a detailed study of the conformational analysis of methyl septanosides **198** and **200**.[124] The related methyl 5-*O*-methyl septanosides **201** and **202**,[125] which have also been reported by us, closely track the conformational preferences observed for the parent septanosides. The 5-*O*-methyl septanosides were used as a model of the reducing-end residue of a septanose-containing disaccharide.[125] As had been done in a furanoside system,[126] the methyl group was a surrogate for the nonreducing end residue of the disaccharide. The intent of this study was to emphasize factors that will probably be diagnostic for the determination of solution state conformations of septanosides in general. The strategy for determining the low-energy conformations of **198** and **200** was a combination of computational chemistry and analysis of $^3J_{H,H}$ coupling constants, both computationally and by ^1H NMR spectroscopy. Because septanosides **198** and **200** were prepared from oxepine **140**, the stereochemistry at C-3—C-6 is already defined. Part of the initial motivation for the conformational analysis was to determine the stereochemistry at C-1 and C-2 resulting from epoxidation of the oxepine and subsequent methanolysis of the intermediate 1,2-anhydro sugar; at the time of the investigation, this had not been established.

A parameter that has been useful for defining the anomeric stereochemistry of glycopyranosides is the chemical shift of the anomeric carbon atom, C-1. Similarly, the ^{13}C NMR shifts of C-1 in methyl septanosides also diagnostic for their anomeric configuration. In general, the δ_{C-1} values for α-septanosides are slightly upfield (99–104 ppm in completely deprotected septanosides) relative to β-septanosides (104–111 ppm) (Table I). This distribution of chemical shifts mirrors the trends for pyranosides; namely, the C-1 chemical shifts for α-pyranosides lie upfield from those of β-pyranosides. This observation suggests that the anomeric effect is operative in

TABLE I
^{13}C Chemical Shifts of C-1 for Selected Septanosides

α Anomer	δ_{C-1} (ppm)	References	β Anomer	δ_{C-1} (ppm)	References
198	106.2	126	200	110.0	126
199	100.6	65	202	109.3	124
201	106.1	124	207	108.0	124
203	100.3	25	208	110.8	74
204	100.7	25	209	110.1	124
205	100.7	25	210	112.2	127
206	106.2	127			

septanose glycosides, as has been illustrated in model seven-membered ring systems.[113] The partitioning of C-1 chemical shifts became evident after data from a number of septanosides prepared in our laboratory were compared. For this conformational analysis exercise, these data were combined with $^3J_{H1,H2}$ data. With these two data points, it was possible to assign tentatively the absolute configuration of the anomeric center via the $\delta^{13}C$ value of C-1 and the relative configuration of the C-2 stereocenter via $^3J_{H1,H2}$. This conclusion assumed that the 1,2-anhydro sugars opened by S_N2 attack to provide the 1,2-*trans* glycosides. A recent computational investigation has calculated $^{13}C-^{13}C$ ^1J coupling constants between C-1 and C-2 in septanose sugars.[129,130] The report uses chair conformations of reducing (see later) septanoses as the starting structures for the computational investigation. Key findings of the paper suggest that trends in $^1J_{C1,C2}$ values may enable assignment of stereochemistry at these centers as well (Fig. 16).

With the C-1 and C-2 stereochemistry established, the other $^3J_{H,H}$ coupling constants determined experimentally were compared to computed $^3J_{H,H}$ values. Experimental $^3J_{H,H}$ values were collected from ^1H NMR spectra and were also simulated to confirm assignments and couplings. The computed $^3J_{H,H}$ values were determined by a stepwise process that included Monte Carlo generation of a large number of conformations for each septanoside (**207** and **209**), which were then minimized using an AMBER mechanics force field. The most stable structures, which were a subset of the original collection of conformers from the Monte Carlo simulation, were then carried on to optimizations at different levels of theory—HF/6-31G* and B3LYP/6-31+G**. Calculations were conducted for both in vacuum and under a continuum solvent model. From these optimizations, a Boltzman distribution of conformers for each

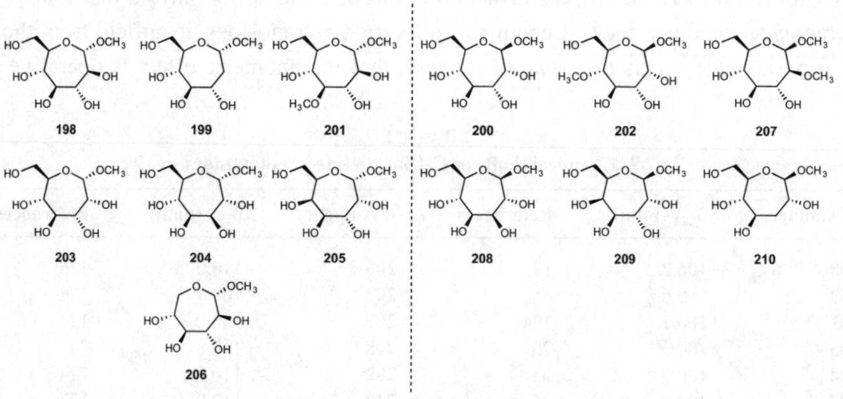

FIG. 16. Methyl septanosides **198–210** whose $\delta^{13}C$ values for C-1 are listed in Table I.

molecule was prepared. Contributions to the calculated $^3J_{H,H}$ values were determined from each conformer's weighting in the Boltzmann distribution. Good agreement between the calculated and observed $^3J_{H,H}$ was observed.

The results showed that for methyl α-D-*glycero*-D-*ido*-septanoside (**209**), the $^{3,4}T_{5,6}$ conformer was the most highly populated; this is especially noteworthy because the same conformation was adopted in the solid state for the related septanoside **210**, as previously discussed. The calculated global low-energy conformer (at the HF/6-31G* level with continuum solvent model) for **209** shown in Fig. 17 represents >73% of the Boltzmann distribution of conformers. Features that governed the conformational preferences for **209** were similarly operative for **210**, namely the anomeric effect, the exo-anomeric effect, and the influence of ring stereocenters. Other conformers that contributed an additional ~23% to the calculated Boltzmann distribution were all in the $^{3,4}T_{5,6}$ conformation but varied in the orientation of hydrogen bonding around the ring and/or the orientation of the C-6—C-7 rotamer. Similar trends were observed for methyl β-D-*glycero*-D-*gulo*-septanoside (**207**); its most populated conformer was the nearby $^{6,O}T_{4,5}$ conformation (Fig. 17).

IV. BIOCHEMICAL AND BIOLOGICAL INVESTIGATIONS USING SEPTANOSE CARBOHYDRATES

Investigations of septanose carbohydrates have been motivated by an interest in exploring their role in biological systems. Inquiries into the biological significance of septanosides range from fundamental biological questions to the utilization of

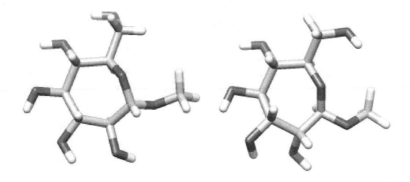

FIG. 17. Computed low-energy conformations of methyl septanosides. Left: Methyl α-D-*glycero*-D-*ido*-septanoside (conf 36); Right: Methyl β-D-*glycero*-D-*gulo*-septanoside (conf 1). (See Color Insert.)

septanoses for designed biological and biochemical applications. Septanoses may be present in biological systems but have heretofore gone unidentified; this includes the proposal suggested earlier that septanoses may be immunogenic. Septanoses could be used in exploratory investigations by replacing pyranoses or furanoses in specific contexts and observing the associated biochemical and/or biochemical consequences. Alternatively, septanoses may be incorporated into designed inhibitors of carbohydrate-processing enzymes or of protein–carbohydrate interactions and assayed for their potency as potential therapeutics. Here, we collect instances of septanose carbohydrates in model biochemical and biological contexts and the resulting effects on the individual systems.

1. Olgonucleotides Containing Septanosyl Nucleotides

There has been a long-standing interest in the synthesis of unnatural nucleoside analogues coupled with the study their biological activities.[131–134] Among them are bicyclic nucleosides,[135] spironucleosides,[136–142] and nucleosides containing sugar rings other than furanoses.[143–148] However, nucleosides incorporating a seven-membered sugar ring are somewhat rare.[149,150] There are also several examples of unnatural oligomers where the nucleoside base has been replaced by other expanded aromatic heterocycles.[2]

This section examines the synthesis of nucleosides that contain seven-membered sugar analogues in place of the deoxyribose component. Nucleosides from the last group have been further incorporated into ONs via solid-phase DNA synthesis. A physical and biochemical investigation of the oligomers thus prepared continues in the next section. The study under review culminated in the assessment of the ability of the oligomers to complex with single-stranded RNA and for the heteroduplexes so formed to serve as substrates of RNAseH.

a. Synthesis of Septanose-Containing Nucleosides and Their Incorporation into Oligomers.—Matsuda *et al.*[151] reported the synthesis of oxepine nucleoside **215** through ring expansion of a furanosyl nucleoside (Scheme 31).[152] Starting from the protected uridine **211**, the TBS group was removed from the 5-hydroxyl group and the resulting alcohol was oxidized to give aldehyde **212** (47% over two steps). Addition onto formaldehyde by an enolate derived from aldehyde **212** and subsequent reduction afforded the bis(hydroxymethyl) derivative **213** (51%). A sequence of protection and deprotection steps, followed by oxidation of the appropriate hydroxymethyl group, provided the 4-α-formyluridine derivative **214** (47% over four steps). Wittig methylenation of **214** and desilylation resulted in the formation of the ring-expanded product **215** in 37% yield. The authors proposed a mechanism for the conversion of **214** into **215** that involved migration of the 3′-*O*-TBS group to

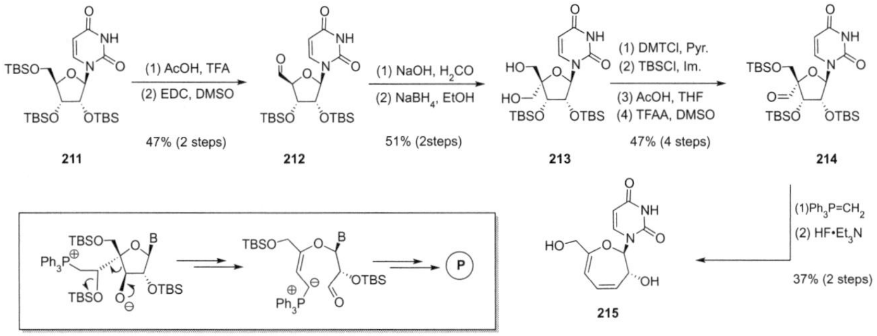

SCHEME 31. Ring-expanded nucleoside analogue **215**.

the oxygen atom on the betaine intermediate, followed by a retro-aldol type of ring cleavage between C-3′ and C-4′ and finally an intramolecular Wittig reaction. The inset of Scheme 31 shows intermediates in the ring-expansion mechanism.

Mandal et al.[135] demonstrated a synthetic route to oxepanyl nucleosides **223** starting from aldehyde **216**, itself prepared from 1,2:5,6-di-*O*-isopropylidene-α-D-glucofuranose (Scheme 32). The fused bicyclic intermediate **217** was obtained in 56% yield[153] through an intramolecular 1,3-dipolar cycloaddition of the nitrone generated *in situ* from the aldehyde **216** and benzylhydroxylamine. Removal of the isopropylidene acetal delivered the hemiacetal **218** (80%). Glycol cleavage and subsequent reduction gave compound **219**, which was then subjected to reductive cleavage of the N—O bond, giving amino-oxepane **220** (44% over three steps). Coupling of **220** with 5-amino-4,6-dichloropyrimidine produced the aminopyrimidinyl oxepane **221** (51%), which was further transformed into chloropurine derivative **222** via imidazole formation. The final nucleoside analogue, **223**, was obtained in 72% yield over the last two steps, by replacing the 6-chloro group by an amino group, using alcoholic ammonia.

Syntheses of ring-expanded nucleosides that are even more closely related to natural nucleosides are shown in Schemes 33 and 34. The target septanose nucleoside **229** positions the purine base, the hydroxymethyl group, and the ring hydroxyl groups in positions reminiscent of natural adenosine (Scheme 33). The synthesis of **229** started with the diacetal **224**, prepared from commercially available D-*glycero*-D-*gulo*-heptono-1,4-lactone following a reported procedure.[149] Reduction of **224** with NaBH$_4$ gave 5-deoxy-2,3:6,7-di-*O*-diethylidene-D-*allo*-heptitol. Stepwise protection of the primary C-1 hydroxyl group as the dimethoxytrityl (DMT) ether and the 4-hydroxyl group as the *p*-chlorobenzyl derivative afforded **225** in 55% yield over three

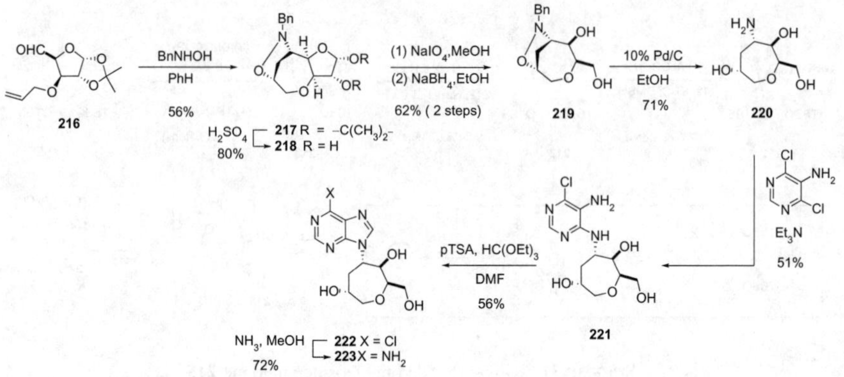

SCHEME 32. Synthetic route to oxepanyl nucleoside analogues.

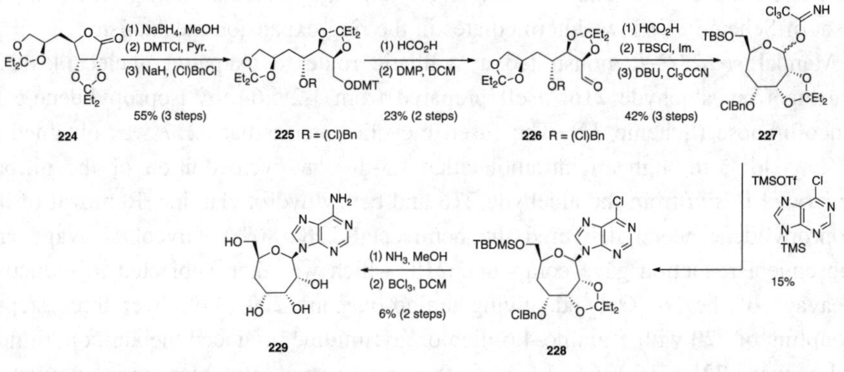

SCHEME 33. Ring-expanded nucleoside **229** closely related to natural adenosine.

steps. Selective removal of the DMT group and oxidation of the newly exposed hydroxyl group with Dess–Martin periodinane produced aldehyde **226** (23% over two steps). Selective deprotection of the C-6,C-7 acetal using formic acid led to the septanose, now in its hemiacetal form, which was protected as the TBS ether at the C-7 hydroxyl group and then further transformed into trichloroacetimidate **227** (42% over three steps). The ability to trap the hemiacetal as the trichloracetimidate reinforces the fact that the septanoses will adopt a seven-membered ring form when other hydroxyl groups are protected. Glycosylation with 6-chloro-9-trimethylsilylpurine, promoted by trimethylsilyl triflate, furnished **228** in 15% yield. Finally, the target nucleoside **229**, essentially a ring-expanded adenosine analogue, was obtained by

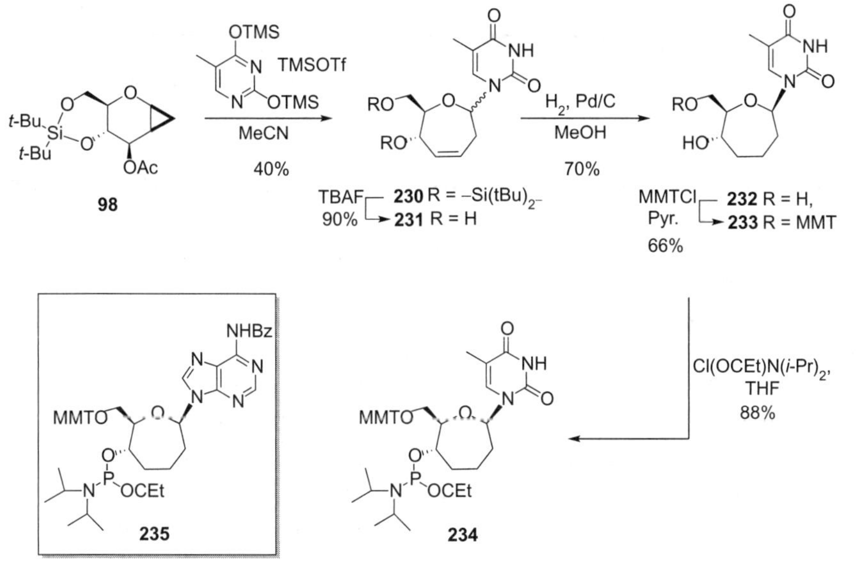

SCHEME 34. synthesis of protected phosphoramidites suitable for solid-phase oligonucleotide synthesis.

reaction with methanolic ammonia and subsequent treatment with BCl₃ (6%). Despite the low yields in the glycosylation reaction and the subsequent deprotection to afford **229**, sufficient material was prepared to assay its potential biological activity.

Damha et al.[150,154] reported the synthesis of substituted nucleoside analogues **234** (designated oT by the authors) and **235** (termed oA). The "o" prefix was used to indicate that the nucleoside contained an oxepanyl sugar, or septanose, while T and A are the standard nucleoside bases thymine and adenine, respectively. The syntheses (Scheme 34) are notable because protected phosphoramidites, appropriate for solid-phase ON synthesis, were prepared. The route to the nucleosides is illustrated by the formation of thymidine analogue **234**; the same sequence delivered **235** in similar yields. The synthesis started from the cyclopropanated glycal **98** originally reported by Hoberg.[54] Vorbruggen-type[155] glycosylation reaction of **98** with silylated thymine furnished the unsaturated oxepine nucleoside **230** (40%) along with a diene product (not shown) that arose by elimination. Ring expansion was concomitant with glycosylic bond formation in this reaction. Glycosylation with the thymine acceptor proceeded with relatively high diastereoselctivity (β:α 10:1), whereas glycosylation with N^6-benzoyladenine resulted in lower selectivity (β:α 2:1). Desilylation (90%) and reduction of the double bond (70%) produced the oxepanyl nucleoside **232**.

Selective functionalization of the 7-hydroxyl group as the monomethoxytrityl (MMT) ether and phosphoramidite formation at O-5 under standard conditions delivered **234** (58% over two steps). This material could be taken through to the solid-phase ON synthesis. This synthesis utilizing **234** and **235** as starting materials delivered homopolymers, oT_{15} and oA_{15}, each of which was 15 nucleoside bases long.

b. Physical and Biochemical Characterization of Septanose-Containing Oligomers.—Studies on modified ONs have provided useful structural and conformational information that could be used to design and synthesize new ONs for therapeutic uses. Synthetic ONs are of high interest, for example, as antisense binding agents.[156,157] Antisense oligonucleotides (AONs) have been used to base-pair with complementary RNAs, and the resultant duplexes were efficiently recognized by ribonuclease H (RNaseH), which cleaves the complementary RNA.[150] Thus, design of stable AONs that are substrates of RNaseH and catalytic in destruction of multiple copies of an RNA transcript is of value in functional genomics and as potential drug targets.[158] Using oligomers that contained a mixture of rigid and flexible nucleotides, Damha *et al.* found that flexibility in the ON backbone played an important role for its antisense activity. Heteroduplexes composed of RNA plus the oligomers containing the flexible analogues were found to enable alignment into the active site of RNAse and subsequently effect accelerated degradation of the target RNA.[150,158] Systematic investigations were carried out with ONs that differed in the ring size of the constituent sugars. It was anticipated that ONs that varied the sugar residue would impart different flexibility to individual AONs and thus provide useful information on the underlying mechanism of substrate recognition by RNaseH.[150] This hypothesis motivated investigation of the oxepanyl ONs.

With the oxepane-based oligomers in hand, the investigation turned to their physical and biochemical characterization. To gain insight about the influence of the oxepane modification on the ONs, low-energy conformations of the oT nucleoside, derived from a conformational analysis, were compared to the known low-energy conformation of dT (Fig. 18).[150] Thymidine (dT) adopts a twist conformation with C-2' above the plane of the ring, and this is referred to as the C-2'-*endo* conformation. An *ab initio* geometry optimization and gradient algorithm calculations on oT showed that a twist conformation ($^{O,6}T_{4,5}$) was the major conformer. Simple inspection of the two low-energy conformations for dT and oT, respectively, shows that they have a similar overall orientation; both sugars dispose the pyrimidine base in an exo-fashion and show related sugar puckering. The largest differences between the two nucleosides is that oT has a larger sugar-ring diameter than dT and the C-3'/C-5' hydroxyl group is oriented in a slightly different manner. Specifically, a shorter O—O distance (7'-OH—5'-OH, 3.57 Å) was noted for oT, in comparison to dT, where the 5'-OH—3'-OH distance is 4.26 Å; this is a difference

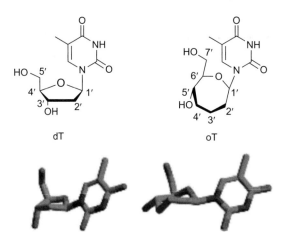

FIG. 18. Thymidine (dT) and an oxepanyl analogue (oT). Structures on the bottom (taken from ref. 149) show minimized conformers for both dT and oT, emphasizing the similarity in the disposition of the base and the C-4/C-6 hydroxymethyl group. (See Color Insert.)

of almost 0.8 Å. Notably, the authors also observed that oT existed in a complex equilibrium between many low-energy states, including boat, twist-boat, chair, and twist forms. The energy barrier for the most favorable transition from a C to a T conformer was calculated to be 0.5–2 kcal/mol, which was considered to be low relative to the furanose form and other analogues. Overall, their analysis indicated that oxepane-modified oligomers should effectively hybridize to both DNA and RNA.[159]

Thermal-denaturation experiments were utilized to collect information on the melting temperatures (T_m) of duplexes that were composed of combinations of the following oligomers: dA_{15}, dT_{15}, rU_{15}, rT_{15}, oT_{15}, and oA_{15} (d=deoxyribose; r=ribose; o=oxepane). T_m values for control duplexes dT_{15}/dA_{15}, rU_{15}/rA_{15}, dT_{15}/rA_{15}, and rU_{15}/dA_{15} ranged from 16 to 37 °C (Table II). The oT_{15}/oA_{15} duplex gave a very weak but distinct T_m value of 12 °C and an associated small change in percent hyperchromicity (Table II and Fig. 19). The data indicated weak hybridization of oxepane-modified ONs. Control T_m experiments were performed with single-stranded oA_{15} and control dA_{15}, which showed only a slight increase in absorbance and was indicative of no recognition between these similar strands. The tentative conclusion, then, was that the oxepanyl oligomers formed a duplex structure.

In order to confirm formation of the oT_{15}/oA_{15} duplex, additional techniques were used to characterize the complex. Stoichiometery analysis (Job's plot binding) supported the formation of a 1:1 complex. CD-assisted melting experiments performed on the

TABLE II
Thermal Melting (T_m) Temperatures and Percentage Hyperchromicity (%H) Changes for Oligonucleotide Duplexes

Duplex	T_m (°C)	%H
Natural		
dT_{15}/dA_{15}	37	24.8
rU_{15}/rA_{15}	25	28.7
dT_{15}/rA_{15}	32	25.4
rU_{15}/dA_{15}	16	20.7
Oxepanyl		
oT_{15}/oA_{15}	12	4.5
oT_{15}/dA_{15}	<5	8.7
oT_{15}/rA_{15}	13	11.3
dT_{15}/oA_{15}	<5	3.3
rU_{15}/oA_{15}	12	3.4

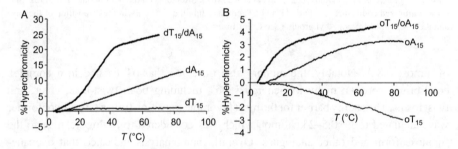

FIG. 19. (Taken from ref. 149) Left: Traces showing change in hyperchromicity as a function of temperature for ssDNAs dT_{15}, and dA_{15} and duplex dT_{15}/dA_{15}. Right: Traces showing change in hyperchromicity as a function of temperature for single-strand oxepanyl oligomers oT_{15} and oA_{15} and duplex oT_{15}/oA_{15}.

oT_{15}/oA_{15} duplex measured molar ellipticity at 254 nm versus temperature and provided a T_m value (12 °C) in agreement with the data obtained from the UV–vis experiment. However, the CD spectra obtained for the oT_{15}/oA_{15} duplex were clearly different than spectra observed for the individual oT_{15} and oA_{15} strands and of the natural duplex (dT_{15}/dA_{15}). The authors suggested that the data reflected different helical structures for the oxepane duplex relative to the DNA duplex. Although a left-handed helix structure of the oT_{15}/oA_{15} duplex was proposed based on the CD profile, no NMR or crystallographic data were available to address the structure of the duplex conclusively.

Oxepanyl oligomers oA15 and oA15 were also evaluated for their ability to hybridize with single-stranded deoxyribosyl or ribosyl oligomers. Binding studies,

Job's plot analyses, T_m values, and temperature-dependent CD experiments confirmed the formation of 1:1 heteroduplexes between the oxepanyl and ribosyl (rA_{15} and rU_{15}) oligomers (Table II). Similar experiments with the deoxyribosyl oligomers dA15 and dT15 showed no significant association. CD spectra of oT_{15}/rA_{15} heteroduplexes were indicative of A-form helices similar to natural dT_{15}/rA_{15} systems. Although oT_{15} alone exhibited a weak CD spectrum that differed significantly from rA_{15} and the oT_{15}/rA_{15} duplex, it was concluded that, because of the greater flexibility of the oxepane ring, the modified oligomer can adjust itself into the helix to become a better fit with the dominant conformation of the RNA strand.

RNAseH is a ribonuclease enzyme that recognizes and cleaves the 3′-O-P bond of the RNA strand in RNA–DNA duplex to produce a 3′-hydroxyl and 5′-phosphate-terminated product. AONs have found major utility as a tool for the treatment of cancer and infectious diseases.[156] Antisense therapy is based on activation of the RNAseH mechanism.[160] The mechanism entails duplex formation between the antisense oligomer strand and cellular RNA—a primary transcript of a specific sequence. Selective hydrolysis of the target RNA strand in the heteroduplex is then catalyzed by RNAseH.[161] Many AONs employed to serve this purpose[162–164] suggest that flexibility of the sugar backbone of the ONs play a key role in the enzyme-recognition event. In principle, RNAse H recognizes an ideal DNA/RNA hybrid. A flexible AON should be able to hybridize effectively with the DNA and be recognized by RNAseH, but unlikely to be a substrate for nucleosidases that may otherwise hydrolyze DNA-based AOns.

Degradation of the RNA strand, mediated by *Escherichia coli* RNAseH for a variety of heteroduplex ONs, was next assayed at three temperatures, 37, 20, and 10 °C. There was essentially no cleavage of the RNA strand of the rA_{15}/oT_{15} heteroduplex at 37 or 20 °C; cleavage was detected at 10 °C, however. Nearly quantitative hydrolysis was observed for the "native" dT_{15}/rA_{15} heteroduplex at all three temperatures. Estimates for RNA hydrolysis were approximately 20% when in the oT_{15}/rA_{15} duplex versus 100% for dT_{15}/rA_{15}. The observation of RNAseH hydrolysis at the lower temperatures tracked well with the duplex stability as expressed by T_m. That is, when the oxepanyl heteroduplex was more highly populated (at the lower temperatures), greater hydrolysis was observed. The authors correlated the conformational flexibility of the deoxyribosyl versus oxepanyl nucleosides involved; according to them, the results reinforced the idea that greater flexibility of the constituent nucleotides equates with greater hydrolysis by RNAseH. This study effectively illustrates how septanose carbohydrates can be used to address fundamental questions in biochemistry, and perhaps even lead to new therapeutics.

2. Protein–Septanose Interactions

a. Lectin-Binding Studies.—Our group has employed methyl septanosides in the study of protein–carbohydrate interactions.[125,128] The jack-bean protein concanavalin A (ConA) was chosen as a model lectin because it is commercially available and there are extensive structural and binding data for its interaction with natural carbohydrates. Isothermal titration calorimetry (ITC) was the primary experimental technique used to measure binding between ConA and the methyl glycosides shown in Fig. 20. The known monosaccharide binders methyl α-D-mannopyranoside (**236**) and methyl α-D-glucopyranoside (**237**) served as positive controls in the experiments and also as reference points for interpreting the results. Association constants for the interaction between ConA and selected monosaccharides are listed in Table III. The methyl α-D-septanosides **198–199** and **205–206** (Fig. 16) showed no evidence of binding as measured by ITC. The methyl β-D-septanosides **200**, **207–208**, and **210**, however, demonstrated variable affinities, ranging from 390 to 840 M^{-1} (Figs. 16 and 21). Exothermic binding was observed for the β-septanoside ligands, paralleling the profiles of **236** and **237**. This result provided preliminary evidence that the binding mode of the β-septanosides was the same as for α-pyranosides, suggesting an inversion in selectivity of anomeric configuration for the septanosides, namely, β septanosides were preferred over α septanosides. This is opposite to the selectivity observed for pyranosides.

Data obtained by ITC experiments were corroborated by saturation transfer difference (STD) ^{1}H NMR spectroscopy. Signals that arise during an STD experiment are a

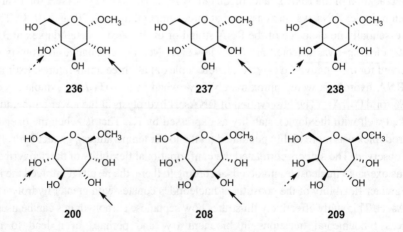

FIG. 20. Methyl glycosides and oxepanes used as ligands for the investigation of ConA/monosaccharide interactions.

TABLE III
Association Constants (K_a) for Interaction Between ConA and Selected Monosaccharides by Isothermal Titration Calorimetry

Ligand	K_a (M^{-1})
236	2000
237	7900
238	NB
200	520
207	450
208	840
209	NB[a]
210	390

NB, no binding as determined by ITC.

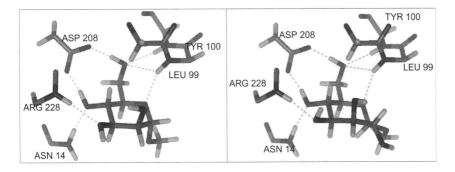

FIG. 21. Detail of the monosaccharide binding pocket with (left) pyranoside ligand **237** and (right) septanoside ligand **208** as determined from the QM/MM computations (taken from ref. 124). (See Color Insert.)

direct measure of binding. Positive STDs were measured for two methyl β-septanosides, compounds **200** and **210**. In an effort to demonstrate that the binding by ConA to β-septanosides was in its monosaccharide binding pocket and not simply a nonspecific interaction, we conducted a competition experiment between β-septanoside **210** and α-pyranoside **237**. As measured by STD, the results indicated that the α-pyranoside displaced the β-septanoside. It was concluded that β-septanosides were being bound by ConA in the ligand binding pocket for the natural α-pyranoside.

A detailed investigation using ITC and the quantum-mechanics/molecular mechanics (QM/MM) computational technique provided further insight into the ConA–septanoside interaction.[125] A major focus of this study was to define how the presence and orientation of functional groups about the monosaccharide ring affected affinity. It has been established that methyl α-D-galactopyranoside (**238**) is not a ligand for ConA, whereas methyl α-D-glucoside and methyl α-D-mannoside are ligands. The stereochemistry at individual carbon atoms of the ring therefore affects affinity. We prepared ring-expanded analogues of D-mannose and D-galactose, namely the methyl septanosides **208** and **209**, from the corresponding oxepines using chemistry previously described. There is a matched stereochemistry (homomorphism) between C-2—C-5 of the pyranosides and C-3—C-6 of the septanosides. This is the portion of the monosaccharides that makes contact with ConA when it is bound. We observed the same rank ordering of affinities for methyl β-septanosides that exists for methyl α-pyranosides. As shown in Table III, the affinity of the D-mannose analogue **208** is within almost a factor of 2 from the affinity of methyl α-D-glucopyranoside (**236**).

QM/MM calculations provided additional support for the model. Relative ΔH values from the calculations correlated with the ΔH values collected by ITC. The correlation allowed inspection of the computed complexes to identify factors important to association (Fig. 21). Three separate comparisons were made; these were between **200** and **208**, **236** and **200**, and **237** and **208**. It was suggested that the C-3—C-7 region of the septanosides was akin to C-2—C-6 of the pyranosides, and thus **208** was therefore a better ligand of ConA relative to **200** in the same way that **237** is better than **236**. The difference of K_a values **236** and **200** was ascribed in the case of **200** to the energetic penalty of reconfiguration for complexation. The magnitude of ΔH for binding of ConA to **208** was greater than for **237**; it was argued that the decreased entropy of the complex therefore made **208** a poorer ligand overall than **237**. The investigation of ConA–monseptanoside interactions revealed that septanosides could, with some qualifications, act as surrogates for pyranosides.

b. Glycosidases.—One design strategy for the development of glyco-enzyme inhibitors relies on finding a close structural relative of the substrate. Imino sugars (such as **9–13**, Fig. 2) are well-studied compounds that inhibit glycosidases through mimicry of the oxacarbenium ion intermediate involved in the mechanism of glycosidase action. Polyhydroxyazepanes **9–12** were also argued to be more conformationally flexible than six-membered ring analogues.[30] Based on the glycosidase inhibitory activity observed for the azepanes, Stütz and Withers asked whether L-idoseptanosides **239** and **240** could act as substrates for D-glucosidases.[165] *p*-Nitrophenyl α-L-idoseptanoside (**239**) and *p*-nitrophenyl β-D-glucoseptanoside (**240**) have the same configuration from C-1—C-4 as the respective β-D-glucopyranoside **241** and α-D-glucopyranoside **242**. Molecular-modeling studies reinforced this mimicry of the 4-nitrophenyl glycopyranosides by the α- and β-L-idoseptanosides. Shown in Fig. 22 are molecular-modeling overlays of **239** with **241** and **240** with **242**. These overlays demonstrate a similar positioning of the C-1, C-2, C-3, and C-4 hydroxyl

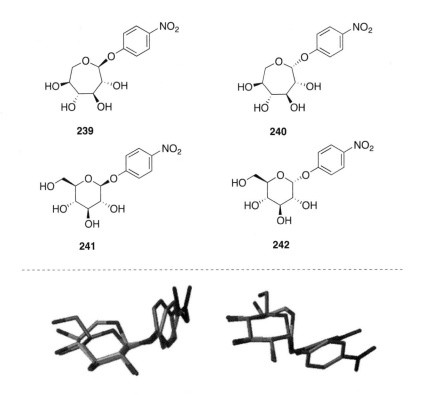

FIG. 22. Top: *p*-Nitrophenyl L-idoseptanosides **239** and **240** used to mimic D-glucopyranosides **241** and **242** as substrates for glycosidases. Bottom: Overlays of minimized structures. Left is the overlay of **239** with **240** and right is the overlay of **240** with **242** (taken from ref. 164). (See Color Insert.)

functionality. Additionally, the 5-hydroxyl groups of the septanoses were close to the C-6 oxygen atoms of the pyranosides.

The *p*-nitrophenyl septanosides **239** and **240** were synthesized from L-idose and then assayed against a collection of α- and β-glycosidases. Compound **239** was found to be a substrate for *Agrobacterium* sp. β-glucosidase, although with low efficiency ($k_{cat}/K_M = 22$ s^{-1} M^{-1}), especially in comparison to the pyranoside substrate **241**, where $k_{cat}/K_M = 2.2 \times 10^6$ s^{-1} M^{-1}). For the *Saccharomyces cerevisiae* α-glucosidase, septanoside **240** was a better substrate, with $k_{cat}/K_M = 24$ s^{-1} M^{-1} as compared to $k_{cat}/K_M = 2.6 \times 10^4$ s^{-1} M^{-1} for pyranoside **242**. The K_M values for **240** (2.9 mM) and **242** (0.27 mM) with the *S. cerevisiae* α-glucosidase suggested that binding was within a factor of 10 of pyranosides; the low efficiency of the septanoses implied that they were unable to be bound in a conformation that was productive for hydrolysis. Nonetheless, the authors convincingly demonstrated specificity for the appropriate anomeric configuration of the substrates by both of these enzymes. They also noted that **239** and **240** seemed to serve as glycosyl acceptors in transglycosidation reactions using *Thermoanaerobacterium saccharolyticum* β-xylosidase and a β-xylopyranoside donor.

In a similar vein, McDonald *et al.* prepared α-D-mannoseptanose mono-, di-, and trisaccharides **243–245** (Fig. 23) as potential inhibitors for α-mannosidases.[166] Monosaccharide **243** and the corresponding donors and acceptors leading to **244** and **245** were prepared via thioglycosides, synthesized themselves from oxepines. Using a jack-bean α-mannosidase, compounds **243–245** were assayed for their ability to inhibit the hydrolysis of *p*-nitrophenyl α-D-mannopyranoside. Akin to the arguments made by Stütz and Withers, the rationale was that a similar orientation the glycosidic bond and the C-2—C-4 hydroxyl groups would facilitate binding to the enzyme. The results showed no differences in K_M or k_{cat} values in the presence of **243–245** when compared to control hydrolysis reactions. The absence of the hydroxymethyl functionality on septanosides **239**, **240**, and **243–245** may contribute to their low activity. Further investigations with additional analogues could clarify this proposition.

3. Additional Reports

There are also a limited number of additional examples where oxepanes and thiepanes were evaluated for activity in biological assays. Septanose nucleoside **229** has been tested for its antiviral activity and assayed against viruses that represent three genera of the ssRNA$^+$ *Flaviviridae* family, namely, *Pestivirus*, *Flavivirus*, and *Hepacivirus*. It has also been studied against *enterovirus*, virus of ssRNA family *Retroviridae*; two representatives of the ssRNA$^-$ family *Paramyxoviridae*, *Pneumovirus* and *Morbillivirus*; and dsDNA virus family *Poxviridae*. No significant antiviral activity was found from the assay against the broad range of the viruses employed, and the nucleoside was not cytotoxic.

5-Thio-β-D-xylopyranosides that contain aromatic aglycons, such as Beciparcil (**246**, Fig. 24), and related analogues have potent antithrombotic activity in

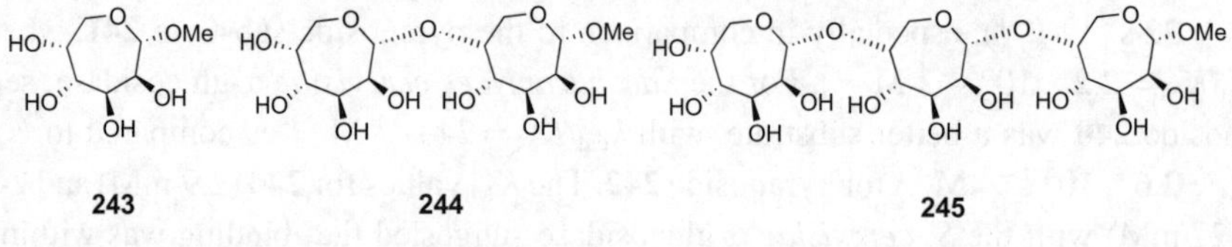

FIG. 23. Methyl α-septanoside mono-, di-, and tri-saccharides **243–245** assayed against jack-bean α-mannosidase.

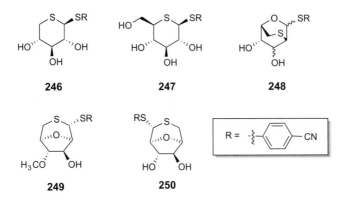

FIG. 24. Carbohydrate-based antithrombotics **246–250**.

mammals.[80,167] Their proposed mechanism of action is based on the ability of the aromatic aglycon to penetrate the plasmic membrane. Extensive research has focused on structure–activity relationships that include modifications made on both the sugar residue and the aglycon. The activity is not solely dependent upon the β-D-xyloside moiety; rather, 1,5-dithio-β-D-glucopyranoside **247** and the bridged thio sugar **248** also exerted significant oral antithrombic activity.[168] With this in mind, the researchers then synthesized and studied the activity of a series of seven-membered thio sugar analogues.[169,170] It was found that variations in the substituents on the *para* position of the aromatic aglycon had little effect on the overall activity. Many of the compounds had significant antithrombin activity in the standard rat model assay. In fact, compounds **249** and **250** exhibited activity that was approximately 10-fold higher than that of the reference compound, **246**.

There are probably many other biological activities of septanose carbohydrates that have yet to be described. Some natural products containing seven-membered cyclic ethers[171] and related analogues[172] have potent cellular activities. In addition to this inherent potential for septanose carbohydrates, the lingering question exists about the possible occurrence of natural septanosides, offering opportunities for additional investigations.

V. OUTLOOK

The chemistry and biochemistry of septanose carbohydrates are at an important turning point. Synthetic methods in the past 10–20 years have enabled the synthesis of a host of septanose carbohydrates. These include the septanose form for several of the natural (and C-6-unsubstituted) aldohexoses as well as a number of C-6-substituted heptoses. The increasing complexity of the structures that can be prepared is significant: it ranges from such oligosaccharides as trisaccharides **124** and **245** to glycoconjugates such as **79**. The next step forward for this nascent field is to identify compelling biological activities of septanose carbohydrates. This could entail discovering as yet unknown activity of natural septanosides, or by demonstrating that a septanoside is important to the activity of a potent small molecule. Additional investigations, on the synthetic, biochemical, and biological fronts, are required to consolidate the importance and usefulness of this class of carbohydrates.

Acknowledgments

The authors thank Bikash Surana for help in preparing a preliminary draft of this article. M. W. P. also thanks NSF (CHE-056311) for support of the research related to septanose carbohydrates in his group.

References

1. S. L. Schreiber, Rethinking relationships between natural products, *Nat. Chem. Biol.*, 3 (2007) 352.
2. A. T. Krueger, H. Lu, A. H. F. Lee, and E. T. Kool, Synthesis and properties of size-expanded DNAs: Toward designed, functional genetic systems, *Acc. Chem. Res.*, 40 (2007) 141–150.
3. R. P. Cheng, S. H. Gellman, and W. F. DeGrado, β-Peptides: From structure to function, *Chem. Rev.*, 101 (2001) 3219–3232.
4. S. H. Gellman, Foldamers: A manifesto, *Acc. Chem. Res.*, 31 (1998) 173–180.
5. J. Gao, H. Liu, and E. T. Kool, Assembly of the complete eight-base artificial genetic helix, xDNA, and its interaction with the natural genetic system, *Angew. Chem. Int. Ed.*, 44 (2005) 3118–3122.
6. A. H. F. Lee and E. T. Kool, A new four-base genetic helix, yDNA, composed of widened benzopyrimidine-purine pairs, *J. Am. Chem. Soc.*, 127 (2005) 3332–3338.
7. H. Liu, J. Gao, and E. T. Kool, Helix-forming properties of size-expanded DNA, an alternative four-base genetic form, *J. Am. Chem. Soc.*, 127 (2005) 1396–1402.
8. H. Liu, J. Gao, and E. T. Kool, Size-expanded analogues of dG and dC: Synthesis and pairing properties in DNA, *J. Org. Chem.*, 70 (2005) 639–647.
9. A. H. F. Lee and E. T. Kool, Novel benzopyrimidines as widened analogues of DNA bases, *J. Org. Chem.*, 70 (2005) 132–140.
10. H. Lu, K. He, and E. T. Kool, yDNA: A new geometry for size-expanded base pairs, *Angew. Chem. Int. Ed.*, 43 (2004) 5834–5836.
11. H. Liu, J. Gao, S. R. Lynch, Y. D. Saito, L. Maynard, and E. T. Kool, A four-base paired genetic helix with expanded size, *Science*, 302 (2003) 868–871.
12. A. R. Hernández and E. T. Kool, The components of xRNA: Synthesis and fluorescence of a full genetic set of size-expanded ribonucleosides, *Org. Lett.*, 13 (2011) 676–679.

13. H. Lu, A. T. Krueger, J. Gao, H. Liu, and E. T. Kool, Toward a designed genetic system with biochemical function: Polymerase synthesis of single and multiple size-expanded DNA base pairs, *Org. Biomol. Chem.*, 8 (2010) 2704–2710.
14. A. C. Gemperli, S. E. Rutledge, A. Maranda, and A. Schepartz, Paralog-selective ligands for Bcl-2 proteins, *J. Am. Chem. Soc.*, 127 (2005) 1596–1597.
15. E. A. Harker and A. Schepartz, Cell-permeable β-peptide inhibitors of p53/hDM2 complexation, *Chembiochem*, 10 (2009) 990–993.
16. J. A. Kritzer, R. Zutshi, M. Cheah, F. A. Ran, R. Webman, T. M. Wongjirad, and A. Schepartz, Miniature protein inhibitors of the p53-hDM2 interaction, *Chembiochem*, 7 (2006) 29–31.
17. O. M. Stephens, S. Kim, B. D. Welch, M. E. Hodsdon, M. S. Kay, and A. Schepartz, Inhibiting HIV fusion with a β-peptide foldamer, *J. Am. Chem. Soc.*, 127 (2005) 13126–13127.
18. Z. Pakulski, Seven membered ring sugars: A decade update, *Pol. J. Chem.*, 80 (2006) 1293–1326.
19. Z. Pakulski, Seven membered ring sugars, *Pol. J. Chem.*, 70 (1996) 667–707.
20. X. Chen and A. Varki, Advances in the biology and chemistry of sialic acids, *ACS Chem. Biol.*, 5 (2010) 163–176.
21. C. Traving and R. Schauer, Structure, function and metabolism of sialic acids, *Cell. Mol. Life Sci.*, 54 (1998) 1330–1349.
22. IUPAC-IUBMB Joint Commission on Biochemical Nomenclature (JCBN), Nomenclature of carbohydrates (Recommendations 1996), http://www.chem.qmul.ac.uk/iupac/2carb/.
23. Z. Pakulski and A. Zamojski, An efficient synthesis of 5-deoxy-D-ribohexose, *Pol. J. Chem.*, 69 (1995) 912–917.
24. E. F. L. J. Anet, Degradation of carbohydrates. Part X. 2,3,4,5-Tetra-*O*-methyl-D-glucose, *Carbohydr. Res.*, 8 (1968) 164–174.
25. S. D. Markad, S. M. Miller, M. Morton, and M. W. Peczuh, Hydroxyl group orientation affects hydrolysis rates of methyl [alpha]-septanosides, *Tetrahedron Lett.*, 51 (2010) 1209–1212.
26. F. Moris-Varas, X.-H. Qian, and C.-H. Wong, Enzymatic/chemical synthesis and biological evaluation of seven-membered iminocyclitols, *J. Am. Chem. Soc.*, 118 (1996) 7647–7652.
27. Q. Xinhua, F. Morís-Varas, and C.-H. Wong, Synthesis of C2-symmetrical polyhydroxyazepanes as inhibitors of glycosidases, *Bioorg. Med. Chem. Lett.*, 6 (1996) 1117–1122.
28. J. C. Y. Wong, P. Lacombe, and C. F. Sturino, A ring closing metathesis-osmylation approach to oxygenated oxepanes as carbohydrate surrogates, *Tetrahedron Lett.*, 40 (1999) 8751–8754.
28a. A. E. Stuetz and T. M. Wrodnigg, Imino sugars and glycosyl hydrolases: historical context, current aspects, emerging trends, *Adv. Carbohydr. Chem. Biochem.*, 66 (2011) 187–298.
29. H. Li, Y. Blériot, C. Chantereau, J. M. Mallet, M. Sollogoub, Y. Zhang, E. Rodríguez-García, P. Vogel, J. Jiménez-Barbero, and P. Sinaÿ, The first synthesis of substituted azepanes mimicking monosaccharides: A new class of potent glycosidase inhibitors, *Org. Biomol. Chem.*, 2 (2004) 1492–1499.
30. F. Marcelo, Y. He, S. A. Yuzwa, L. Nieto, J. Jimènez-Barbero, M. Sollogoub, D. J. Vocadlo, G. D. Davies, and Y. Blériot, Molecular basis for inhibition of GH84 glycoside hydrolases by substituted azepanes: Conformational flexibility enables probing of substrate distortion, *J. Am. Chem. Soc.*, 131 (2009) 5390–5392.
31. M. L. Candenas, F. M. Pinto, C. G. Cintado, E. Q. Morales, I. Brouard, M. T. Diaz, M. Rico, E. Rodriguez, R. M. Rodriguez, R. Perez, R. L. Perez, and J. D. Martin, Synthesis and biological studies of flexible brevetoxin/ciguatoxin models with marked conformational preference, *Tetrahedron*, 58 (2002) 1921–1942.
32. F. Micheel and F. Sückfull, Eine neue Klasse von Zuckerderivaten, *Justus Liebigs Ann. Chem.*, 502 (1933) 85–98.
33. F. Micheel and W. Spruck, Über eine weitere Dar-stellungsmethode für Zucker-Derivate mit siebengliedrigem Ringe (Septanosen), *Ber. Dtsch. Chem. Ges. B*, 67B (1934) 1665–1667.

34. J. D. Stevens, Large scale isolation of 1,2:3,4-di-*O*-isopropylidene-α-D-glucoseptanose and 2,3:4,5-di-*O*-isopropylidene-β-D-glucoseptanose, *Carbohydr. Res.*, 346 (2011) 689–690.
35. T. Q. Tran and J. D. Stevens, Septanose carbohydrates. VII. Preparation of mono-*O*-isopropylidene derivatives of methyl β-D-glucoseptanoside and preparation of methyl α-L-idoseptanoside and its derivatives, *Aust. J. Chem.*, 55 (2002) 171–178.
36. G. E. Driver and J. D. Stevens, Preparation of 1,2-*O*-isopropylidene derivatives of α-D-galactoseptanose, β-L-altroseptanose, and 3-*O*-methyl-α-D-guloseptanose, *Carbohydr. Res.*, 334 (2001) 81–89.
37. C. J. Ng, D. C. Craig, and J. D. Stevens, Preparation of methyl β-idoseptanoside and its derivatives, *Carbohydr. Res.*, 284 (1996) 249–263.
38. C. J. Ng and J. D. Stevens, Preparation of mono-*O*-isopropylidene derivatives of methyl α-D-glucoseptanoside, *Carbohydr. Res.*, 284 (1996) 241–248.
39. G. E. Driver and J. D. Stevens, Septanose carbohydrates. III. Oxidation-reduction products from 1,2-3,4-di-*O*-isopropylidene-α-D-glucoseptanose: Preparation of L-idose derivatives, *Aust. J. Chem.*, 43 (1990) 2063–2081.
40. C. J. Ng and J. D. Stevens, Methyl α- and β-D-glucoseptanosides, *Methods Carbohydr. Chem.*, 7 (1976) 7–14.
41. J. D. Stevens, Septanose carbohydrates. I. Acid-catalysed reaction of D-glucose with acetone, *Aust. J. Chem.*, 28 (1975) 525–557.
42. J. D. Stevens, Two septanose diacetals of D-glucose, *J. Chem. Soc. D* (1969) 1140–1141.
43. J. D. Stevens, Isolation of aldehydo and septanoside derivatives from the acid-catalyzed reaction of D-glucose and some of its derivatives with acetone-methanol, *Carbohydr. Res.*, 21 (1972) 490–492.
44. J. C. McAuliffe and O. Hindsgaul, Use of acyclic glycosyl donors for furanoside synthesis, *J. Org. Chem.*, 62 (1997) 1234–1239.
45. M. L. Wolfrom, D. I. Weisblat, and A. R. Hanze, The reactivity of the monothioacetals of glucose and galactose in relation to furanoside synthesis. I, *J. Am. Chem. Soc.*, 66 (1944) 2065–2068.
46. M. L. Wolfrom, D. I. Weisblat, and A. R. Hanze, D-Glucose *S*-ethyl *O*-methyl monothioacetal, *J. Am. Chem. Soc.*, 62 (1940) 3246–3250.
47. M. L. Wolfrom and D. I. Weisblat, Monothioacetals of galactose, *J. Am. Chem. Soc.*, 62 (1940) 878–880.
48. J. C. McAuliffe and O. Hindsgaul, The synthesis of disaccharides terminating in D-septanosyl residues using acyclic intermediates, *Synlett* (1998) 307–309.
49. A. Köver, M. I. Matheu, Y. Díaz, and S. Castillón, A study of the oxepane synthesis by a 7-endo electrophile-induced cyclization reaction of alkenylsulfides. An approach towards the synthesis of septanosides, *Arkivoc* (2007) 364–379.
50. J. Saha and M. W. Peczuh, Access to ring-expanded analogues of 2-amino sugars, *Org. Lett.*, 11 (2009) 4482–4484.
51. K. N. Kirschner and R. J. Woods, Solvent interactions determine carbohydrate conformation, *Proc. Natl. Acad. Sci. U.S.A.*, 98 (2001) 10541–10545.
52. J. Saha and M. W. Peczuh, Expanding the scope of aminosugars: synthesis of 2-amino septanosyl glycoconjugates using septanosyl fluoride donors, *Chem. Eur. J.*, 17 (2011) 7357–7365.
53. J. O. Hoberg and J. J. Bozell, Cyclopropanation and ring-expansion of unsaturated sugars, *Tetrahedron Lett.*, 36 (1995) 6831–6834.
54. J. O. Hoberg, Formation of seven-membered oxacycles through ring expansion of cyclopropanated carbohydrates, *J. Org. Chem.*, 62 (1997) 6615–6618.
55. R. Batchelor, J. E. Harvey, P. T. Northcote, P. Teesdale-Spittle, and J. O. Hoberg, Heptanosides from galactose-derived oxepenes via stereoselective addition reactions, *J. Org. Chem.*, 74 (2009) 7627–7632.
56. C. V. Ramana, R. Murali, and M. Nagarajan, Synthesis and reactions of 1,2-cyclopropanated sugars, *J. Org. Chem.*, 62 (1997) 7694–7703.

57. N. V. Ganesh and N. Jayaraman, Synthesis of septanosides through an oxyglycal route, *J. Org. Chem.*, 72 (2007) 5500–5504.
58. G. S. Jones and W. J. Scott, Oxidative addition of palladium(0) to the anomeric center of carbohydrate electrophiles, *J. Am. Chem. Soc.*, 114 (1992) 1491–1492.
59. O. Varela, G. M. de Fina, and R. M. de Lederkremer, The reaction of 2-hydroxyglycal esters with alcohols in the presence of *N*-iodosuccinimide, stereoselective synthesis of α anomers of alkyl 3-deoxyhex-2-enopyranosides and 3,4-dideoxyhex-3-enopyranosid-2-uloses, *Carbohydr. Res.*, 167 (1987) 187–196.
60. N. V. Ganesh and N. Jayaraman, Synthesis of aryl, glycosyl, and azido septanosides through ring expansion of 1,2-cyclopropanated sugars, *J. Org. Chem.*, 74 (2009) 739–746.
61. N. V. Ganesh, S. Raghothama, R. Sonti, and N. Jayaraman, Ring expansion of oxyglycals. Synthesis and conformational analysis of septanoside-containing trisaccharides, *J. Org. Chem.*, 75 (2010) 215–218.
62. H. Ovaa, M. A. Leeuwenburgh, H. S. Overkleeft, G. A. van der Marel, and J. H. van Boom, An expeditious route to the synthesis of highly functionalized chiral oxepines from monosaccharides, *Tetrahedron Lett.*, 39 (1998) 3025–3028.
63. M. W. Peczuh and N. L. Snyder, Carbohydrate-based oxepines: Ring expanded glycals for the synthesis of septanose saccharides, *Tetrahedron Lett.*, 44 (2003) 4057–4061.
64. J. D. Rainier and S. P. Allwein, A highly efficient iterative approach to fused ether ring systems, *Tetrahedron Lett.*, 39 (1998) 9601–9604.
65. E. J. Corey, M. C. Kang, M. C. Desai, A. K. Ghosh, and I. N. Houpis, Total synthesis of (±)-ginkgolide B, *J. Am. Chem. Soc.*, 110 (1988) 649–651.
66. S. Castro and M. W. Peczuh, Sequential cyclization-elimination route to carbohydrate-based oxepines, *J. Org. Chem.*, 70 (2005) 3312–3315.
67. E. Alcazar, J. M. Pletcher, and F. E. McDonald, Synthesis of seven-membered ring glycals via endo-selective alkynol cyclo-isomerization, *Org. Lett.*, 6 (2004) 3877–3880.
68. F. E. McDonald, Alkynol endo-cycloisomerizations and conceptually related transformations, *Chem. Eur. J.*, 5 (1999) 3103–3106.
69. B. Koo and F. E. McDonald, Fischer carbene Catalysis of alkynol cycloisomerization: Application to the synthesis of the altromycin B disaccharide, *Org. Lett.*, 9 (2007) 1737–1740.
70. S. Castro, C. S. Johnson, B. Surana, and M. W. Peczuh, An oxepinone route to carbohydrate based oxepines, *Tetrahedron*, 65 (2009) 7921–7926.
71. W. Pitsch, A. Russel, M. Zabel, and B. König, Synthesis of a functionalized cyclohepten-one from erythronic acid-4-lactone, *Tetrahedron*, 57 (2001) 2345–2347.
72. L. Alberch, G. Cheng, S.-K. Seo, X. Li, F. P. Boulineau, and A. Wei, Stereoelectronic factors in the stereoselective epoxidation of glycals and 4-deoxypentenosides, *J. Org. Chem.*, 76 (2011) 2532–2547.
73. G. Cheng, F. P. Boulineau, S.-T. Liew, Q. Shi, P. G. Wenthold, and A. Wei, Stereoselective epoxidation of 4-deoxypentenosides: A polarized-π model, *Org. Lett.*, 8 (2006) 4545–4548.
74. A. M. Orendt, S. W. Roberts, and J. D. Rainier, The role of asynchronous bond formation in the diastereoselective epoxidation of cyclic enol ethers: A density functional theory study, *J. Org. Chem.*, 71 (2006) 5565–5573.
75. S. D. Markad, S. Xia, N. L. Snyder, B. Surana, M. D. Morton, C. M. Hadad, and M. W. Peczuh, Stereoselectivity in the epoxidation of carbohydrate-based oxepines, *J. Org. Chem.*, 73 (2008) 6341–6354.
76. W. S. Fyvie, M. Morton, and M. W. Peczuh, Synthesis of 2-iodo-2-deoxy septanosides from a D-xylose-based oxepine: Intramolecular cyclization in the absence of a glycosyl acceptor, *Carbohydr. Res.*, 339 (2004) 2363–2370.
77. M. W. Peczuh, N. L. Snyder, and W. S. Fyvie, Synthesis, crystal structure, and reactivity of a D-xylose based oxepine, *Carbohydr. Res.*, 339 (2004) 1163–1171.

78. F. P. Boulineau and A. Wei, Synthesis of l-sugars from 4-deoxypentenosides, *Org. Lett.*, 4 (2002) 2281–2283.
79. M. A. Boone, F. E. McDonald, J. Lichter, S. Lutz, R. Cao, and K. I. Hardcastle, 1,5-α-D-Mannoseptanosides, ring-size isomers that are impervious to a-mannosidase-catalyzed hydrolysis, *Org. Lett.*, 11 (2009) 851–854.
80. S. Castro, W. S. Fyvie, S. A. Hatcher, and M. W. Peczuh, Synthesis of α-D-idoseptanosyl glycosides using an S-phenyl septanoside donor, *Org. Lett.*, 7 (2005) 4709–4712.
81. H. M. I. Osborn and A. Turkson, Synthesis and NMR spectroscopic analysis of 3-nitro-pyranoside, 3-nitro-septanoside and 4-nitro-septanoside derivatives by condensation of the anion of nitromethane with glycoside dialdehydes, *Tetrahedron: Asymmetry*, 20 (2009) 2162–2166.
82. M. L. Wolfrom, U. G. Nayak, and T. Radford, A novel reaction of a nitro sugar with alcohols, *Proc. Natl. Acad. Sci. U.S.A.*, 58 (1967) 1848–1851.
83. M. E. Butcher, J. C. Ireson, J. B. Lee, and M. J. Tyler, Seven and eight membered ring sugars and related systems: The synthesis of some septanose rings from dioxepans, *Tetrahedron*, 33 (1977) 1501–1507.
84. M. E. Butcher and J. B. Lee, Seven-membered ring sugars: factors influencing the formation of branched-chain 3-deoxy-3-nitro-septanosides, *J. Chem. Soc., Chem. Commun.* (1974) 1010–1011.
85. K. C. Nicolaou, H. J. Mitchell, F. L. van Delft, F. Rübsam, and R. M. Rodríguez, Expeditious routes to evernitrose and vancosamine derivatives and synthesis of a model vancomycin aryl glycoside, *Angew. Chem. Int. Ed.*, 37 (1998) 1871–1874.
86. D. E. Wright, The orthosomycins, a new family of antibiotics, *Tetrahedron*, 35 (1979) 1207–1237.
87. K. Sato and J. Yoshimura, Branched-chain sugars. XII. The stereoselectivities in the reaction of methyl 4,6-O-benzylidene-α- and -β-D-hexopyranosid-3-uloses with diazomethane, *Bull. Chem. Soc. Jpn.*, 51 (1978) 2116–2121.
88. B. Flaherty, S. Nahar, W. G. Overend, and N. R. Williams, Branched-chain sugars. XIV. Reactions of glycosulose derivatives with diazomethane. Ring expansion of glycosulose derivatives, *J. Chem. Soc., Perkin Trans. 1* (1973) 632–638.
89. B. Flaherty, W. G. Overend, and N. R. Williams, Ring expansion of a glycoside, *J. Chem. Soc., Chem. Commun.* (1966) 434–436.
90. R. L. Whistler and C. S. Campbell, Synthesis of septanose derivatives of 6-thio-D-galactose, *J. Org. Chem.*, 31 (1966) 816–818.
91. R. Batchelor, J. E. Harvey, P. Teesdale-Spittle, and J. O. Hoberg, Mechanistic studies of rearrangements during the ring expansions of cyclopropanated carbohydrates, *Tetrahedron Lett.*, 50 (2009) 7283–7285.
92. Y. Le Merrer, M. Fuzier, I. Dosbaa, M.-J. Foglietti, and J.-C. Depezay, Synthesis of thiosugars as weak inhibitors of glycosidases, *Tetrahedron*, 53 (1997) 16731–16746.
93. M. Fuzier, Y. Le Merrer, and J.-C. Depezay, Thiosugars from D-mannitol, *Tetrahedron Lett.*, 36 (1995) 6443–6446.
94. V. Spiwok, B. Králová, and I. Tvaroška, Modelling of β-D-glucopyranose ring distortion in different force fields: a metadynamics study, *Carbohydr. Res.*, 345 (2010) 530–537.
95. G. Favini, Conformational analysis of seven-membered cyclic systems, *THEOCHEM*, 93 (1983) 139–152.
96. D. F. Bocian and H. L. Strauss, Conformational structure and energy of cycloheptane and some related oxepanes, *J. Am. Chem. Soc.*, 99 (1977) 2876–2882.
97. J. B. Hendrickson, Molecular geometry. VII. Modes of interconversion in the medium rings, *J. Am. Chem. Soc.*, 89 (1967) 7047–7061.
98. J. B. Hendrickson, Molecular geometry. VI. Methyl-substituted cycloalkanes, *J. Am. Chem. Soc.*, 89 (1967) 7043–7046.

99. D. F. Bocian, H. M. Pickett, T. C. Rounds, and H. L. Strauss, Conformations of cycloheptane, *J. Am. Chem. Soc.*, 97 (1975) 687–695.
100. A. Espinosa, M. A. Gallo, A. Entrena, and J. A. Gómez, Theoretical conformational analysis of seven-membered rings. V. MM2 and MM3 study of oxepane, *J. Mol. Struct.*, 323 (1994) 247–256.
101. A. Entrena, J. Campos, J. A. Gómez, M. A. Gallo, and A. Espinosa, A new systematization of the conformational behavior of seven-membered rings. Isoclinal anomeric and related orientations. 1, *J. Org. Chem.*, 62 (1997) 337–349.
102. A. Espinosa, A. Entrena, M. A. Gallo, J. Campos, J. F. Dominguez, E. Camacho, and R. Garrido, Conformational analysis of some 1,4-dioxepane systems. 2. Methoxy-1,4-dioxepanes, *J. Org. Chem.*, 55 (1990) 6018–6023.
103. J. F. Stoddart and W. A. Szarek, Conformational studies on 1,3-dioxepans. Part III. 2,5-*O*-methylene-D-mannitol and some related compounds, *J. Chem. Soc. B* (1971) 437–442.
104. J. F. Stoddart, Stereochemistry of Carbohydrates (1971) Wiley & Sons, New York.
105. J. B. Hendrickson, Molecular geometry. IV. The medium rings, *J. Am. Chem. Soc.*, 86 (1964) 4854–4866.
106. C. Altona and M. Sundaralingam, Conformational analysis of the sugar ring in nucleosides and nucleotides. New description using the concept of pseudorotation, *J. Am. Chem. Soc.*, 94 (1972) 8205–8212.
107. D. Cremer and J. A. Pople, General definition of ring puckering coordinates, *J. Am. Chem. Soc.*, 97 (1975) 1354–1358.
108. G.R. J. Thatcher, (Ed.), *In* The Anomeric Effect and Associated Stereoelectronic Effects, American Chemical Society, Washington, DC, 1993.
109. P. Deslongchamps, Stereoelectronic Effects in Organic Chemistry, (1983) Wiley, New York.
110. W. A. Szarek, and D. Horton, (Eds.) *In* Anomeric Effect: Origins and Consequences, *ACS Symp. Ser.*, Vol. 87, pp. 1–127.
111. A. J. Kirby, The Anomeric Effect and Related Stereoelectronic Effects at Oxygen, (1983) Springer Verlag, New York.
112. S. Desilets and M. St. Jacques, The anomeric effect in thio derivatives of benzocycloheptene, *Can. J. Chem.*, 70 (1992) 2650–2658.
113. S. Desilets and M. St. Jacques, Anomeric effect in seven-membered rings: A conformational study of 2-oxa derivatives of benzocycloheptene by NMR, *J. Am. Chem. Soc.*, 109 (1987) 1641–1648.
114. H. Booth, K. A. Khedhair, and S. A. Readshaw, Experimental studies of the anomeric effect: Part I. 2-substituted tetrahydropyrans, *Tetrahedron*, 43 (1987) 4699–4723.
115. H. Booth, J. Mark Dixon, and R. Simon, Experimental studies of the anomeric effect. Part V. The influence of some solvents on the conformational equilibria in 2-methoxy- and 2-(2′,2′,2′-trifluoroethoxy)-tetrahydropyran, *Tetrahedron*, 48 (1992) 6151–6160.
116. U. Salzner and P.v.R. Schleyer, *Ab initio* examination of anomeric effects in tetrahydropyrans, 1,3-dioxanes, and glucose, *J. Org. Chem.*, 59 (1994) 2138–2155.
117. C. B. Anderson and D. T. Sepp, Conformation and the anomeric effect in 2-halotetrahydropyrans, *J. Org. Chem.*, 32 (1967) 607–611.
118. J.-P. Praly and R. U. Lemieux, Influence of solvent on the magnitude of the anomeric effect, *Can. J. Chem.*, 65 (1987) 213–223.
119. R. U. Lemieux, A. A. Pavia, J. C. Martin, and K. A. Watanabe, Solvation effects on conformational equilibria. Studies related to the conformational properties of 2-methoxytetrahydropyran and related methyl glycopyranosides, *Can. J. Chem.*, 47 (1969) 4427–4439.
120. C. Mayato, R. Dorta, and J. Vázquez, Experimental evidence on the hydroxymethyl group conformation in alkyl β-D-mannopyranosides, *Tetrahedron: Asymmetry*, 15 (2004) 2385–2397.
121. M. Kraszni, Z. Szakács, and B. Noszál, Determination of rotamer populations and related parameters from NMR coupling constants: A critical review, *Anal. Bioanal. Chem.*, 378 (2004) 1449–1463.

122. R. Stenutz, I. Carmichael, G. Widmalm, and A. S. Serianni, Hydroxymethyl group conformation in saccharides: Structural dependencies of 2JHH, 3JHH, and 1JCH spin-spin coupling constants, *J. Org. Chem.*, 67 (2002) 949–958.
123. C. Nóbrega and J. T. Vázquez, Conformational study of the hydroxymethyl group in α-D-mannose derivatives, *Tetrahedron: Asymmetry*, 14 (2003) 2793–2801.
124. M. P. DeMatteo, N. L. Snyder, M. Morton, D. M. Baldisseri, C. M. Hadad, and M. W. Peczuh, Septanose carbohydrates: Synthesis and conformational studies of methyl α-D-*glycero*-D-idoseptanoside and methyl β-D-*glycero*-D-guloseptanoside, *J. Org. Chem.*, 70 (2005) 24–38.
125. M. P. DeMatteo, S. Mei, R. Fenton, M. Morton, D. M. Baldisseri, C. M. Hadad, and M. W. Peczuh, Conformational analysis of methyl 5-*O*-methyl septanosides: effect of glycosylation on conformer populations, *Carbohydr. Res.*, 341 (2006) 2927–2945.
126. V. A. Danilova, N. V. Istomina, and L. B. Krivdin, 13C-13C Spin-spin coupling constants in structural studies: XXXVII. Rotational conformations of hydroxy groups in pyranose, furanose, and septanose rings, *Russ. J. Org. Chem.*, 40 (2004) 1194–1199.
127. M. R. Duff, Jr.,, W. S. Fyvie, S. D. Markad, A. E. Frankel, C. V. Kumar, J. A. Gascón, and M. W. Peczuh, Computational and experimental investigations of mono-septanoside binding by Concanavalin A: Correlation of ligand stereochemistry to enthalpies of binding, *Org. Biomol. Chem.*, 9 (2011) 154–164.
128. J. B. Houseknecht, P. R. McCarren, T. L. Lowary, and C. M. Hadad, Conformational studies of methyl 3-*O*-methyl-α-D-arabinofuranoside: An approach for studying the conformation of furanose rings, *J. Am. Chem. Soc.*, 123 (2001) 8811–8824.
129. V. A. Danilova and L. B. Krivdin, 13C-13C spin-spin coupling constants in structural studies: XXXVI. Stereochemical study of the septanose ring, *Russ. J. Org. Chem.*, 40 (2004) 57–62.
130. S. Castro, M. Duff, N. L. Snyder, M. Morton, C. V. Kumar, and M. W. Peczuh, Recognition of septanose carbohydrates by concanavalin A, *Org. Biomol. Chem.*, 3 (2005) 3869–3872.
131. A. A. Koshkin, P. Nielsen, M. Meldgaard, V. K. Rajwanshi, S. K. Singh, and J. Wengel, LNA (locked nucleic acid): An RNA mimic forming exceedingly stable LNA:LNA duplexes, *J. Am. Chem. Soc.*, 120 (1998) 13252–13253.
132. V. K. Rajwanshi, A. E. Hakansson, M. D. Sorensen, S. Pitsch, S. K. Singh, R. Kumar, P. Nielsen, and J. Wengel, The eight stereoisomers of LNA (Locked nucleic acid): A remarkable family of strong RNA binding molecules, *Angew. Chem. Int. Ed.*, 39 (2000) 1656–1659.
133. S. Hildbrand and C. Leumann, Enhancing DNA triple helix stability at neutral pH by the use of oligonucleotides containing a more basic deoxycytidine analog, *Angew. Chem. Int. Ed.*, 35 (1996) 1968–1970.
134. R. Steffens and C. J. Leumann, Synthesis and thermodynamic and biophysical properties of tricyclo-DNA, *J. Am. Chem. Soc.*, 121 (1999) 3249–3255.
135. S. Tripathi, B. G. Roy, M. G. B. Drew, B. Achari, and S. B. Mandal, Synthesis of oxepane ring containing monocyclic, conformationally restricted bicyclic and spirocyclic nucleosides from D-glucose: A cycloaddition approach, *J. Org. Chem.*, 72 (2007) 7427–7430.
136. A. Kittaka, T. Asakura, T. Kuze, H. Tanaka, N. Yamada, K. T. Nakamura, and T. Miyasaka, Cyclization reactions of nucleoside anomeric radical with olefin tethered on base: Factors that induce anomeric stereochemistry, *J. Org. Chem.*, 64 (1999) 7081–7093.
137. B. R. Babu, L. Keinicke, M. Petersen, C. Nielsen, and J. Wengel, 2'-Spiro ribo- and arabinonucleosides: Synthesis, molecular modelling and incorporation into oligodeoxynucleotides, *Org. Biomol. Chem.*, 1 (2003) 3514–3526.
138. S. Z. Dong and L. A. Paquette, Stereoselective synthesis of conformationally constrained 2'-deoxy-4'-thia beta-anomeric spirocyclic nucleosides featuring either hydroxyl configuration at C5', *J. Org. Chem.*, 70 (2005) 1580–1596.

139. R. Hartung and L. A. Paquette, Practical synthesis of enantiopure spiro 4.4 nonane C-(2′-deoxy) ribonucleosides, *J. Org. Chem.*, 70 (2005) 1597–1604.
140. L. A. Paquette and S. Z. Dong, Stereoselective synthesis of beta-anomeric 4′-thiaspirocyclic ribonucleosides carrying the full complement of RNA-level hydroxyl substitution, *J. Org. Chem.*, 70 (2005) 5655–5664.
141. A. Roy, B. Achari, and S. B. Mandal, An easy access to spiroannulated glyco-oxetane, -thietane and -azetane rings: Synthesis of spironucleosides, *Tetrahedron Lett.*, 47 (2006) 3875–3879.
142. S. Sahabuddin, A. Roy, M. G. B. Drew, B. G. Roy, B. Achari, and S. B. Mandal, Sequential ring-closing metathesis and nitrone cycloaddition on glucose-derived substrates: A divergent approach to analogues of spiroannulated carbanucleosides and conformationally locked nucleosides, *J. Org. Chem.*, 71 (2006) 5980–5992.
143. A. Eschenmoser, Searching for nucleic acid alternatives, *Chimia*, 59 (2005) 836–850.
144. M. O. Ebert, A. Luther, H. K. Huynh, R. Krishnamurthy, A. Eschenmoser, and B. Jaun, NMR solution structure of the duplex formed by self-pairing of α-L-arabinopyranosyl-(4′,2′)-(CGAATTCG), *Helv. Chim. Acta*, 85 (2002) 4055–4073.
145. A. Eschenmoser and R. Krishnamurthy, Chemical etiology of nucleic acid structure, *Pure Appl. Chem.*, 72 (2000) 343–345.
146. K. U. Schoning, P. Scholz, S. Guntha, X. Wu, R. Krishnamurthy, and A. Eschenmoser, Chemical etiology of nucleic acid structure: The α-threofuranosyl-(3′,2′) oligonucleotide system, *Science*, 290 (2000) 1347–1351.
147. S. Pitsch, S. Wendeborn, R. Krishnamurthy, A. Holzner, M. Minton, M. Bolli, C. Miculca, N. Windhab, R. Micura, M. Stanek, B. Jaun, and A. Eschenmoser, Pentopyranosyl oligonucleotide systems. 9th Communication, *Helv. Chim. Acta*, 86 (2003) 4270–4363.
148. M. Egli, P. S. Pallan, R. Pattanayek, C. J. Wilds, P. Lubini, G. Minasov, M. Dobler, C. J. Leumann, and A. Eschenmoser, Crystal structure of homo-DNA and Nature's choice of pentose over hexose in the genetic system, *J. Am. Chem. Soc.*, 128 (2006) 10847–10856.
149. G. Sizun, D. Dukhan, J. F. Griffon, L. Griffe, J. C. Meillon, F. Leroy, R. Storer, J. P. Sommadossi, and G. Gosselin, Synthesis of the first example of a nucleoside analogue bearing a 5′-deoxy-β-D-*allo*-septanose as a seven-membered ring sugar moiety, *Carbohydr. Res.*, 344 (2009) 448–453.
150. D. Sabatino and M. J. Damha, Oxepane nucleic acids: Synthesis, characterization, and properties of oligonucleotides bearing a seven-membered carbohydrate ring, *J. Am. Chem. Soc.*, 129 (2007) 8259–8270.
151. M. Nomura, K. Endo, S. Shuto, and A. Matsuda, Nucleosides and nucleotides. 191. Ring expansion reaction of 1-[2,3,5-tri-O-TBS-4 alpha-formyl-beta-D-*ribo*-pentofuranosyl] uracil by treating with (methylene)triphenylphosphorane to give a new nucleoside containing dihydrooxepine ring at the sugar moiety, *Tetrahedron*, 55 (1999) 14847–14854.
152. M. Nomura, S. Shuto, M. Tanaka, T. Sasaki, S. Mori, S. Shigeta, and A. Matsuda, Nucleosides and Nucleotides. 185. Synthesis and biological activities of 4′-α-C-branched-chain sugar pyrimidine nucleosides, *J. Med. Chem.*, 42 (1999) 2901–2908.
153. A. Bhattacharjee, S. Datta, P. Chattopadhyay, N. Ghoshal, A. P. Kundu, A. Pal, R. Mukhopadhyay, S. Chowdhury, A. Bhattacharjya, and A. Patra, Synthesis of chiral oxepanes and pyrans by 3-O-allyl-carbohydrate nitrone cycloaddition (3-OACNC), *Tetrahedron*, 59 (2003) 4623–4639.
154. D. Sabatino and M. J. Damha, Synthesis and properties of oligonucleotides containing a 7-membered (oxepane) sugar ring, *Nucleosides Nucleotides*, 26 (2007) 1185–1188.
155. B. Bennuaskalmowski, K. Krolikiewicz, and H. Vorbruggen, A new simple nucleoside synthesis, *Tetrahedron Lett.*, 36 (1995) 7845–7848.
156. A. De Mesmaeker, R. Haener, P. Martin, and H. E. Moser, Antisense oligonucleotides, *Acc. Chem. Res.*, 28 (1995) 366–374.

157. E. Uhlmann and A. Peyman, Antisense oligonucleotides: A new therapeutic principle, *Chem. Rev.*, 90 (1990) 543–584.
158. M. M. Mangos, K. L. Min, E. Viazovkina, A. Galarneau, M. I. Elzagheid, M. A. Parniak, and M. J. Damha, Efficient RNase H-directed cleavage of RNA promoted by antisense DNA or 2' F-ANA constructs containing acyclic nucleotide inserts, *J. Am. Chem. Soc.*, 125 (2003) 654–661.
159. A. Gelbin, B. Schneider, L. Clowney, S. H. Hsieh, W. K. Olson, and H. M. Berman, Geometric parameters in nucleic acids: Sugar and phosphate constituents, *J. Am. Chem. Soc.*, 118 (1996) 519–529.
160. A. Noy, A. Perez, M. Marquez, F. J. Luque, and M. Orozco, Structure, recognition properties, and flexibility of the DNA-RNA hybrid, *J. Am. Chem. Soc.*, 127 (2005) 4910–4920.
161. M. J. Damha, C. J. Wilds, A. Noronha, I. Brukner, G. Borkow, D. Arion, and M. A. Parniak, Hybrids of RNA and arabinonucleic acids (ANA and 2' F-ANA) are substrates of ribonuclease H, *J. Am. Chem. Soc.*, 120 (1998) 12976–12977.
162. P. Nielsen, H. M. Pfundheller, and J. Wengel, A novel class of conformationally restricted oligonucleotide analogues: Synthesis of 2',3'-bridged monomers and RNA-selective hybridisation, *Chem. Commun.* (1997) 825–826.
163. J. Li and R. M. Wartell, RNase H1 can catalyze RNA/DNA hybrid formation and cleavage with stable hairpin or duplex DNA oligomers, *Biochemistry*, 37 (1998) 5154–5161.
164. Y. Oda, S. Iwai, E. Ohtsuka, M. Ishikawa, M. Ikehara, and H. Nakamura, Binding of nucleic acids to E. coli RNase HI observed by NMR and CD spectroscopy, *Nucleic Acids Res.*, 21 (1993) 4690–4695.
165. A. Tauss, A. J. Steiner, A. E. Stütz, C. A. Tarling, S. G. Withers, and T. M. Wrodnigg, 1-Idoseptanosides: substrates of D-glucosidases?*Tetrahedron: Asymmetry*, 17 (2006) 234–239.
166. Y. Li, D. Horton, V. Barberousse, F. Bellamy, P. Renaut, and S. Samreth, Chemical interconversions of 4-cyanophenyl 1,5-dithio-β-D-xylopyranoside (Beciparcil): Structural modification at the C-4 position, *Carbohydr. Res.*, 314 (1998) 161–167.
167. F. Bellamy, D. Horton, J. Millet, F. Picart, S. Samreth, and J. B. Chazan, Glycosylated derivatives of benzophenone, benzhydrol and benzhydril as potential venous antithrombotic agents, *J. Med. Chem.*, 36 (1993) 898–903.
168. E. Bozó, A. Medgyes, S. Boros, and J. Kuszmann, Synthesis of 4-substituted phenyl 2,5-anhydro-1,6-dithio-α-gluco- and -α-guloseptanosides possessing antithrombotic activity, *Carbohydr. Res.*, 329 (2000) 25–40.
169. E. Bozó, S. Boros, L. Párkányi, and J. Kuszmann, Synthesis of 4-cyano- and 4-nitrophenyl 2,5-anhydro-1,6-dithio-α-D-gluco- and -α-L-guloseptanosides carrying different substituents at C-3 and C-4, *Carbohydr. Res.*, 329 (2000) 269–286.
170. E. Bozó, T. Gáti, Á. Demeter, and J. Kuszmann, Synthesis of 4-cyano and 4-nitrophenyl 1,6-dithio–manno-, -ido- and -glucoseptanosides possessing antithrombotic activity, *Carbohydr. Res.*, 337 (2002) 1351–1365.
171. D. Berger, L. E. Overman, and P. A. Renhowe, Total synthesis of (+)-isolaurepinnacin. Use of acetal-alkene cyclizations to prepare highly functionalized seven-membered cyclic ethers, *J. Am. Chem. Soc.*, 119 (1997) 2446–2452.
172. S. Basu, B. Ellinger, S. Rizzo, C.l. Deraeve, M. Schürmann, H. Preut, H.-D. Arndt, and H. Waldmann, Biology-oriented synthesis of a natural-product inspired oxepane. Collection yields a small-molecule activator of the Wnt-pathway, *Proc. Natl. Acad. Sci. USA.*, 108 (2011) 6805–6810.

IMINO SUGARS AND GLYCOSYL HYDROLASES: HISTORICAL CONTEXT, CURRENT ASPECTS, EMERGING TRENDS

Arnold E. Stütz and Tanja M. Wrodnigg

Glycogroup, Institut für Organische Chemie, Technische Universität Graz, Graz, Austria

I. Introduction	188
1. Discovery of Imino Sugars as Natural Products	189
2. Early Synthetic Efforts	192
II. Glycoside Hydrolases	193
1. General Features and Means of Classification	193
2. Historical Background and Development	194
III. Glycon Structure, Functional Groups, and Catalysis	197
1. β-Glucosidases and Nonnatural Substrates	197
2. β-Galactosidases and Nonnatural Substrates	199
3. α-Glucosidases and Nonnatural Substrates	200
4. α-Galactosidases and Nonnatural Substrates	202
5. α-Mannosidases and Nonnatural Substrates	202
IV. Glycosidase Inhibitors	203
1. High-Molecular-Weight Inhibitors	203
2. Small Molecules: Covalent ("Irreversible"), Competitive (Reversible), and Noncompetitive (Allosteric) Inhibitors	203
3. Nonsugar Inhibitors	204
4. Carbohydrate-Related Inhibitors	206
V. Imino Sugars	211
1. pK_a Value and State of Ionization Approaching the Active Site	212
2. Stereochemical Considerations Concerning the Catalytic Protonation Process	214
3. Significance of Individual Functional Groups to Recognition and Binding	216
4. Deoxyfluoro Derivatives as Active-Site Probes	217
5. Imino Sugars as Active-Site Ligands for Structure and Catalysis Visualization	224
6. "Non-Glycon" Binding Sites and Inhibitor Activity	246
VI. Conclusions	256
VII. Paradigm Changes and Emerging Topics	258
VIII. Table of PDB-Entries: Enzyme–Inhibitor Complexes of Imino Sugars and Selected Other Ligands	259
Acknowledgment	276
References	276

Abbreviations

AH-Sepharose, 6-aminohexyl-Sepharose; CAZy, Carbohydrate-Active EnZymes database; DADMe-immucillin, 4'-deaza-1'-aza-2'-deoxy-1'-(9-methylene)-immucillin; DAST, diethylaminosulfur trifluoride; DMDP, 2,5-dihydroxymethyl-3,4-dihydroxypyrrolidine; ER, endoplasmatic reticulum; GH, glycohydrolase; Glc, D-glucose; GlcNAc, N-acetyl-D-glucosamine; HexA, HexB, human N-acetylhexosaminidase; Man, D-mannose; MCO, mode of co-substrate orientation; MTAN, 5'-methylthioadenosine nucleosidase; NOEV, N-octyl-1-epi-valienamine; PAGE, polyacrylamide gel electrophoresis; PDB, protein data bank; http://www.pdb.org/pdb/home/home.do; PNP, purine nucleoside phosphorylase; PUGNAc, O-(2-acetamido-2-deoxy-D-glucopyranosylidene)amino N-phenyl carbamate; RSCB, research collaboratory for structural bioinformatics

I. Introduction

The 1960s and 1970s were characterized by significant advances in general organic chemistry as well as in carbohydrate chemistry. Cahn, Ingold, and Prelog had published[1] their "Specification of Molecular Chirality," providing a new paradigm in the understanding of organic stereochemistry. Corey and his group gained synthetic access to (at that time) "difficult" natural products, such as the prostaglandins[2] and set the first standards of computer-assisted (retro-) synthetic analysis and design,[3] and Corey and Seebach reported the much acclaimed 1,3-dithiane method[4] for the synthesis of unsymmetrical ketones employing the Umpolung concept. The Woodward–Hoffmann rules which govern the orbital symmetry control of chemical reactions[5] were established, and Kishi was on his way to the total synthesis of tetrodotoxin.[6] These mention just a very few of the many achievements and highlights of this period.

In the 1980s, both natural-products discovery and biological applications created a thriving environment for synthetic chemists, offering a large variety of new structures to be chosen from for synthesis. Research on antibiotics, a huge and very rewarding field of this period, was seemingly about to win the fight against pathogenic bacteria, and led to a distinctly decreasing interest in the area, both in academia and in pharmaceutical companies. The companies considerably downsized, if not terminated, their research on antibiotics. Sophisticated contributions by numerous groups, including Ingolf Dyong and coworkers,[7,8] Horton and his group,[9,10] Bock et al.,[11,12] as well as Thiem and his team,[13,14] just to mention a few, are readily remembered in this context, with synthetic efforts addressing the chemistry of sugar-containing antibiotics in the 1980s. The areas of amino sugars and deoxy sugars as constituents of antibiotic substances and anticancer drugs provided solid knowledge about the properties of highly functionalized compounds featuring one or more amino groups, together with a number of hydroxyl functionalities. Chemical routes thus became available to provide novel compounds for a totally new research area in carbohydrate chemistry, namely the imino sugars as glycosidase inhibitors. With the stage nicely set for both synthesis and biochemistry, this field attracted a significant number of scientists who had been waiting for new challenges with such well-established and reliable synthetic tools at hand.

1. Discovery of Imino Sugars as Natural Products

Largely unnoticed by the general chemical community and many carbohydrate chemists alike, 5-amino-5-deoxy D glucose (**1**, Scheme 1), was discovered as a metabolite of *Streptomyces roseochromogenes* R-468. It acts as an antibiotic substance and is an unusual sugar featuring nitrogen, a basic function, instead of oxygen in the pyranose ring. It was isolated from the fermentation broth of a Japanese *Streptomyces* species, *S. nojiriensis*,[15] characterized and termed nojirimycin after its source and its antibacterial activity.[16,17] In 1968, its structure and synthesis were reported in a paper that would become frequently cited in the first three decades of imino sugar research.[18]

1,5-Dideoxy-1,5-imino-D-glucitol (**2**), which was termed 1-deoxynojirimycin, was obtained by intramolecular reductive amination of nojirimycin. The corresponding 2-deoxy derivative, fagomine (1,2,5-trideoxy-1,5-imino-D-*arabino*-hexitol, **3**), was isolated from buckwheat (*Fagopyrum esculentum*) seeds in 1974.[19]

As early as 1962/1963, the impressive impact of amino sugar-containing antibiotics on science and health-care systems alike had sparked the general curiosity of the carbohydrate community in exploring the "scope and limitations" of sugar synthesis and properties of new compounds and had triggered two sentinel carbohydrate chemists, Hanessian[20] and Paulsen,[21] into synthesizing such compounds, starting

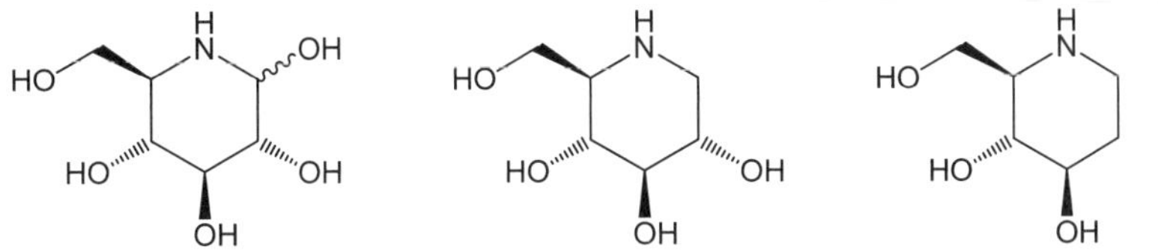

SCHEME 1.

with the more readily accessible pentopyranoses. The time was also ripe for another type of "heterose," featuring sulfur instead of oxygen in the ring. Notwithstanding the efforts of earlier workers in this area, for example, Adley and Owen[22] who prepared, among others, 5-thio-L-idopyranose (**4**, Scheme 2), or Schwarz and Yule[23] (5-thio-D-xylopyranose, **5**), this field usually arouses associations with Whistler's synthesis of 5-thio-D-glucose (**6**) and epimers thereof.[24,25]

In 1976, 1-deoxynojirimycin (**2**) was discovered as a natural product in a *Morus* species and coined moranoline.[26] Incidentally, in the same year, a novel type of alkaloid was found in the liana plant, *Derris elliptica*, a native of Borneo but widespread in the tropics, and it was identified as 2,5-dideoxy-2,5-imino-D-mannitol (**7**, Scheme 3), also known as DMDP (2,5-dihydroxymethyl-3,4-dihydroxypyrrolidine).[27]

This compound exhibits the same relationship to β-D-fructofuranose as 1-deoxynojirimcin does to D-glucopyranose. In 1979, a polyhydroxyindolizidine alkaloid was isolated from the poisonous fruit of *Castanospermum australe*, a handsome Australian indigenous tree, and coined castanospermine (**8**) after its source.[28] A compound subsequently isolated from the seeds of this plant was the pyrrolizidine australine (**9**),[29] also found in *Alexa leiopetala*,[30] along with other compounds. These discoveries increasingly supported the hypothesis that imino sugars and their structural relatives might be a fairly common family of natural products, and their widespread

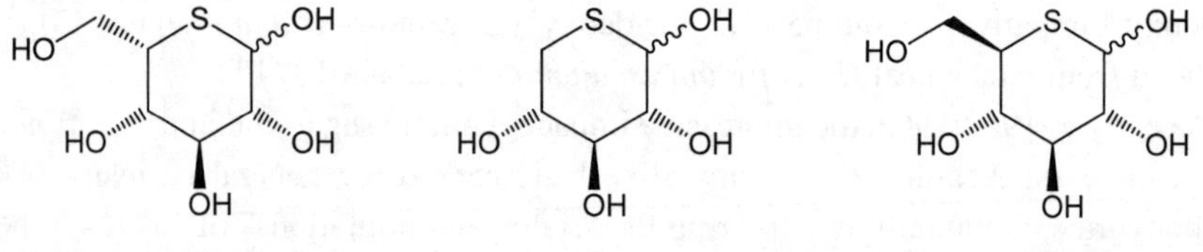

SCHEME 2.

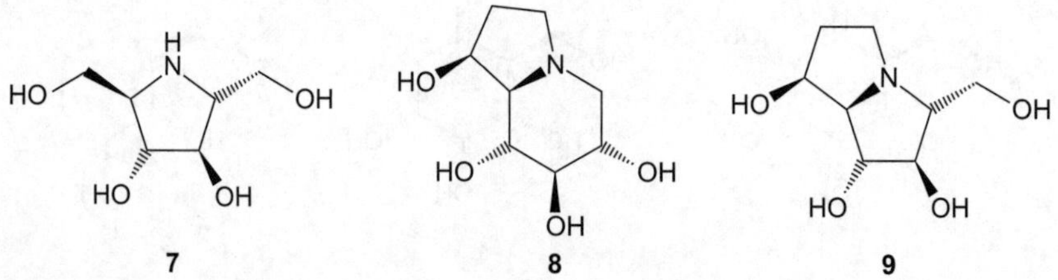

SCHEME 3.

occurrence in plants in various climate zones might have to do with biological effects exerted by these compounds.

The year 1979 was also the year of the first true breakthrough of the imino sugars. This was initiated by a medical researcher with Bayer company, Dr. W. Puls, who suggested a novel therapeutic concept for the treatment of diabetes type 2 symptoms after carbohydrate uptake; this involved the inhibition of intestinal α-glucosidases with various compounds found in fermentation broths, among them 1,5-dideoxy-1,5-imino-D-glucitol, **2**. The same research group also found a "pseudotetrasaccharide," acarbose, featuring a branched cyclohexene ring, valienamine (**10**, Scheme 4). The latter triggered a strong interest in the "pseudosugars," a term coined in 1966 by McCasland, the pioneer in this area.[31] Later, this class of branched-chain cyclitols (and aminocyclitols) was more appropriately and generally termed carba sugars.[32]

1,5-Dideoxy-1,5-imino-D-mannitol (**11**), the C-2 epimer of compound **2**, was found in the legumes *Lonchocarpus sericeus* and *Lonchocarpus costaricensis*.[33]

In 1980, Paulsen and coworkers prepared 1,5-dideoxy-1,5-imino-D-galactitol (**12**),[34] and in 1981, a another important indolizidine alkaloid and powerful mannosidase inhibitor, swainsonine (**13**, Scheme 5),[35] was found in the poisonous Australian locoweed (*Swainsona canescens*), which was known to be dangerous for cattle. Synthetic 1,5-dideoxy-1,5-imino-D-mannitol (**11**) was mentioned by Kinast and Schedel in a footnote to their publication on a four-step *Gluconobacter oxidans*-based chemo-enzymatic approach to 1,5-dideoxy-1,5-imino-D-glucitol (**1**).[36] The first published synthesis came in 1982 from Bengt Lindberg's laboratory, where Karin Leontein had synthesized it, together with the parent sugar 5-amino-5-deoxy-D-mannose, in the context of her work on the capsular polysaccharide of a *Streptococcus pneumoniae* species.[37] Subsequently, this compound was shown to be the first example of an inhibitor blocking the conversion of high-mannose to complex-type oligosaccharides *in vivo*, operating by interfering with mannosidase IA/B.[38]

In 1985, 1,4-dideoxy-1,4-imino-D-arabinitol (**14**) was independently discovered in a legume, *Angylocalyx boutiqueanus*[39] and a fern, *Arachniodes standishii*.[40]

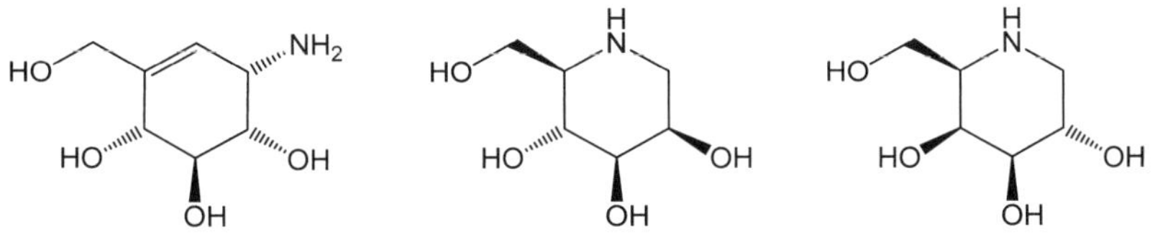

SCHEME 4.

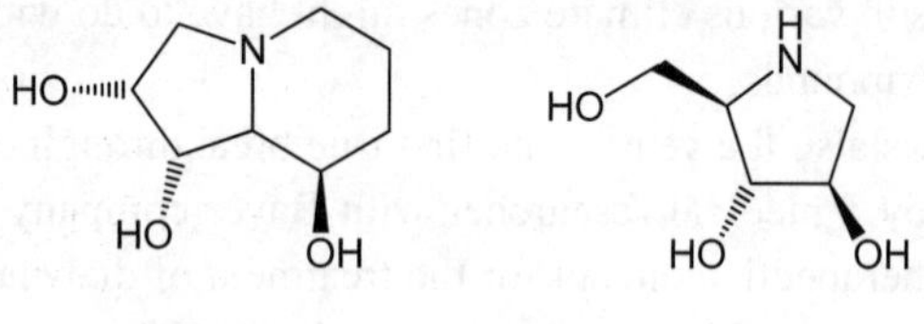

SCHEME 5.

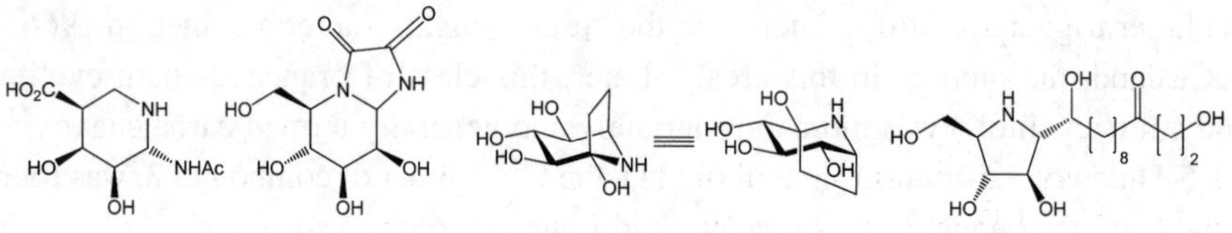

SCHEME 6.

Related but more "unusual" structures found were the sialidase inhibitors siastatins (siastatin B, **15**, Scheme 6)[41] isolated in 1974, as well as kifunensine (**16**),[42,43] an immunomodulating agent and powerful inhibitor of the glycoprotein-processing mannosidase I.[44] Further discoveries included epimers of polyhydroxypyrrolizidine australine,[45] the *nor*-tropane alkaloids calystegines (calystegine B$_2$, **17**),[46] as well as the pyrrolidine alkaloids broussonetines (broussonetine E, **18**) which are structurally related to 2,5-dideoxy-2,5-imino-D-mannitol (**7**).[47,48]

2. Early Synthetic Efforts

Researchers at the Royal Botanical Gardens, Kew, made particularly significant contributions to developments in the field from 1984 onwards. Their extremely fruitful collaboration with George Fleet, a key player in the synthesis section of the young and highly promising imino sugar area, resulted in some of the most-cited reports in imino sugar characterization and synthesis of the subsequent decade.

Selected from among their contributions were the first example of a synthetic furanoid iminoalditol, 1,4-dideoxy-1,4-imino-D-mannitol (**19**, Scheme 7),[49] the first synthesis of the powerful L-fucosidase inhibitor 1,5-dideoxy-1,5-imino-L-fucitol (**20**),[50] the synthesis and structure confirmation of 1,4-dideoxy-1,4-imino-D-arabinitol

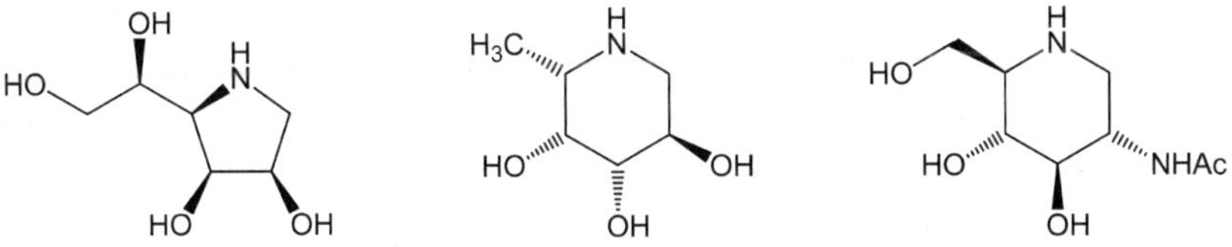

19 **20** **21**

SCHEME 7.

(**14**),[51] and the synthesis of the potent hexosaminidase inhibitor 2-acetamido-1,2,5-trideoxy-1,5-imino-D-glucitol (**21**),[52] the imino analogue of N-acetyl-D-glucosamine.

Parallel in time, Legler and coworkers made a strong impact on the field. In addition to important enzymological studies employing 5-amino-5-deoxy-D-mannose (mannojirimycin) and 1,5-dideoxy-1,5-imino-D-mannitol (1-deoxymannojirimycin or 1-deoxymannonojirimycin, **11**),[53] this group pioneered the application of imino alditols as affinity ligands for enzyme purification purposes. Compound **11**, by immobilization on AH-Sepharose 4B, became the first example of an affinity ligand in this series of compounds.[54] Applying this new technique, lysosomal glucosylceramidases of healthy human placenta as well as of a Gaucher disease-affected cell line could be purified by employing 1-deoxynojirimycin.[55] Following these early studies, this method was employed for the purification protocols of a wide range of important glycosidases.[56]

Several additional groups joined the field during the second half of the 1980s, when the "imino sugar concept" had already fallen on fertile ground. This exciting period in time also witnessed the first total synthesis of palytoxin,[57] with the Nobel Prize award to E. J. Corey underscoring the significance of organic synthesis in contemporary science, and the introduction of generally applicable catalysts for ring-closing metathesis,[58] as well as the discovery of the buckminsterfullerenes.[59]

Excellent and comprehensive reviews on imino sugars as natural products and synthetic methodologies are available.[60–68]

II. GLYCOSIDE HYDROLASES

1. General Features and Means of Classification

Glycoside hydrolases (glycosidases) are essential and consequently widely abundant enzymes in all living systems that rely on the processing of carbohydrates. From the degradation of such polysaccharides as starch, cellulose, or chitin to the highly sophisticated deglycosylation–reglycosylation sequence in the maturation and functionalization of glycoproteins, a wide range of glycosidases catalyze the selective release of a structural plethora of aglycons from their glycon partners. Glycosidases cleaving O-glycosides and thioglycosides belong to the class EC 3.2.1, whereas proteins breaking N-glycosylic bonds such as the nucleoside hydrolases are classified in EC 3.2.2. Their molecular masses vary considerably, between about 20 kDa and well over 100 kDa for the monomeric structures, with the majority of examples being in the range between 40 and 70 kDa. These monomers may form homo- or heterodimers, or higher aggregates such as tetramers, as is the case with the *Escherichia coli* β-galactosidase.[69]

Most glycosidases work best around neutral pH, although such acid enzymes as lysosomal glycosidases may favor the range between pH 3.5 and 5, and pH optima may be as low as pH 2.2.[70,71]

There is a clear distinction between exo- and endo-glycosidases; the former act at the nonreducing end of a glycan chain, whereas the latter cleave at some point along an oligosaccharide chain, either at random or in a "chemical environment"-specific manner.

Interestingly, despite the fact that glycosidases are essentially catabolic, degrading enzymes, the α-glucosidases and α-mannosidases in particular also play key roles in an early phase of essential biochemical glycoprotein construction and maturation, namely, the posttranslational trimming process of juvenile N-linked glycoproteins in the endoplasmatic reticulum and the Golgi.

2. Historical Background and Development

The classification of glycoside hydrolases and understanding their modes of action have provided a powerful incentive for decades of experimental work and contemplation. From early considerations on the mechanistic principles of glycoside hydrolysis by such pioneers as Kuhn,[72] Veibel,[73] Pigman,[74,75] and Shafizadeh,[76] the picture has become increasingly more clear with Koshland's contributions,[77–79] which are now considered to be the spark that has shed new light on our modern understanding of glycosidase catalysis. Supported by the first X-ray diffraction study of a glycosidase by Phillips and coworkers,[80] who solved the structure of hen-egg white lysozyme in the first half of the 1960s, the main objectives of glycosidase research have been to understand the detailed mechanism of the reaction pathway. This was considered to progress via "rapid and reversible protonation of the anomeric oxygen, followed by a rate-determining heterolysis giving a carbonium ion,"[81] as also outlined by Vernon.[82]

Glycosidases had been distinguished by their configurational as well as anomeric specificity, and by the stereochemical outcome of the hydrolysis reaction.[83–88] There are two most significant groups, the inverting enzymes and the retaining enzymes. The inverting glycosidases first generate the oxocarbenium ion then and quench this transition state under inversion of configuration at the anomeric center (Fig. 1). Retaining glycosidases act via initial formation under inversion of configuration of a covalent enzyme–glycon bond—as initially concluded from kinetic data,[89] which in turn is hydrolyzed in a second inversion step, leading to overall retention of configuration at the anomeric carbon (Fig. 2).

The key structural difference between inverting and retaining enzymes is the spatial distance between the two catalytic carboxyl groups in the active site. Inverting enzymes show distances between the general acid–base pair at the active site in the range of 6–12 Å, whereas representatives of the retaining type rather generally have the two catalytic residues at a distance of approximately 5 Å.[85]

The concept of retaining versus inverting glycohydrolases (GHs) was extended by Hehre based on observations made in his group as well as by others concerning the stereochemical outcome of protonation of glycals and exo glycals by glycosidases as well as unexpected findings with the hydrolysis of glycosyl fluorides of "wrong" configuration by several enzymes. In his important critical review,[91] a range of crystal structures of glycosidases and glycosyl phosphorylases was thoroughly commented on and the terms 1-MCO (1 mode of co-substrate orientation toward the active site) and 2-MCO (2 modes of co-substrate orientation, i.e., inorganic co-substrates such as water or phosphate approach from a different direction to the trajectory of organic co-substrates) were introduced. 1-MCO glycosidases are equivalent to the retaining enzymes, 2-MCO represent the group of inverting enzymes, and the concept was proposed to be extended to the glycosyl transferases due to the fact that "the glycosyl moiety of a substrate replaces a proton of a co-substrate and is itself replaced by a proton."

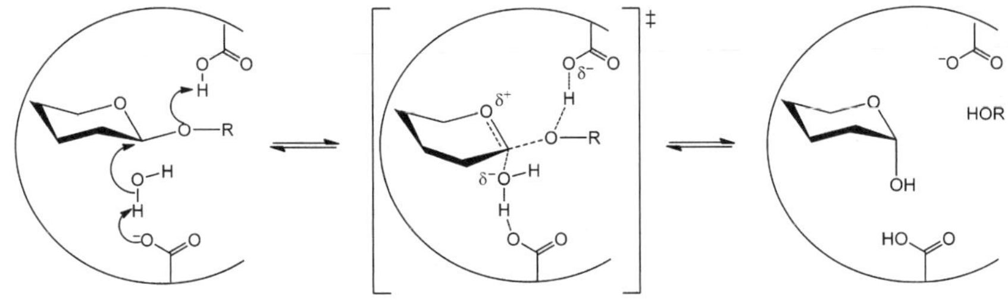

FIG. 1. Mechanism of inverting glycosidases.[90]

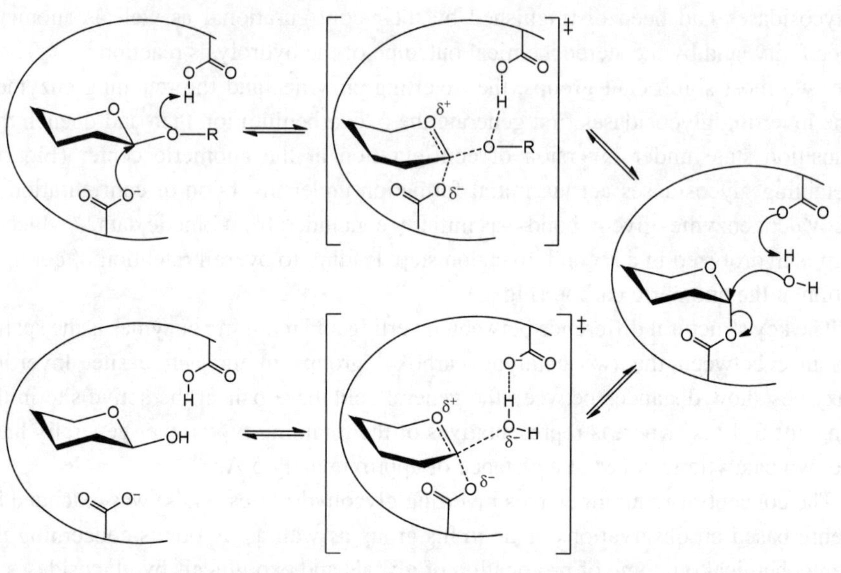

FIG. 2. Mechanism of retaining glycosidases.[90]

An alternative mechanism for the enzymatic hydrolysis was suggested by Post and Karplus,[92] and also by Fleet,[93] that invoked protonation of the endocyclic O-5 instead of the anomeric O-1, thus leading to an open-chain transition state, as was also discussed in Franck's interesting account.[94]

Some retaining N-acetylhexosaminidases have been demonstrated to follow a different catalytic pathway, with initial formation by anchimeric assistance from the N-acetyl group of a bicyclic α-configured oxazoline, which in turn is hydrolyzed from the β face to release free N-acetylglucosamine as the product.[95] This pathway had been found operative for the spontaneous hydrolysis of 3- as well as 4-nitrophenyl N-acetylhexosaminides at various pH values[96] and was suggested[96,97] as early as 1967 as an alternative to the generally accepted route via the covalent enzyme–glycon adduct, proposed by Koshland.[79]

In the absence of structural data for many interesting glycosidases, comparison of their sequences has revealed important structural relationships and similarities of catalytic properties. The database CAZy (Carbohydrate-Active EnZymes) is a sequence-based collection of currently around 100 GH families (currently GH 1 to GH 117, with around 15 families deleted), which are further grouped in 14 clans. It provides family-typical structural and mechanistic details and has become an invaluable tool in GH research.[98]

III. Glycon Structure, Functional Groups, and Catalysis

The significance of individual hydroxyl groups in the glycon part of glycosides for recognition/binding and for the stabilization of the transition state(s) of the enzymatic glycoside hydrolysis has constituted a major focus of research. Each hydroxyl group of the glycon was soon recognized to contribute to varying degrees to the recognition of the substrate by noncovalent interactions with the residues at the active site in the glycosidase, exerting stabilizing effects on the transition states of the glycoside hydrolysis reaction and thus aiding the turnover of the correct substrate. Modifications of the functional groups around the ring, and the biochemical consequences thereof, became an important tool for the characterization of a wide range of glycosidases.

In particular, substrate analogues having a decreased number of hydroxyl groups, such as deoxygenated (2-deoxy, 3-deoxy, 4-deoxy, and 6-deoxy sugars), or compounds O-methylated at various positions, along with the corresponding deoxyfluoro derivatives, have helped to gain key items of evidence and information as to the minimal requirements for substrate recognition, binding, and processing.

1. β-Glucosidases and Nonnatural Substrates

In early experiments, it was shown that 4-nitrophenyl 6-deoxy-β-D-glucopyranoside (**22**, Scheme 8), resembling aryl 6-O-methyl-β-D-glucopyranosides (**23**), is a slow substrate of the β-glucosidase from *Stachybotrys atra*.[99] Legler found that the hydrolysis of phenyl 6-deoxy-β-D-glucopyranoside (**24**), by the family 3 (GH 3) β-glucosidase A₃ of *Aspergillus wentii*, proceeded at a rate of about one-tenth of that of the corresponding β-glucoside.[100] The same deoxy sugar was also noted to be a substrate, albeit a poor one, of the β-glucosidase B₁ from *Aspergillus oryzae*.[101] The lack of a 2-OH group in 2-deoxyglucopyranosides slowed the action of the β-glucosidase from sweet almonds by factors ranging between 700 and 2000.[102]

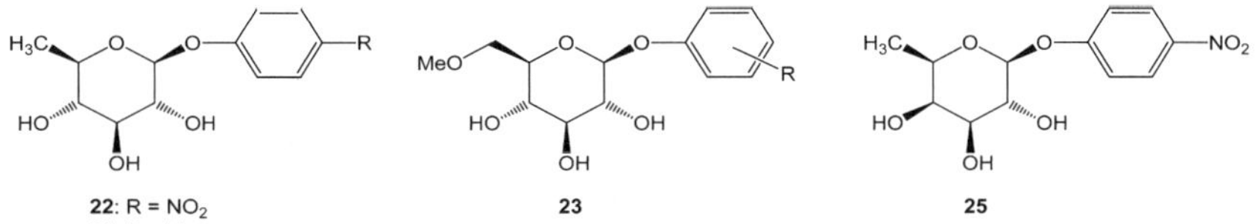

22: R = NO₂
24: R = H

23

25

Scheme 8.

Walker and Axelrod discovered that 4-nitrophenyl 6-deoxy-β-D-galactopyranoside (**25**) was a better substrate of the "β-glucosidase–β-galactosidase" of almond emulsin than the corresponding galactosides and glucosides.[103] Subsequent systematic investigations by Legler and Roeser in 1981 compared the kinetic effects of removing, from 4-methylumbelliferyl β-D-glucopyranoside, of the hydroxyl groups at C-2 (**26**, Scheme 9), C-4 (**27**), as well as of the hydroxymethyl group at C-5, of in hydrolysis by the β-glucosidase from *Aspergillus wentii*.[104]

In these studies, an even more pronounced effect, of four to five orders of magnitude in the rate constants, was found for glycosylation and deglycosylation with the 2-deoxy-D-*arabino*-hexoside (**26**) when compared to the parent compound, corresponding to a 32 kJ/mol contribution of 2-OH to the stabilization of the transition state. For the contribution of the 4-OH group, a value of 21 kJ/mol was calculated in a comparative study of *Agrobacterium* β-glucosidase with a comprehensive range of deoxygenated as well as deoxyfluoro-modified substrates.[105] Mega and Matsushima obtained similar results for the β-glucosidase of *Aspergillus oryzae*, with the strongest activation exerted by 2-OH and the lowest contributed by 6-OH.[106] The β-glucosidase of *Aspergillus niger* was investigated by Bock and coworkers, who monitored the enzymatic hydrolysis of cellobiose and methyl cellobioside as well as analogues monodeoxygenated in the nonreducing ring.[107] From the results, they concluded that all secondary hydroxyl groups of the nonreducing moiety are essential for acceptance by the enzyme, while the 6′-deoxy derivative was hydrolyzed, albeit slowly. The 6′-bromo-6′-deoxy analogue of cellobiose was not a substrate, possibly because of the large size of the 6-substituent and consequent steric clashes. Withers and coworkers compared the relative rates of spontaneous hydrolysis of dinitrophenyl β-D-glucopyranosides in an attempt to distinguish between the contributions of individual hydroxyl groups to the enzymatic and the spontaneous hydrolysis.[108]

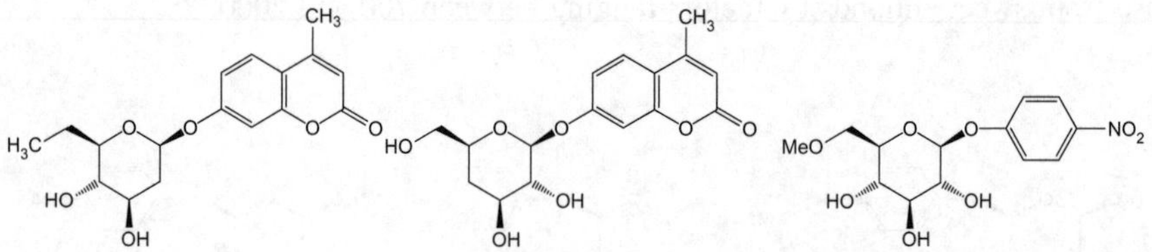

SCHEME 9.

Decreasing the number of hydroxyl groups, in particular removal of the 2-OH group, generally diminished the interaction with the β-glucosidase.

Since 2-deoxy-D-*arabino*-hexopyranosides ("2-deoxy glucosides") are more prone to spontaneous hydrolysis than the parent compounds, the 2-OH group apparently plays a most significant role in the enzymatic binding event and the conformational pathway toward the transition state, despite its role in electronically destabilizing the transition state. This was also supported by the finding that 2-deoxyfluoro glucosides are poor substrates, pointing to 2-OH as a vital hydrogen-bond donor toward a charged partner group of the glucosidase.[105]

Recently, a selection of fungal 1,4-β-glucosidases was probed with mono-*O*-methyl derivatives of 4-nitrophenyl β-D-glucopyranoside, and it was discovered that only mono-*O*-methylation of O-6 (**28**) was tolerated to a varying extent, whereas none of the substrate analogues *O*-methylated at a secondary position was a substrate of these enzymes.[109] Steric clashes with the active site were invoked to rationalize these findings.

2. β-Galactosidases and Nonnatural Substrates

Akin to the findings with β-glucosidases, a strong contribution of 2-OH was also observed in the β-galactosidase series. Employing 2′-, 3′-, 4′-, as well as 6′-monodeoxy methyl β-D-lactosides, Bock and Adelhorst demonstrated the vital role of 2-OH for the enzymatic process. The 3′-OH group and 4′-OH were also found to be essential, whereas 6-OH of the glucose moiety was important but not necessary for hydrolysis.[110] In a subsequent in-depth investigation exploiting a series of deoxy (**29–32**, Scheme 10) as well as deoxyfluoro 2,4-dinitrophenyl β-D-galactosides, the *E. coli (lacZ)* β-galactosidase (GH 2) was shown to exert a more than 33.5 kJ/mol interaction with 2-OH, with contributions of 3-OH, 4-OH, and 6-OH in the range of 15 (4-deoxy, compound **31**) to nearly 24 kJ/mol (3-F, compound **33**).[111]

Small-intestine lactase (GH 1), the absence of which is associated with lactose intolerance in a large proportion of the adult population, was probed with monodeoxy as well as selected mono-*O*-methyl derivatives in the galactose moiety of methyl lactosides.[112] This enzyme required hydroxyl groups at C-2 and also C-3, while neither the presence of a 4-OH group nor a specific configuration at C-4 was prerequisites for successful hydrolysis. A 6′-deoxy lactoside was readily hydrolyzed, but the absence of 6-OH in the *gluco* moiety caused it to be a poor substrate.

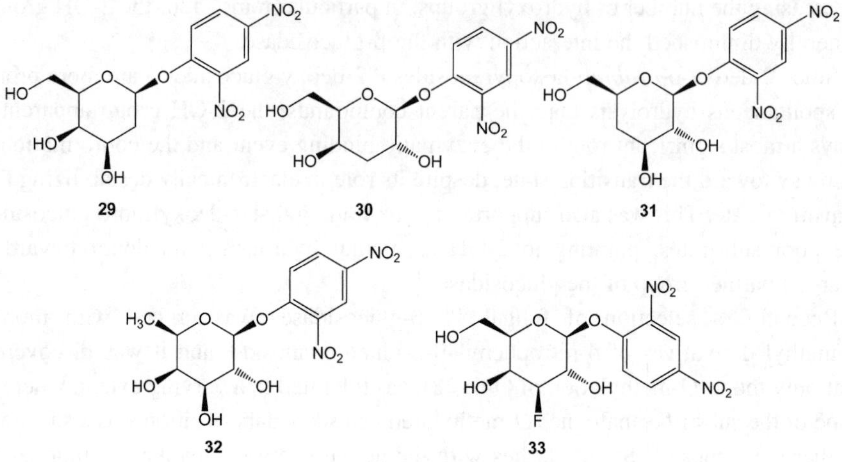

SCHEME 10.

3. α-Glucosidases and Nonnatural Substrates

Probing α-glucosidases with various substrate analogues, Bock and Pedersen investigated hydrolysis, by the inverting amyloglucosidase (glucoamylase, EC 3.2.1.3; GH 15) from *Aspergillus niger*, of all eight possible monodeoxy derivatives of methyl β-maltoside. Hydroxyl groups at positions 4' and 6' (in the nonreducing moiety) as well as 3-OH were found essential for processing by the enzyme. 6-Deoxy and 2'-deoxy derivatives were hydrolyzed faster than the corresponding fully hydroxylated compound, with transition states of similar energy but weaker substrate binding as compared to the parent compound.[113] Such derivatives as the 6-deoxyfluoro analogue were also found to be substrates, but charged groups (6-aminodeoxy as well as 5-carboxy) were not tolerated. Again, 6'-OH was found essential for successful hydrolysis.[114] In-depth kinetic comparison revealed powerful binding of 4'-OH as well as of 6'-OH to charged partners of the enzyme's active site, providing 17–19 kJ/mol each to the transition-state interactions.[115,116] With monodeoxy and mono-*O*-methyl modified methyl α-isomaltosides, a similarly pronounced significance of 4'-OH as well as 6'-OH was observed.[117,118] The retaining, high-isoelectric-point α-glucosidase from barley malt (EC 3.2.1.20, GH 31) was investigated by Svensson and coworkers. Again, a strong contribution by 4'-OH (18.9 kJ/mol) as well as by 6'-OH (18.7 kJ/mol) was found, with significantly smaller effects of 2'-OH and 3'-OH, amounting to around 10 kJ/mol each, suggesting charged partners for the former two hydroxyl groups and neutral hydrogen bonds between 2′-OH, 3′-OH, and the enzyme.[119] For the α-glucosidase of *Aspergillus niger*, the involvement of 2-OH as well as 3-OH in the ionization of the carboxyl groups located in the active site was demonstrated, employing 2-deoxy and 3-deoxy analogues of methyl β-maltoside.[120] Rice α-glucosidase was probed with all four monodeoxy derivatives (**34–37**, Scheme 11) of 4-nitrophenyl α-D-glucopyranoside.

This enzyme did not hydrolyze the 3-, 4-, or 6-deoxy derivatives but readily processed the 2-deoxy derivative.[121] In an extended study, the same group probed these substrates, plus the four corresponding mono-*O*-methyl derivatives of 4-nitrophenyl α-D-glucopyranoside, with a panel of seven α-glucosidases from six different sources, namely the GH 13 enzymes from *Saccharomyces cerevisiae*, *Bacillus stearothermophilus*, and two α-glucosidases from the honey bee (honey bee I and III) along with the family 31 α-glucosidases from sugar beet, flint corn, and *Aspergillus niger*. It turned out that the enzymes from sugar beet, flint corn, and *Aspergillus niger* hydrolyzed the 2-deoxy glucosides with a higher rate than the parent α-glucoside. The 3-deoxy glucosides were hydrolyzed by the flint corn and the *Aspergillus niger* enzyme of this GH family, whereas neither the 4-deoxy nor the 6-deoxy sugar served as substrates. The GH 13 glucosidases invariably required the intact D-*gluco* configuration, with all hydroxyl groups present. In line with the observations with the α-glucosidase from rice, no hydrolytic activity was observed with any of the mono-*O*-methyl glucopyranosides, and steric hindrance plus hydrophobic repulsions were

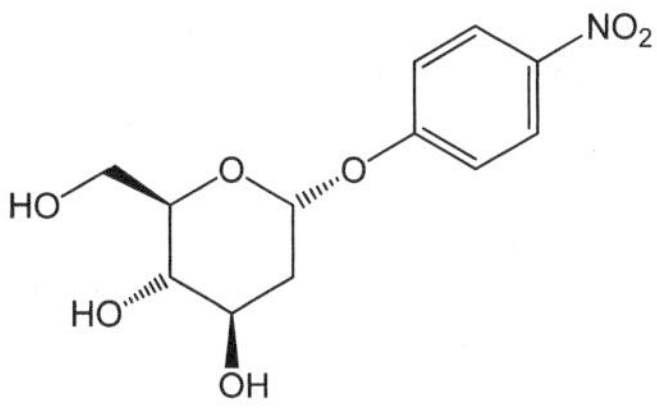

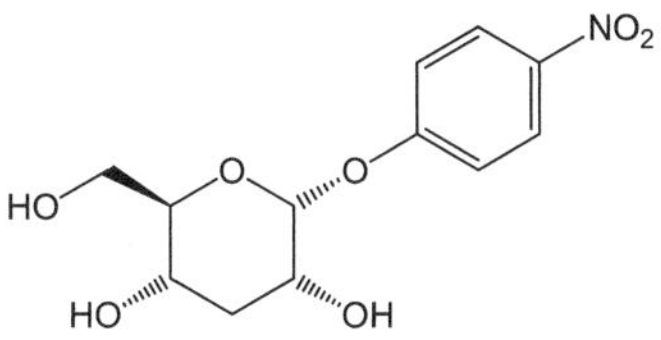

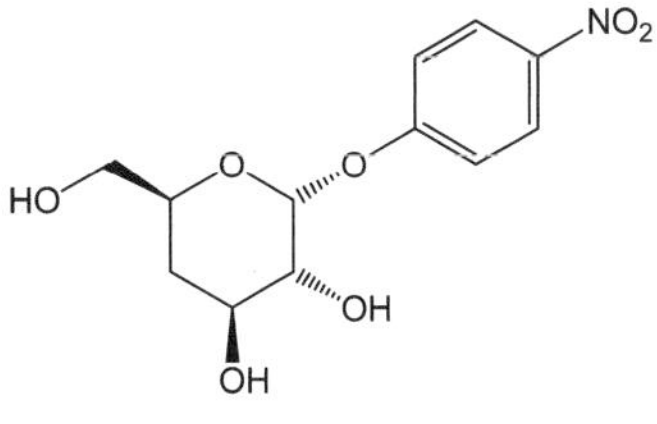

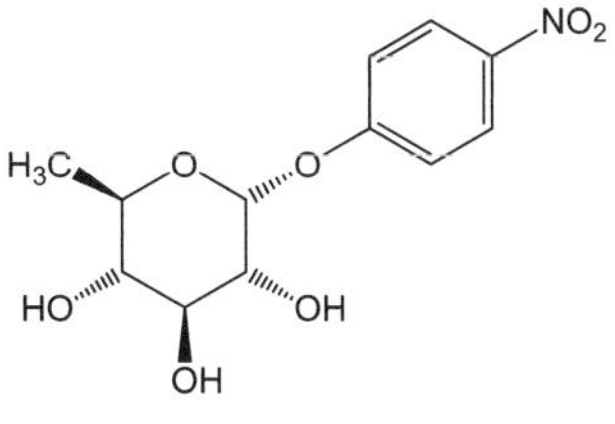

SCHEME 11.

postulated as the factors responsible for this nonreactivity. It was therefore concluded that 2-OH is not involved in interactions of the family 31 α-glucosidases with their substrates.[122]

The processing α-glucosidases I and II from rat liver were also probed with the monodeoxygenated 4-nitrophenyl α-glucosides (**34–37**, Scheme 11). Whereas α-glucosidase II readily hydrolyzed the 2-deoxy glucoside at a higher rate than the intact substrate, all of the deoxy derivatives inhibited α-glucosidase I.[123]

4. α-Galactosidases and Nonnatural Substrates

Investigation of the hydrolytic activities of GH 27 α-galactosidases (EC 3.2.1.22) from green coffee bean, *Mortierella vinacea*, and *Aspergillus niger* with the four possible monodeoxy derivatives of 4-nitrophenyl α-D-galactopyranoside provided the following picture: the *Aspergillus niger* α-galactosidase only hydrolyzed the 2-deoxy substrate. The enzymes from green coffee bean and *Mortierella* also accepted the 2-deoxygalactoside, albeit at substantially lower rates than the parent substrate, but additionally showed some activity for the 6-deoxygalactoside. None of the enzymes in this study was able to utilize either the 3-deoxy or the 4-deoxy sugar, pointing to the vital role of these two functional groups in the enzymatic hydrolysis of α-galactosides.[124] The α-galactosidase A from human liver hydrolyzed the 2-deoxy derivative of 4-nitrophenyl α-D-galactopyranoside but failed with the 6-deoxy analogue.[125]

5. α-Mannosidases and Nonnatural Substrates

In the mannosidase series, two GH 38 enzymes, that of jack bean and also almond α-mannosidase, were evaluated with deoxy derivatives of 4-nitrophenyl α-D-mannopyranoside.[126] For both enzymes, it was found that scarcely any hydrolysis took place with the 2-deoxy, 3-deoxy, or 4-deoxy sugars, but the 6-deoxymannopyranoside was turned over at about the same rate as the parent compound, indicating that a strict conservation of functional groups is necessary for productive enzymatic action.

In conclusion, modified substrates have served as excellent tools for the characterization and categorization of glycosidases, demonstrating the subtle network of polar and nonpolar forces and interactions involved in recognition, binding, and catalytic processing of glycosides. In particular, the contributions of individual hydroxyl groups to productive processing could be observed and measured. The essential role of the substituent at C-2 for the catalytic machinery of most glycosidases is clearly evident.

IV. GLYCOSIDASE INHIBITORS

In addition to natural substrates and structurally modified analogues, chemical compounds acting as inhibitors of GHs have served as valuable probes for the characterization of glycosidases. In turn, such compounds have been characterized by their effects on these enzymes.

1. High-Molecular-Weight Inhibitors

Seemingly, nature frequently exploits glycosidase inhibitors of high molecular weight that are usually themselves proteins. They may serve as "silencers" of the respective glycosidases under "normal" conditions but may dissociate upon exposure to biochemical triggers and release the active enzyme. A report as early as 1933 described "sistoamylase" a water-insoluble protein from buckwheat malts, which was found to inhibit malt alpha amylase.[127] Subsequently, several high-molecular-weight inhibitors of alpha amylases have been found in various plants and were shown to be proteins or glycoproteins, as nicely reviewed by Truscheit and coworkers.[128] Some molecular masses were demonstrated to be in the same range as those of the inhibited enzyme, as in the case of phaseolamin ($\sim$49 kDa), a powerful noncompetitive inhibitor of the alpha amylases of snail, hog, and humans.[129] Others, for example, *Streptomyces*-derived compounds, range between 10^3 and 10^4 mass units.[128,130,131] For comparison, some acarbose homologues as well as trestatins containing nine (trestatin A)[132] or more sugar/carba sugar components also fit into this mass range.

2. Small Molecules: Covalent ("Irreversible"), Competitive (Reversible), and Noncompetitive (Allosteric) Inhibitors

For investigations of glycosidase properties and mechanisms, chemists have conveniently employed mainly inhibitors of low molecular weight that are available in a molecular-mass range of 100[133] to around 1000 mass units.[128] Fitting into this range is a vast number of available covalent as well as competitive and noncompetitive inhibitors that have served as invaluable tools in GH research. Irreversible inhibitors form covalent bonds at (or very close to) the active site, thus blocking any catalytic activity. Reversible, competitive inhibitors are bound by noncovalent interactions and compete with the natural substrate for the binding site. Consequently, they may be removed by an excess of substrate or even by changes in the pH to release the active enzyme. Noncompetitive inhibitors act by binding to locations away from the catalytic center, effecting conformational changes or other means of sterically hindering access to the active site.

3. Nonsugar Inhibitors

Many glycosidase inhibitors are structurally related to natural substrates, closely resembling carbohydrates. However, several exceptions to this generalization are known. These include such diverse structures as the diterpene andrographolide and its analogues (**38**, Scheme 12),[134] 3,21-di-*O*-acetylcichoridiol (**39**, a triterpenoid),[135] acridone alkaloids, and oriciacridone (**40**).[136]

Other examples include the macrolide penarolide (**41**, Scheme 13),[137] flavonoids (**42**, **43**),[138] and also chalcones (**44**).[139]

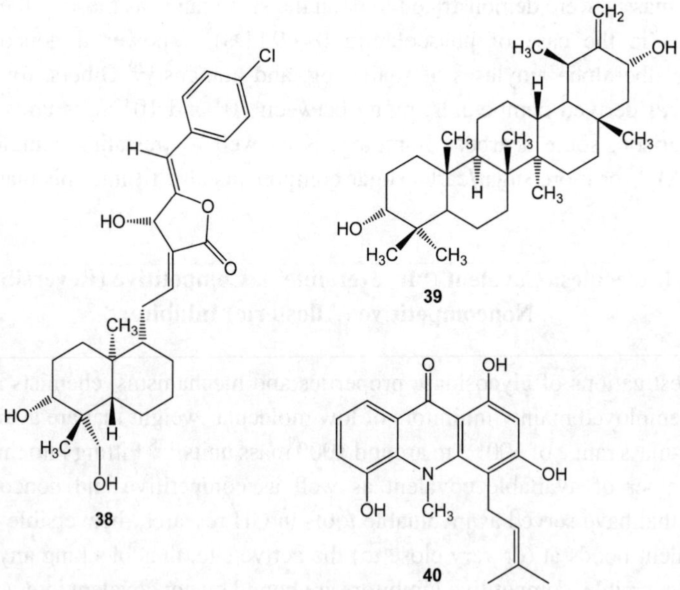

SCHEME 12.

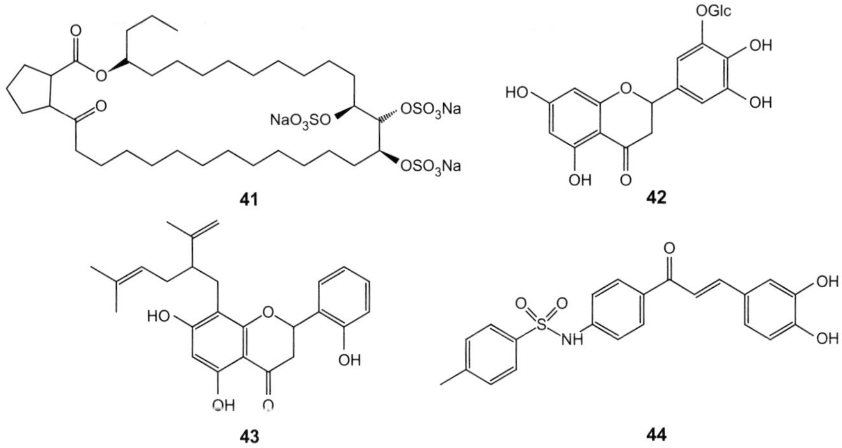

SCHEME 13.

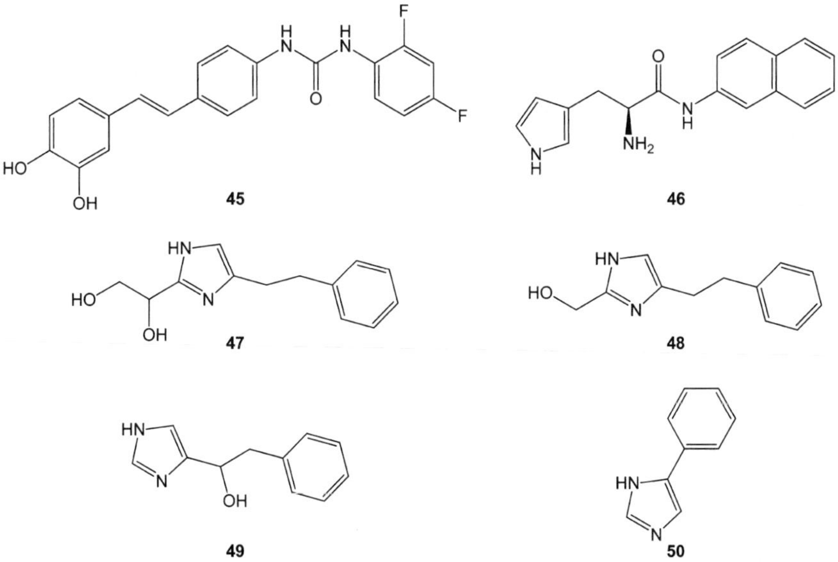

SCHEME 14.

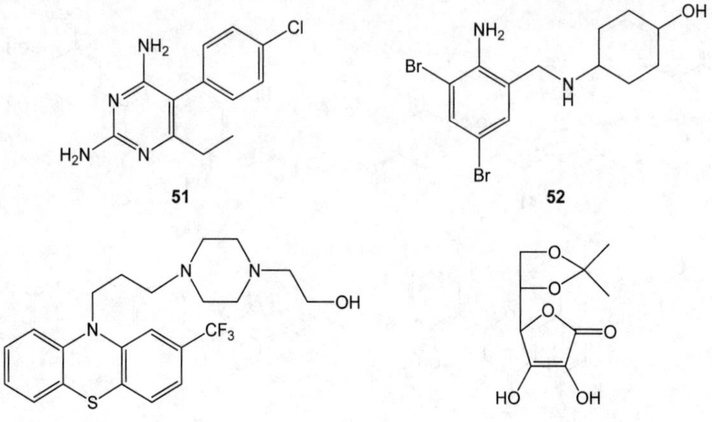

SCHEME 15.

In addition, flavones,[140] aromatic urea derivatives (**45**, Scheme 14),[141] lipophilic histidines (**46**), and histamines and imidazoles (**47–50**) have been reported as glycosidase inhibitors.[142–144]

Compounds also found active include (among many others) tannins,[145] the pharmaceutical products pyrimethamine (**51**, Scheme 15),[146] ambroxol (**52**),[147] the secretolytic ingredient in mucosolvan, or fluphenazine (**53**),[147] an antipsychotic substance, as well as—and closer to "real" sugars—ascorbic acid derivatives (**54**).[148]

4. Carbohydrate-Related Inhibitors

Carbohydrate-related inhibitors can be grouped into three important classes: the carbocycles, the sugar analogues having sulfur in the ring, and also the large class of imino sugars, including iminoalditols, bicyclic analogues, and related polyhydroxylated alkaloids.

a. Substrate-Related Covalent ("Irreversible") Inhibitors.—Such covalent inhibitors as bromoconduritols (**55**, Scheme 16) and anhydroconduritols (**56**)[149] or cyclophellitol (**57**),[150,151] along with corresponding aziridines, are substrate-related, highly functionalized cyclohexane derivatives. They undergo chemical reactions (nucleophilic substitution processes, including oxirane- and aziridine-ring opening, addition,

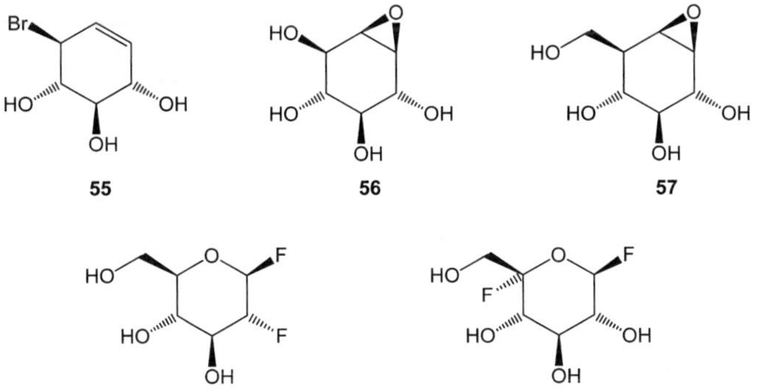

SCHEME 16.

Michael addition, or allylic rearrangements) at (or very close to) the active site, resulting in covalent and, hence, catalytically incompetent enzyme–inhibitor adducts.

These adducts can be further investigated, either as such or subsequent to partial digestion, to provide clues as to the location of the active site as well as—from the product deriving from the inhibitor—the anomeric specificity of the enzyme. Cyclophellitol (**57**)[152] is a powerful and selective irreversible inhibitor of β-glucosidases.[153] The inhibition of β-glucosidase from *Thermotoga maritima* was recently shown by X-ray crystallography to be the result of nucleophilic oxirane opening by one of the catalytic glutamate residues with formation of a covalent ester bond.[154] The type of inhibition by the corresponding aziridine or epimers thereof has not been reported,[155,156] and it was suggested "that the inhibition mechanisms of the epoxide and the aziridine analogues are different."[156] However, results obtained by Tong and Ganem[157] for a related compound having an endocyclic aziridine nitrogen suggest carboxylate-induced opening of the aziridine ring with formation of an ester bond between the sugar analogue and the nucleophile.

Several other electrophilic structures that may also react readily with a carboxylate group at the active site have been reported[158–160] and reviewed.[83] More recently, 2-deoxy-2-fluoro- (**58**) together with 5-deoxy-5-fluoroglycosyl fluorides (**59**), which form analytically stable covalent glycosyl esters in the active site, have been introduced and employed by Withers and coworkers in elegant studies of active-site mapping.[161] (Unmodified glycosyl fluorides and glycals bind to the active site, also forming an ester bond initially. However, the lack of sufficiently stabilizing electron-withdrawing substituents at C-2 usually renders this bond prone to subsequent hydrolysis).

The area of covalent glycosidase inhibitors has recently been highlighted in an excellent review that also includes compounds containing such reporter groups as photolabile functional groups as photoaffinity probes, and others.[161]

b. Competitive Inhibitors.—*(i) Structural Types of Amine-Based Inhibitors.* A key question in looking at basic glycosidase inhibitors has been the significance of the position of the basic or cationic center (Fig. 3).[162] This basic center, or more specifically, the ring nitrogen atom, may be situated in replacement of O-5 as in 1,5-imino sugars (**I**), O-4 as in furanoid imino sugars (**K**), at the anomeric position providing "isoimino sugars" (**J, L**), or exocyclically, replacing O-1 as in glycosyl-amines (**A, E**) or aminocyclitols (**B–D, F–H**).

This positioning in the context of ring size and conformation may have major influence on enthalpic and entropic terms of the inhibitory interaction and, consequently, on the relative power of inhibition of α- and β-specific enzymes, as distinguished by more or less subtle differences in the geometry of their active sites.[83]

Most competitive inhibitors form noncovalent complexes, usually ion pairs, as with the imino sugars,[83] amine-substituted cyclopentanes[163] and cyclohexanes[83] (furanoid and pyranoid carba sugars), and charged thio sugar derivatives, for example, salacinol (**60**, Scheme 17)[164–166] and its relatives, just to mention the most frequent examples in the literature. Depending on the number and the "quality" of their interactions, these inhibitors compete with the substrate and the complexes formed may have substantial half-lives, but the inhibitor is eventually released in the presence of an excess of

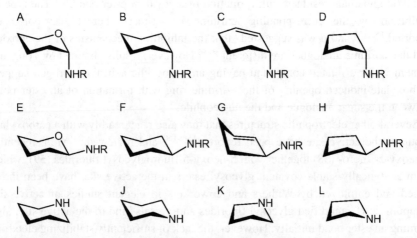

FIG. 3. Position of the nitrogen atom in various basic sugar analogues of proven glycosidase inhibitory activity (disregarding conformational freedom as well as other ring sizes).

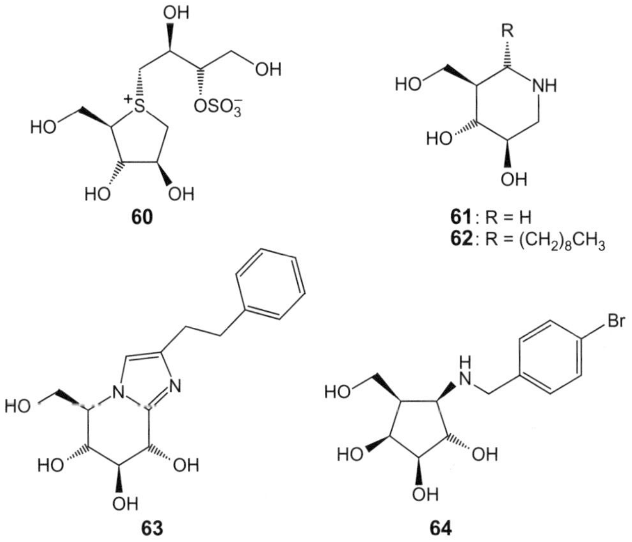

SCHEME 17.

substrate or by changes of the pH value. Their inhibitory activity is usually given as IC$_{50}$ (concentration of inhibitor that results in decrease of the enzyme activity to 50%, or "functional power"), or as the K_i value (dissociation constant of the enzyme–inhibitor complex, or "binding affinity").

From the rate enhancement of enzymatic versus spontaneous glycoside hydrolysis, it has been concluded that the transition state is the enzymatically most stabilized species on the way from substrate to product, and that reversible inhibitors closely mimicking in theory its geometry and charge distribution may bind with K_i values in the atto-(10^{-18}) to zepto-molar (10^{-21}) range.[167,168] Known glycosidase inhibitors have not even come close to such values (as yet) in telling the difference between "ideal transition-state analogue" and "real life" structural ramifications of mimicking the "perfect" oxocarbenium ion–enzyme complex. Of the many powerful inhibitors known to date, those having values in the low nanomolar range are already quite exceptional.

For example, some of the most powerful β-glucosidase inhibitors of their times, such as isofagomine **61**[169] (1994, K_i 110 nM, β-glucosidase almonds, pH 6.8; K_i 25 nM,[170] human β-glucocerebrosidase), C-nonylisofagomine **62** (2005, IC$_{50}$ 0.6 nM, human β-glucocerebrosidase),[170] glucoimidazole **63** (2000, K_i 1.2 nM, β-glucosidase

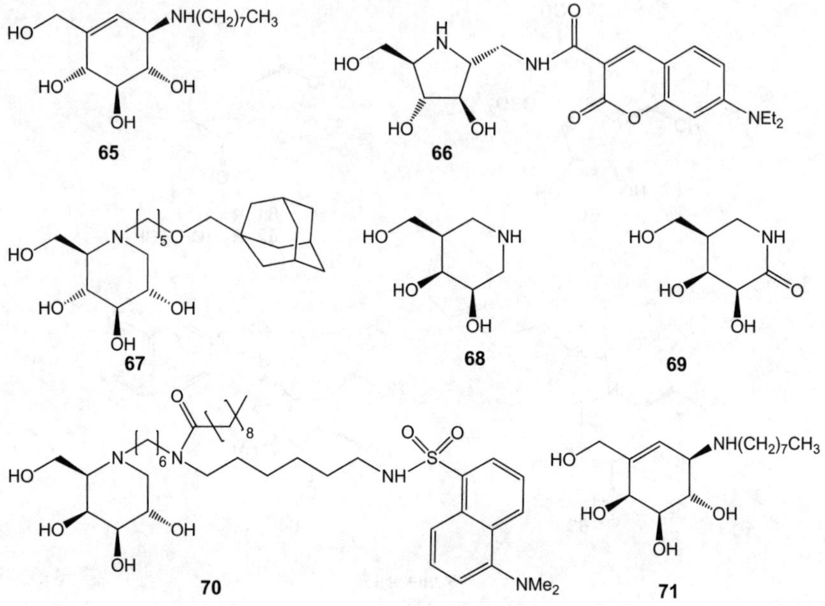

SCHEME 18.

almonds, pH 6.8; K_i 0.11 nM, β-glucosidase *Caldocellum saccharolyticum*),[171] castanospermine **8** (1984, K_i 10 μM, β-glucosidase almonds, pH 6.5),[172] cyclopentane derivative **64** (2001, K_i 17 nM, β-glucosidase almonds, 120 nM *C. saccharolyticum*),[173] *N*-octyl-β-valienamine **65** (1996, IC$_{50}$ 30 nM, mouse liver β-glucocerebrosidase, pH 5.5; Scheme 18),[174] DMDP derivative **66** (2001, K_i 1.2 nM, β-glucosidase *Agrobacterium* sp.),[175] and adamantly substituted *N*-alkyl-1-deoxynojirimycin **67** (2009, K_i 1 nM, membrane-bound β-glucosidase 2)[176] all range between K_i 10 μM and 100 pM.

The β-galactosidase inhibitors 4-*epi*-isofagomine **68** (1995, K_i 4.1 nM, *Asp. oryzae*, pH 6.8),[177] the related lactam **69** (2001, K_i 18 nM, *Asp. oryzae*), the cyclopentane **64** (2001, K_i 0.6 nM, β-galactosidase from *E. coli*; K_i 4 nM, *Aspergillus oryzae*; Scheme 17),[173] *N*-alkyl-1-deoxy-D-galactonojirimycin **70** (2010, K_i 47 nM, β-gal *E. coli*),[178] and NOEV **71** (2002, K_i 300 nM, human β-galactosidase)[179] can be found roughly in the same range, with the imino sugars certainly trailing the furanoid carbacycle.

In a class of their own are the noteworthy examples of imino sugar nucleoside analogues as inhibitors, provided by Tyler, Schramm, and their coworkers, based on

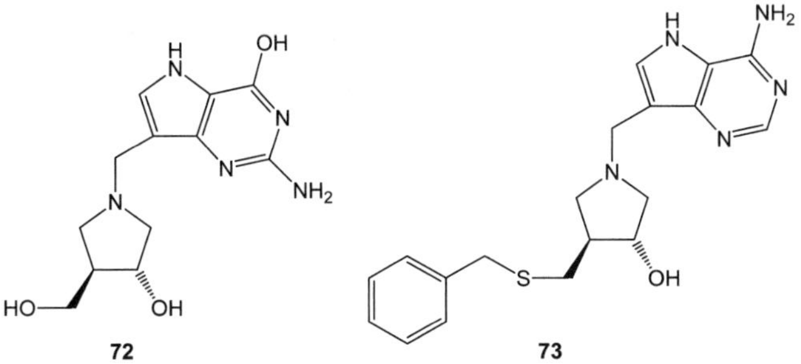

SCHEME 19.

the observation of kinetic isotope effects during enzymatic nucleoside hydrolysis and rational design of transition-state analogues. Purine nucleoside phosphorylase (PNP) inhibitors, such as compound **72** (K_i 7 pM, human PNP, Scheme 19), or 5′-methylthioadenosine nucleosidase (MTAN) inhibitors such as **73** (K_i 0.047 pM), are by far the most potent imino sugar-based inhibitors, to date. The topic and its implications for the treatment of a variety of diseases have recently been excellently reviewed.[180]

V. IMINO SUGARS

Among the considerable number of glycosidase-inhibitor taxa, imino sugars in the widest sense have become the largest, most versatile, and most readily available in terms of synthetic availability, structural variability, and inhibitory properties. With practically any configuration, ring size, substitution pattern, as well as a wide pK_a range readily accessible, these compounds provide invaluable support for glycosidase research.

Imino sugars feature a ring nitrogen atom formally replacing the corresponding ring oxygen, which is O-4 in furanoid compounds or O-5 in pyranoid structures, and a few examples of synthetic polyhydroxyazepanes are also known. The term "aza-sugars" as frequently (but incorrectly) employed by some authors, properly refers to those compounds in which a carbon atom has been replaced by nitrogen (see IUPAC 2-Carb-34.1). In this sense, glucooxazine **74**[181–183] (Scheme 20) is an aza sugar (formal replacement of C-1 by nitrogen in the ring, O-5 as ring oxygen still present), whereas isofagomine (**61**, Scheme 17) or noeuromycin (**75**)[184] is not.

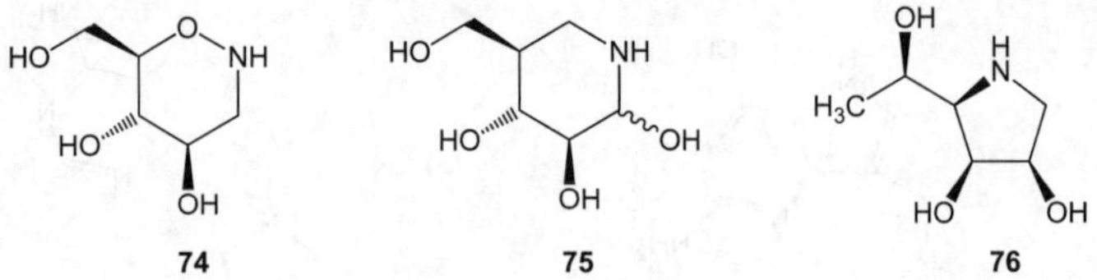

SCHEME 20.

Such imino sugars as free nojirimycin (**1**, 5-amino-5-deoxy-D-glucose) and stable derivatives thereof, iminoalditols in particular, have proved of great value due to the availability of several of them as natural products and because of the many excellent synthetic approaches that allow access to practically any configuration, relevant ring size, and mode of derivatization. Many of these polyhydroxy compounds featuring a ring nitrogen and a sugar-like configuration of the functional groups around a pyranoid ring have been shown to be excellent inhibitors of glycosidases specific for the corresponding configuration. For example, 1-deoxynojirimycin (**2**, Scheme 1) is a good D-glucosidase inhibitor, the D-*manno* epimer **11** (Scheme 4) inhibits mannosidases, the 6-deoxy-L-*manno*-configured compound **20** (Scheme 7) is a very powerful inhibitor of α-L-fucosidases, and the corresponding GlcNAc analogue **21** interacts strongly with hexosaminidases.

Accordingly, one of the key guidelines to design good imino sugar-type inhibitors of a particular glycoside hydrolase is to mimic closely the configuration pattern of its natural substrate(s), although this may not be as straightforward as it sounds. Beyond the carbon backbone and the obvious charge at the nitrogen atom, substrates and inhibitors are clusters of donors and acceptors of hydrogen bonds, with polar as well as nonpolar interactions. Slight differences of intermolecular atomic distances may still fit the glycosidase's "expectation" of its substrate, even when the chemist sees a completely different structure. This can be exemplified by 2,5-dideoxy-2,5-imino-D-mannitol (**7**), a good β-glucosidase inhibitor with obvious fructofuranose shape, as well as 1,4,6-trideoxy-1,4-imino-D-mannitol (**76**, Scheme 20) and swainsonine (**13**, Scheme 5), two powerful α-mannosidase inhibitors.

1. pK_a Value and State of Ionization Approaching the Active Site

An important parameter of glycosidase inhibition by basic sugar analogues is certainly the pK_a value of the respective potential inhibitor in relation to the state of protonation of the active site under consideration, and the pH optimum of the enzyme.

Considerable work has been conducted concerning the relationships between the pK_a value of a given glycosidase inhibitor and its activity toward particular enzymes.

Legler has considered two mechanisms for the ion-pair formation between basic inhibitor and active-site carboxyl group[162]:

(a) The basic inhibitor approaches in equilibrium with its protonated form depending on its pK_a, and the protonated form associates with a carboxylate in the active site, or
(b) the inhibitor is bound as the free base and becomes protonated by an active-site carboxyl group.

As an example of a "cation-binder," the β-glucosidase A_3 from *Aspergillus wentii* (GH 3) was strongly inhibited by β-D glucopyranosylpyridinium cations,[185] whereas, for example, the β-glucosidase from almonds (GH 1) and the β-galactosidase of *E. coli* (GH 2) are not significantly affected by cationic substrate analogues.

The removal of hydroxyl groups in proximity to the basic center may increase the basicity of imino sugars, whereas the introduction of strongly electronegative substituents, in particular fluorine atoms, may cause the free base to prevail, or at least be present in a comparably higher proportion in the equilibrium at the same pH value.

Bols and coworkers compared the stereoelectronic effects of substituents at individual ring positions on the basicity of the ring nitrogen of various inhibitors. They had observed that the pK_a of 4-*epi*-isofagomine **68** (Scheme 18), featuring one axial hydroxyl group, was measurably higher (pK_a 8.8) than that of the corresponding parent compound, isofagomine **61** (Scheme 17) (pK_a 8.4).[186] Similar findings were subsequently made with other compounds, demonstrating that an axial hydroxyl group in the γ-position to the ring nitrogen led to a higher pK_a value than was measured with the corresponding equatorially substituted imine.[187] An in-depth investigation employing a set of around 60 functionalized piperidines bearing different substituents supported the theory that equatorial hydroxyl groups in the 4- or 3-position of these compounds and related structures are generally (up to three times) more electron-withdrawing than axial ones, thus influencing the basicity of the ring nitrogen.[188] This was also assumed to be a reason for conformational changes observed upon protonation, a phenomenon previously observed with mono- as well as difluoropiperidines.[189,190] A versatile and useful simple formula for the prediction of pK_a values in poly-heterosubstituted piperidines was deduced in the course of these investigations, providing estimates as close as 0.1 pH unit to the values measured with comparative data for fluorine substituents also available.[191] Conformational inversion of piperidine chairs from an all-equatorial to an all-axial substitution pattern resulted in a pK_a increase of as much as two pH units.[192] These observations may also have implications in regard to the ring conformation of protonated and unprotonated inhibitors as they approach the active site of glycosidases, as well as the possibility of conformational changes of unprotonated inhibitors upon their protonation in the active site. Prediction of the pK_a value is a very useful feature for the design of active-site probes prior to undertaking elaborate synthetic efforts and provides an option to narrow down the number of interesting candidate molecules to be used.

2. Stereochemical Considerations Concerning the Catalytic Protonation Process

In a most valuable contribution concerning the stereochemical features of the protonation pathway for glycosides in the active site, Vasella and his group have designed highly sophisticated tools to distinguish, among glycosidases, between syn- and anti-protonators (Fig. 4).

In the first group, the catalytic acid is syn to the ring oxygen atom relative to the plane defined by O-1, C-1, and H-1, whereas in the second group, the proton approaches anti with respect to the ring oxygen atom.[193] While the protonation in anti-protonators occurs in a glycoside conformation favored by the exo-anomeric effect, syn-protonators require distortion of the sugar ring for the aglycon to attain a quasiaxial orientation.

This notion affords another criterion for categorizing glycosidases and, additionally, is a very meaningful piece of information regarding inhibitor design and selectivity. With the inhibitory data of such glycoimidazoles as compound **77** (Scheme 21), 1,2,4-triazoles (**78**), and tetrazoles (**79**) in various glycon configurations, it is possible to design and explore powerful inhibitors as well as sensitive probes for the protonation trajectory.[194–197]

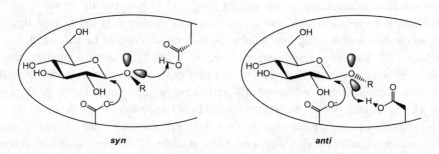

FIG. 4. syn- and anti-protonation, respectively, of O-1 as a means of classification of glycosidases.[193]

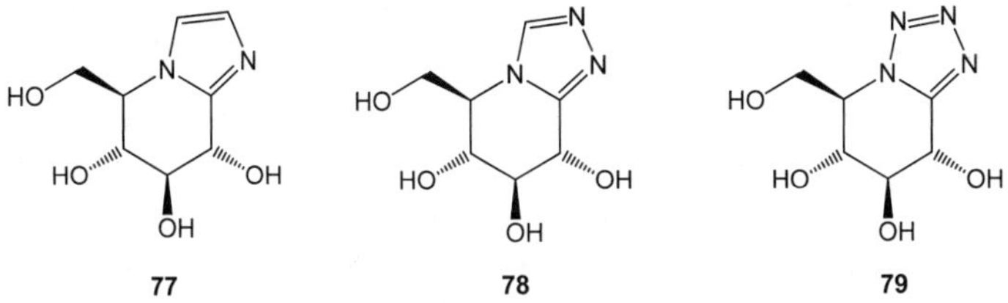

SCHEME 21.

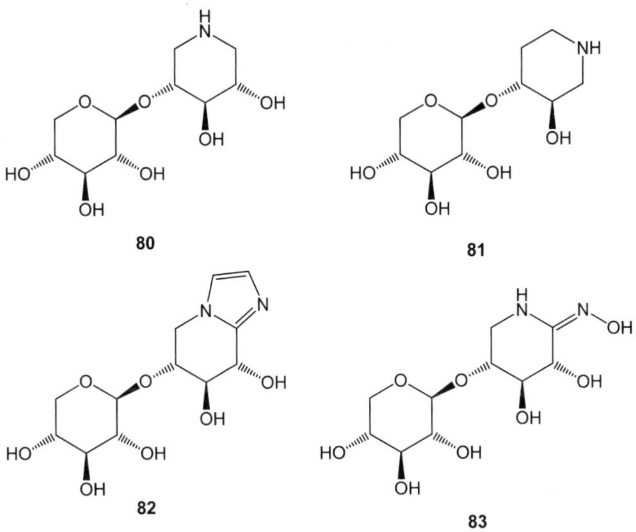

SCHEME 22.

This important concept was rapidly absorbed by the research community, providing "fine-tuning" of both glycosidase categorization and inhibitor design. In particular, subtle differences of xylosidase and xylanase specificities were probed, and they provided useful insights into the complex and finely tuned enzymatic hydrolysis of glycosides.

Family 10 xylanase Cex from *Cellulomonas fimi* and family 11 xylanase Bcx from *Bacillus circulans*[198,199] (the first being an anti-, the second a syn-protonator) were exposed to deoxynojirimycin-related (**80**, Scheme 22) and isofagomine analogous (**81**) xylosidase inhibitors, along with the anti-protonation-selective xylobiosyl imidazole (**82**) and the lactam oxime (**83**). Whereas these inhibitors inhibited Cex, it was found that Bcx was less prone to inhibition by the compounds probed.

In the context of structure determination of glycosidases by X-ray diffraction and, consequently, the availability of an increasing number of structural data for enzyme–inhibitor complexes, the concept has become a most valuable tool. A further refinement of the syn/anti concept was offered by Nerinckx and coworkers,[200] who combined Vasella's approach with Vyas's[201] A–B half-space nomenclature and introduced space quadrants (syn-A, syn-B, anti-A, anti-B). Investigating the liganded glycoside hydrolase structures then known, they noted that an electron-rich functional group is nearly always (with very few exceptions) positioned at a distance close enough "to be able to interact with the substrate-ring oxygen $2p_z$ orbital at the transition state," facilitating stabilization of the transition state. These workers found that syn-protonators of β-glycoside hydrolases (equatorial leaving group) feature their proton donor consistently at the syn-A, and syn-α enzymes (axial leaving group) at the syn-B quadrant, thus providing an "extra interaction toward the transition state's ring oxygen," concluding that "they should be in catalytic advantage over anti-protonators." With anti-protonators, where this interaction with the ring oxygen $2p_z$ orbital is not available, compensating alternative interactions with electron-rich components such as the second oxygen atom of the carboxylate or a hydroxyl group of the ligand itself, which are located in the syn-A or syn-B quadrants, were inferred. Such involvement of substrate atoms has also been found, for example, by Collins and coworkers.[202]

From this work, glycoside hydrolases may be subdivided into eight classes according to their action as retaining or inverting enzymes, axial or equatorial orientation of the aglycon, as well as syn- or anti-protonation. A very interesting view on the evolutionary development of glycosidases from "very primordial" anti-protonators also provides interesting reading.

3. Significance of Individual Functional Groups to Recognition and Binding

Akin to the modified substrates mentioned, generally monodeoxy, mono-*O*-methyl, as well as monodeoxyfluoro derivatives, the contribution of the number and location of hydroxyl groups in glycosidase inhibitors is of significance in the characterization of enzymes and inhibitors. Natural products, such as monodeoxy derivatives of known inhibitors, typically in the pair 1-deoxynojirimycin (**2**) and fagomine (**3**, Scheme 1), along with synthetic deoxy and deoxyfluoro analogues, have been frequently investigated. For example, the parent compound **2** is a considerably stronger

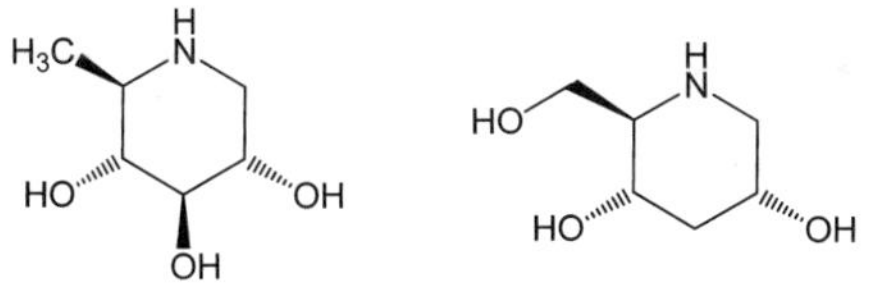

SCHEME 23.

glucosidase inhibitor as compared with fagomine (**3**) across a range of α- and β-glucosidases,[203,207] such as the α-glucosidase from rice (IC_{50} 0.05[204] vs. 320 μM),[205] yeast (IC_{50} 880 vs. 190 μM),[206] and the maltase from rat intestine (IC_{50} 0.63 vs. 820 μM),[207] but noteworthy exceptions being with the β-galactosidases from rat intestine (IC_{50} 26 vs. 15 μM) and bovine liver cytosol (no inhibition vs. IC_{50} 38 μM), albeit in a different group of glycosidases.[207] In the D-mannosidase series, and considering fagomine as a 2-deoxy derivative of 1-deoxymannojirimycin (**11**, Scheme 4), a total loss of activity is again observed when it is compared with the parent compound.[207] This result can be attributed to the lack of the important transition state-stabilizing interaction of 2-OH with the active site, as well as by the increase of basicity caused by the absence of the electron-withdrawing hydroxyl group.[104,105] The same loss of inhibitory activity was found for the 6-deoxy derivative **84** (Scheme 23) of compound **2** prepared by Wong and coworkers,[208] with the α-glucosidase from brewer's yeast showing an activity loss of two orders of magnitude (K_i 1.5 mM vs. 8.7 μM at pH 6.5), as well as with β-glucosidase from almonds where a 40-fold reduction of inhibitory activity was observed (780 vs. 18 μM). The 3-deoxy derivative (**85**) of 1-deoxynojirimycin was not an inhibitor.[209]

Similarly, pronounced decreases of inhibitory activity were observed with pyranoid inhibitors of other configurations upon deoxygenation, in conformity with the general picture provided by the modified substrates as mentioned earlier.

Interestingly, the furanoid derivative 1,4,6-trideoxy-1,4-imino-D-mannitol (**76**, Scheme 20) was found a potent inhibitor of human α-mannosidases and was comparable to the parent compound **19** (Scheme 7).[210]

4. Deoxyfluoro Derivatives as Active-Site Probes

The substitution of hydroxyl groups by fluorine atoms has also been an option, considering that this alteration can be regarded a "polar" deoxygenation and that the fluorine atom may still serve as a hydrogen-bond acceptor or by exerting dipole and

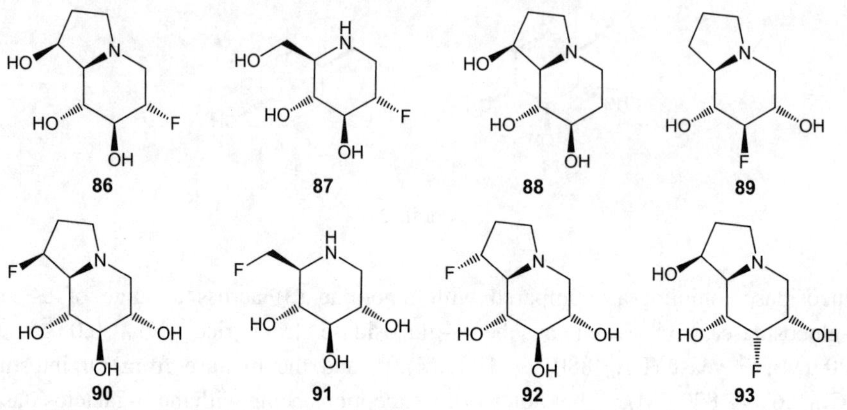

SCHEME 24.

field effects.[111] This consideration raised the question as to whether replacement of hydroxyl functions in glycosidase inhibitors by fluorine atoms would provide additional items of information on the quality as well as the selectivity of glycosidase inhibition.

The first published deoxyfluoro derivative of an imino sugar was Vogel's 6-deoxyfluorocastanospermine (**86**, Scheme 24), which was synthesized in 1990 by a *de novo* route exploiting tartaric acid-derived ketene equivalents.[211,212] This compound is the structural analogue of 1,2,5-trideoxy-2-fluoro-1,5-imino-D-glucitol (**87**) in the deoxynojirimycin series, with a pK_a of 5.1 [Castanospermine (**8**, Scheme 3), pK_a 6.0; 6-deoxycastanospermine (**88**), pK_a 7.3].

1,7-Dideoxy-7-fluorocastanospermine (**89**) was synthesized by Lee and coworkers.[213] From the natural product castanospermine (**8**, Scheme 3), Furneaux *et al.* prepared, among numerous other interesting derivatives, 1-deoxyfluorocastanospermine (**90**) corresponding to the 6-deoxyfluoro derivative (**91**) of compound **2**, as well as the 1-*epi*-F-derivative (**92**),[214] the 6-deoxyfluoro derivative (**86**),[215] the 7-deoxy-7-*epi*-F analogue (**93**),[214] and both 8-deoxyfluoro epimers (**94, 95**, Scheme 25).[216,217]

In the deoxynojirimycin series, the 6-deoxyfluoro derivative (**91**; pK_a 5.85) was the first published example of a deoxyfluoro analogue.[218] This compound was also made available by Szarek and coworkers from L-sorbose.[219] The 2-deoxyfluoro derivative (**87**),[220] and subsequently 1,4-dideoxy-4-fluoronojirimycin (**96**), was also reported.[221]

The 3-deoxyfluoro compound **97** (pK_a 5.75) was published by Lee and coworkers[222] who also synthesized the corresponding 3,6-difluoro derivative (**98**) of 1-deoxynojirimycin.[223,224]

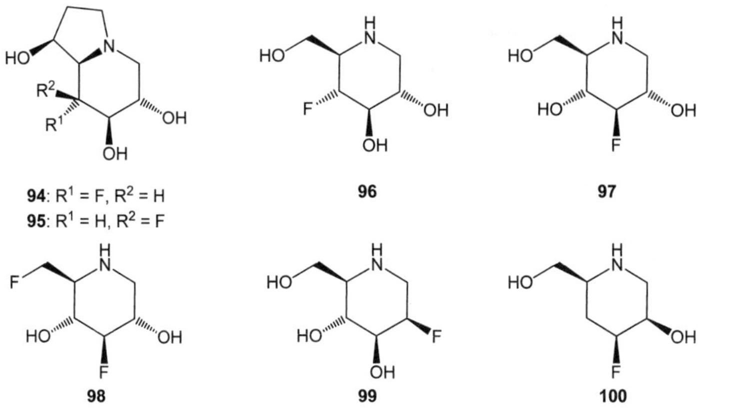

IMINO SUGARS AND GLYCOSYL HYDROLASES 219

94: R¹ = F, R² = H
95: R¹ = H, R² = F

96

97

98

99

100

SCHEME 25.

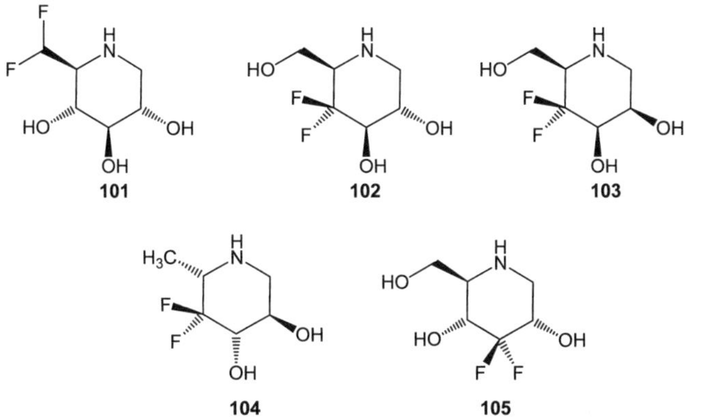

101

102

103

104

105

SCHEME 26.

The D-*manno*-configured 2-deoxyfluoro derivative **99** was prepared by Wong and his group, who exploited an aldolase-catalyzed carbon–carbon bond formation reaction[208] and a 3,4-dideoxy-3-fluoro derivative (**100**, Scheme 26) of compound **11** was provided by Resnati and coworkers.[225]

gem-Difluoro derivatives of 1-deoxynojirimycin were synthesized by Szarek and coworkers, who prepared 1,6-dideoxy-6,6-difluoronojirimycin (**101**)[226] by oxidation of a suitably protected L-sorbose derivative at C-1 and reaction of the resulting aldehyde with DAST (diethylaminosulfur trifluoride), followed by introduction of

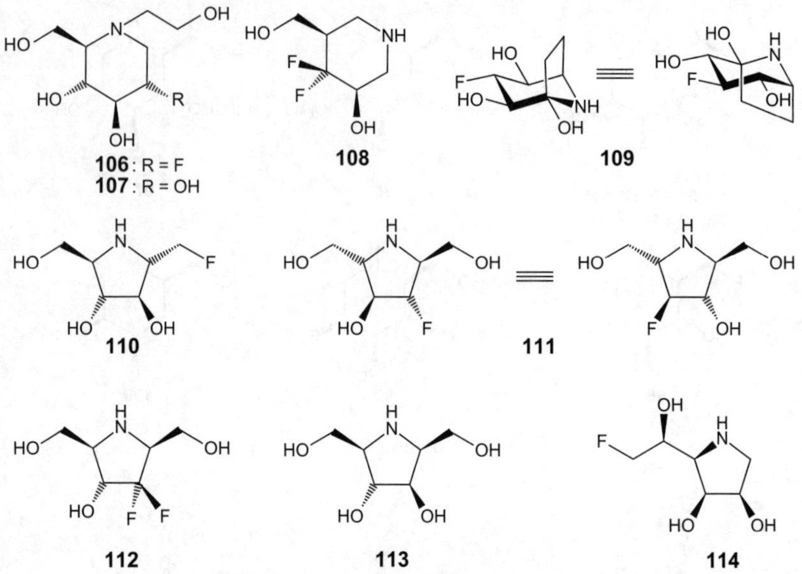

SCHEME 27.

the amine at C-6 and intramolecular reductive amination of the free sugar. 4,4-Difluoro derivatives in the "D-*gluco*" (**102**), the "D-*manno*" (**103**), as well as the "6-deoxy-L-*gluco*" series (**104**) were recently reported by Qing and coworkers.[227] The corresponding 3,3-difluoro derivative **105** was prepared by Csuk et al.[228]

Examples of other fluorinated inhibitors, mainly of D-glucosidases, are the 2-deoxy-2-fluoro derivative **106** (Scheme 27) of miglitol (*N*-hydroxyethyl-1-deoxynojirimycin, **107**),[229] 4-deoxy-4,4-difluoroisofagomine (**108**),[230] 3-deoxy-3-fluoro-calystegin B$_2$ (**109**),[231] the 1-deoxyfluoro derivative (**110**) of 2,5-dideoxy-2,5-imino-D-mannitol,[232] as well as the 3-deoxyfluoro analogue (**111**) of nonnatural L-DMDP (2,5-dideoxy-2,5-imino-L-mannitol)[233] and the 3-deoxy-3,3-difluoro derivative (**112**) of 2,5-dideoxy-2,5-imino-D-glucitol (**113**).[234] All of these are weaker inhibitors than the parent compounds. 1,4,6-Trideoxy-6-fluoro-1,4-imino-D-mannitol (**114**) was prepared by Winchester and coworkers.[210]

No direct comparison of the fluorinated castanospermine derivatives with the corresponding 1-deoxynojirimycin analogues has yet been conducted. In the castanospermine series, the fluorinated compounds exhibited low inhibitory activities. Only the 6-deoxyfluoro derivative **86** (corresponding to the 2-deoxyfluoro compound

87 in the 1-deoxynojirimycin series, Scheme 24) was found to show "only very slightly reduced" activities with lysosomal as well as processing α-glucosidases.[235]

In the 1-deoxynojirimycin series, the 2- (**87**), the 3- (**97**), and the 6-deoxyfluoro compound **91** (Scheme 24) were compared with the parent compound against a set of standard glucosidases.[232] 3-Deoxyfluoro (**97**, K_i 0.35 μM) as well as 6-deoxyfluoro (**91**, 0.4 μM) derivatives exhibited a 40-fold decreased power with the α-glucosidase from rice (GH 31) as compared with the parent compound **2** (K_i 0.01 μM at pH 6). Fagomine (1,2-dideoxynojirimycin, **3**) was reported to have an IC_{50} of 320 μM with this enzyme.[205]

With the β-glucosidase from almonds (GH 1), the 2- and the 3-deoxyfluoro compounds were totally inactive (which is also true for fagomine),[206] whereas the 6-deoxyfluoro analogue had a K_i value of 600 μM at pH 6 (1-deoxynojirimycin **2**, 38 μM; 1,6-dideoxynojirimycin **84**, 780 μM; Scheme 23). Similar losses of activity were found with the enzymes from *Asp. wentii* (GH 3) and bovine kidney lysosomes. Interestingly, the 6-deoxyfluoro derivative **91** (K_i 19 μM) was as active as compound **2** (K_i 25 μM) with the α-glucosidase from yeast, but the 2- (**87**) and 6-modified compounds (**91**) were less active by two orders of magnitude (K_i 2000 and 2500 μM, respectively; fagomine **2**[206]: 880 μM; compound **84**: 1560 μM). The largest relative losses of activity were found for the 2-deoxyfluoro derivative **87** with the β-glucosidases from almonds (10^3-fold) and from *Asp. wentii* (10^5-fold), reflecting the previously observed decreases in bond-cleaving rates with aryl 2-deoxy-D-glucopyranosides. Both enzymes are known to require an intact hydroxyl group at C-2 for efficient hydrolysis but not for substrate binding.[104,185] Neither the 3-deoxy-3-fluoro- (**97**) nor the 6-deoxy-6-fluoro compound **91** inhibited pig liver α-glucosidase II.[236]

3-Deoxy-3-fluoro-calystegin B$_2$ (**109**, Scheme 27), which relates to both 1-deoxynojirimycin (**2**) and isofagomine (**3**), exhibited no activity with the α-glucosidase from yeast but was, interestingly, only 15-fold less active than the parent compound with almond β-glucosidase (82 vs. 5.9 μM).[231] In the light of the fairly high activities of 4,4-difluoro isofagomine (**108**), this calystegine derivative seems to act as an isofagomine analogue rather than in the "1-deoxynojirimycin-mode."

Not unexpectedly by comparison with the monofluorinated 1-deoxynojirimycin derivatives, the 4,4-difluoro derivative **102** (Scheme 26) was found to be inactive with almond β-glucosidase and the α-glucosidase from yeast.[227] Equally inactive with these enzymes was the 3,3-difluoro compound **105** (Scheme 26).[228] It is noteworthy that the 4,4-difluoro derivative of 1-deoxymannojirimycin, **103** (pK_a 5.3) turned out a good inhibitor of the β-glucosidase from almonds (K_i 45 μM, at pH 6.8; **2**: 47 μM) and remained unprotonated and active at pH 5 (K_i 92 μM; **2**: K_i 300 μM).[227]

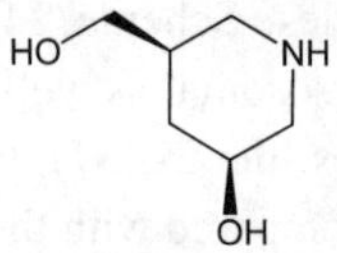

115

SCHEME 28.

In the isofagomine series, the 4,4-difluoro analogue **108** (pK_a 6.86; Scheme 27) exhibited good activity (K_i 1.2 μM, pH 6.8) against the β-glucosidase from almonds when compared with the parent compound **61** (K_i 0.1 μM, pK_a 8.4; Scheme 17) as well as the corresponding 4-deoxy derivative **115** (K_i 2.8 μM, pK_a 9.0; Scheme 28), yet it was practically inactive against α-glucosidases (baker's yeast, *B. stearothermophilus*), β-galactosidases (bovine liver, *S. fragilis*, *A. oryzae*), and jack-bean α-mannosidase.[230]

Among furanoid iminoalditols, the 1-deoxyfluoro derivative **110** (Scheme 27) of 2,5-dideoxy-2,5-imino-D-mannitol was found to be a glucosidase inhibitor having activity comparable to that of 1-deoxynojirimycin (α-glucosidase from yeast, K_i 57 vs. 25 μM; almond β-glucosidase 260 vs. 300 μM; and β-glucosidase from *Agrobacterium faecalis* 30 vs. 12 μM), but less active than its parent compound.[232] 1,4,6-Trideoxy-6-fluoro-1,4-imino-D-mannitol (**114**, Scheme 27) was found a powerful inhibitor of α-mannosidases: human lysosomal (K_i 1.5 μM), neutral (IC$_{50}$ 25 μM), as well as Golgi II (IC$_{50}$ 30 μM), and was more active than the parent compound **19** (13, 200, and 100 μM; Scheme 7) or the corresponding 6-deoxy derivative **76** (1.3, 300, and 60 μM; Scheme 20).[210] Not unexpectedly, the 3-deoxy-3-fluoro derivative **111** (Scheme 27) of nonnatural 2,5-dideoxy-2,5-imino-L-mannitol, the enantiomer of inhibitor **7** (Scheme 3), was not a glycosidase inhibitor.[233]

Examples of fluorine-containing imino sugars as partial structures of PNPs have been reported. The 5'-deoxy-5'-fluoro compound **116** ("5'-F-Imucillin H," Scheme 29) was found to be a nanomolar inhibitor of human PNP (K_i 7 nM) as well as that of *Plasmodium falciparum* (K_i 60 nM).[237] 2'-Deoxy-2',2'-difluoro derivative **117** ("2',2'-diF-immucillin-H") exhibited slightly more inhibitory power against the microbial enzyme and an enhanced selectivity of more than three orders of magnitude (K_i 1.4 nM vs. 15 μM).[238] Another type of inhibitor, the 3-fluoro compound **118** ("F-DADMe-immucillin-H") was shown a potent picomolar inhibitor of the human enzyme (K_i* 32 pM), with some selectivity over the *Plasmodium* phosphorylase (2.6 nM), whereas the enantiomer **119** was 50- to 100-fold less active.[239,240]

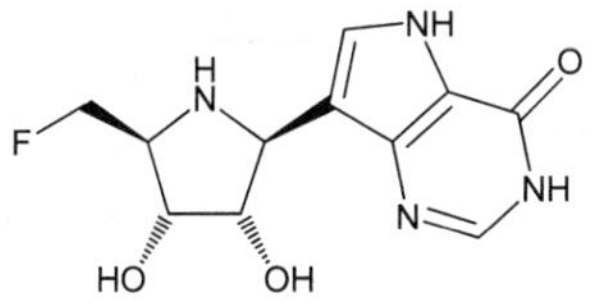

SCHEME 29.

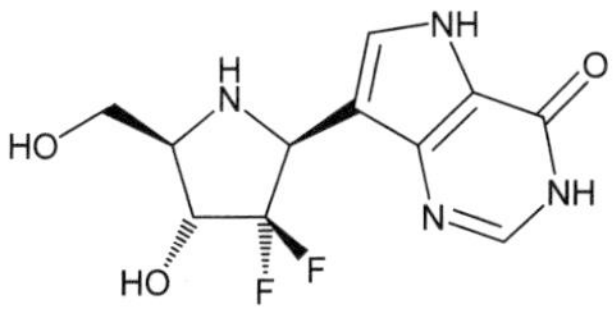

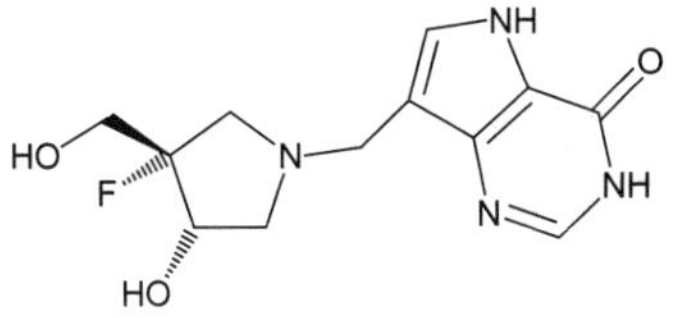

From the available items of information, it can be concluded that deoxyfluoro derivatives of imino sugars and relatives are valuable tools for characterization of glycosidases as they pertain to the significance of individual functional groups around the ring, as well as considerations concerning the pH dependency of glycosidase and inhibitor activity. Their potential as leads toward pH-dependent, selective inhibitors has yet, however, to be realized in particular applications.

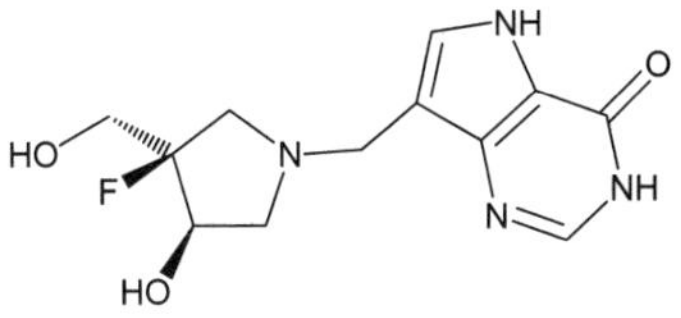

Exceptional in their activities are the fluorine-containing immucillins. Clearly, the additional structural features of the "aglycon" contribute strongly to their inhibitory power.

Noteworthy are the reportedly high activities of 6-deoxy-6-fluorocastanospermine (**86**, Scheme 24) and 1,2,5-trideoxy-1-fluoro-2,5-imino-D-mannitol (**110**, Scheme 27) in comparison with the structurally related 1,2-dideoxy-2-fluoronojirimycin (**87**) and the lack of inhibitory properties observed with 3-deoxyfluoro-calystegin (**109**). In the case of compound **109**, this "double-faced" molecule may bind as an isofagomine (**61**, Scheme 17) or noeuromycin (**75**, Scheme 20) analogue, as opposed to the "1-deoxynojirimycin-type binding," or as a mixture of both binding modes.

Similarly, a "laterally inverted" binding event may take place in the case of 6-deoxy-6-fluorocastanospermine (**86**), with O-1 in the position of O-6 and the fluorine atom in the area where O-1 is located in the parent alkaloid **8**, thus, compound **86** now being a bicyclic mimetic of 1,6-dideoxy-6-fluoronojirimycin (**91**, Scheme 24).[241] Interestingly, due to the impressive progress in X-ray structure determination and the steadily increasing number of well-resolved enzyme–inhibitor complexes, this concept of laterally inverted binding of castanospermine when compared to

1-deoxynojirimycin (**2**) has been substantiated by the structures of trehalulose synthase MutB from *Pseudomonas mesoacidophila* MX-45 liganded by 1-deoxynojirimycin (**2**, PDB 2PWD) and castanospermine (**8**, PDB 2PWG), respectively (Fig. 7).[242] This GH 13 enzyme catalyzes the isomerization of sucrose to trehalulose and isomaltulose and is inhibited by castanospermine (K_i 15 µM) as well as by 1-deoxynojirimycin (K_i 40 µM). Inspection of the bound ligands reveals that 1-deoxynojirimycin is bound in the $^4C_1$ conformation as usually found for this inhibitor. Interestingly, castanospermine binds with the pyranoid ring as a boat form with the pyrrolidine ring "up" so that the lone pair on nitrogen is *trans* to the C-1—C-8a bond. Furthermore, O-6 takes the position of C-6 in 1-deoxynojirimycin, and O-7 along with O-8 is nicely superimposed with O-4 and superimposed fairly well with O-3 of the latter inhibitor. The O-1 and O-2 of castanospermine are not matched with 1-deoxynojirimycin but are located in proximity of the same amino acid residues in the active site. This "proof of concept" is an additional challenge for inhibitor design.

Introduction of strategically positioned fluorine atoms into inhibitors still appears to be limited to just a few examples. Further efforts seem desirable to synthesize new structures while avoiding the removal of vital hydroxyl groups, thereby providing compounds with new and telling properties as molecular probes.

5. Imino Sugars as Active-Site Ligands for Structure and Catalysis Visualization

Structures of numerous glycosidases with inhibitors bound to them have become available via the RCSB Protein Data Base (compare Section VIII).[243] From this increasing number of crystal structures of enzyme–ligand complexes, important conclusions can be drawn concerning conformational changes of substrates and inhibitors at the active sites, protonation states approaching active sites, protonation trajectories, and also structural features of transition-state analogues versus "fortuitous binders." In synergistic synopsis with CAZy,[98] the *C*arbohydrate-*A*ctive En*Z*ymes sequence database, a powerful set of tools has become available for data mining and rational design of inhibitors.

The various conformational itineraries of substrates (or inhibitors) on their way to the transition state(s) and, subsequently to the respective products, have been nicely outlined by Moremen and coworkers[244] (Fig. 5) and may serve as visual support in the following survey.

a. D-Glucosidase Inhibitors

(i) 1-Deoxynojirimycin Versus Isofagomine; D-gluco- and D-xylo-Configured Derivatives. The first examples of the exploitation of iminoalditols as ligands for crystal-structure determinations of carbohydrate-modifying enzymes were reported in

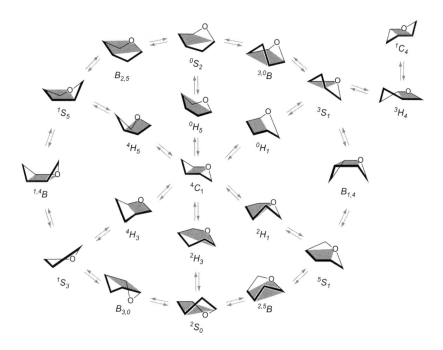

FIG. 5. View at "GPS," the "Globe of Pyranoid Sugar conformations."[244] (See Color Insert.)

the early 1990s by Blow, Fleet, and coworkers, who investigated the catalytic mechanism of D-xylose isomerase by studying the enzyme–inhibitor adducts of 1-deoxynojirimycin (**2**, PDB 1DIE) and 2,5-dideoxy-2,5-imino-D-glucitol (**113**, Scheme 27; PDB 1DID), respectively,[245] and by Honzatko, who studied glucoamylase from *Aspergillus awamori* in free form as well as in complex with 1-deoxynojirimycin (K_i 96 μM; PDB 1DOG).[246] As with free 1-deoxynojirimycin,[247] a $^4C_1$ chair conformation could be inferred for the enzyme-bound inhibitor. This conformation was also found in the complex with the cyclodextrin glucanotransferase from alkalophilic *Bacillus* sp. 1011 by Harata and coworkers (PDB 1I75)[248] Direct comparison[249] of 1-deoxynojirimycin **2** (K_i 9 μM) and isofagomine **61** (Scheme 17; K_i 23 nM) in complex with β-glucosidase from *T. maritima* (GH 1) revealed a flattened $^4C_1$ chair conformation for 1-deoxynojirimycin (PDB 1OIM), whereas isofagomine (PDB 1OIF) was bound in a $^4C_1$ chair in a tight electrostatic interaction with a presumably doubly ionized active site, as also observed, for example, with a cellobiose-related 4-

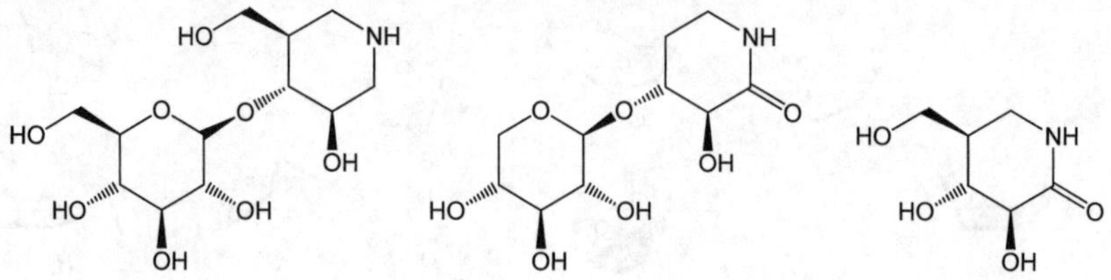

SCHEME 30.

O-β-D-glucopyranosyl-isofagomine **120** (Scheme 30; K_i 700 nM) in complex (PDB 1OCQ) with endocellulase Cel5A from *Bacillus agaradhaerens* (GH 5).[250] The difference of inhibitory activities was attributed to the more-pronounced entropic effects provided by the displacement of water molecules in the binding of isofagomine. Such solvent effects were also inferred as a reason for the comparably weak binding of the structurally closely related D-*gluco* configured tetrahydrooxazine **74** (Scheme 20) to the *T. maritima* enzyme (PDB 1W3J)[251] as well as for the difference in inhibition (K_i 40 vs. 0.48 μM) of the xylobiose-related 1,5-dideoxy-1,5-iminoxylitol **80** (Scheme 22) at pH 5.8 (PDB 1V0K) and pH 7.5 (PDB 1V0M), and the corresponding dehydroxymethyl-isofagomine **81** (Scheme 22; PDB 1V0L and 1V0N, respectively). Both of these were found in the $^4C_1$ chair conformation in their complexes with xylanase Xyn10A from *Strepomyces lividans*, a retaining endoxylanase (GH 10).[252] It is noteworthy that the complex with the corresponding lactam **121** (Scheme 30; PDB 1OD8) was discovered to feature the acid–base catalytic residue Glu128 in the deprotonated state, whereas the nucleophile Glu 236 was found in protonated form.[253] In keeping with Hehre's conclusions concerning enzymatic variability in the course of protonation of nonnatural substrate analogues,[91] this might suggest that inhibitor-dependent "reversed" protonation states could be exploited for glycosidase characterization as well as selective pH-dependent inhibition. Another interesting facet of conformation-dependent glycosidase inhibition was found with the isofagomine lactam **122** (Scheme 30), a good inhibitor of GH 1 β-glucosidases from *T. maritima* (K_i 130 nM) and sweet almond (29 μM), and which had also been found to be a potent β-mannosidase inhibitor with the enzyme from snail (K_i 9 μM).[254] Its structure in its complex with the *Thermotoga* enzyme (PDB 1UZ1) and the *Cellvibrio mixtus* exo-β-mannanase *Cm*Man5 (GH 26) revealed the inhibitor in the $^{2,5}B$ conformation with 3-OH in quasiaxial orientation in the latter complex (PDB 1UZ4) and in a $^4H_5$ half-chair in the former,[255] obviously providing sufficient flexibility and structural details to satisfy both types of glycosidases.

Interestingly, in an inverting cellulase from *Humacola insolens*, the cellobiose-related isofagomine **120** (K_i 400 nM; Scheme 30) was found in what for this compound was a quite unusual, highly distorted conformation close to a $^{2,5}B$ boat (PDB 1OCN).[256] Such a conformation was subsequently also observed in the complex (PDB 1UP2) with another GH 6 β-glucosidase, a cellulase from *Mycobacterium tuberculosis*, Cel6 (GH 6).[257] Compound **120** (K_i 5 μM; PDB 2OYK) was also elegantly exploited by Withers, Strynadka, and collaborators in a comparative investigation into the structural parameters responsible for the potent inhibition of many glycosidases by furanoid inhibitors, employing endoglycoceramidase II (GH 5) from *Rhodococcus* sp. as a model system.[258] Likewise, these authors also utilized the corresponding 4-*O*-glucosylated glucoimidazole **123** (K_i 0.5 μM; Scheme 31; PDB 2OYL).

Hydroximolactams **124** (K_i 1 μM; PDB 1UWU) and **125** (K_i 1 μM; PDB 1UWT), structurally related to D-glucono-1,5-lactam (**126**) and D-galactono-1,5-lactam (**127**), respectively, were employed as probes for the GH 1 β-glucosidase from *Sulfolobus solfataricus*.[259]

Despite the trigonal anomeric center and the consequential $^4H_3$ conformation, both inhibitors are too weak to constitute transition-state analogues, making them good examples among glycosidases as protonation trajectory-selective probes for in-plane

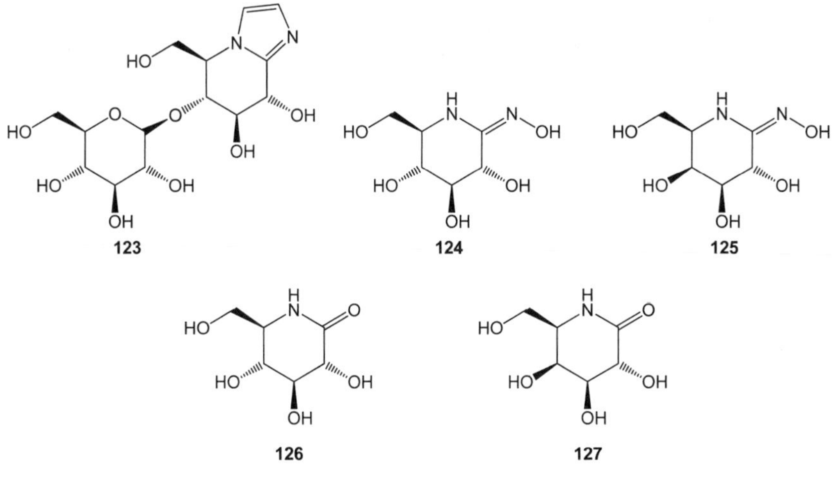

SCHEME 31.

syn- and anti-protonators.[193] In the series of strong "glycoimidazole" inhibitors first synthesized by Streith[260] as well as Tatsuta et al.[261] and investigated in depth by Vasella,[193,262,263] comparison of enzyme–inhibitor complexes of unsubstituted (77, Scheme 21; K_i 74 and 138 nM, respectively; PDB 2CES and 2CEQ) and 2-phenyl-ethyl derivative 63 (Scheme 17; K_i 2.8 and 7.5 nM, respectively; PDB 2CET and 2CER) constituting telling probes for anti-protonating glucosidases, with GH 1 β-glucosidases from T. maritima and S. solfataricus, revealed both inhibitors in a ^4E envelope, with no apparent interaction of the phenylethyl substituent of compound 63 with these enzymes.[196] The significant difference of inhibitory power of compounds 63 and 77 was attributed to the displacement[250] or binding of water molecules at the enzyme by the bulky substituent, thus highlighting the complicated effects exerted by changes in the solvent sphere on glycosidase activity and inhibition. In contrast to this lack of specific interaction of the lipophilic substituent, barley β-D-glucan glucohydrolase (GH 3) in complex (PDB 1LQ2) with structurally closely related phenyl substituted glucoimidazole 128 (Scheme 32; K_i 1.8 nM) exhibited localized hydrophobic interactions with the phenyl rings of Trp 286 and Trp 434, restricting the conformational mobility of the inhibitor's aromatic substituent.[264,265]

In 2007, a noteworthy comparison of 18 glucosidase inhibitors as ligands of the β-glucosidase from T. maritima was reported.[266] In contrast to the results with the GH 3 enzyme from barley, it was shown that the potent glucoimidazole-derived lipophilic inhibitors 63 (Scheme 17; PDB 2CET) and 129 (Scheme 32; PDB 2J7C) did not make any particular contact with the active site through their aromatic substituents. Isofagomine 61 (Scheme 17; PDB 1OIF), noeuromycin 75 (Scheme 20; PDB 2J75), tetrahydrooxazine 74 (PDB 1W3J), and also "azafagomine" 130 (PDB 2J7H) were found to bind in $^4C_1$ conformation. 1-Deoxynojirimycin (PDB 2J77) was

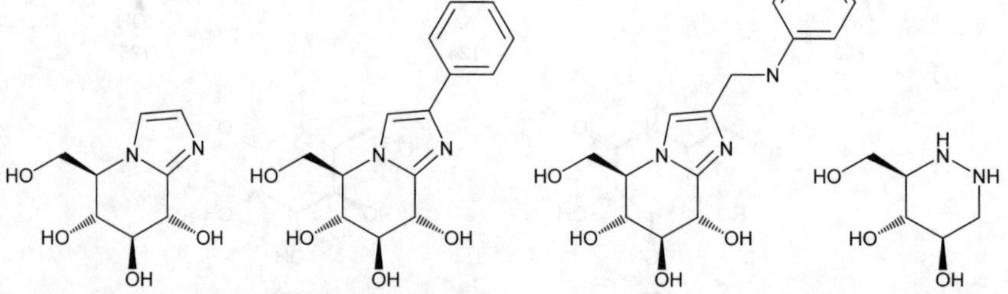

SCHEME 32.

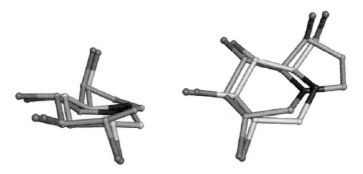

FIG. 6. Superposition of ligands 1-deoxynojirimycin **2** (yellow, PDB 2J77) and castanospermine **8** (green, PDB 2CBU) bound to *Thermotoga maritima* β-glucosidase. The ring nitrogen atoms, all hydroxyl groups in the pyranoid rings, as well as O-1 (**8**) and the primary O 6 (**2**), respectively, are closely matched (See Color Insert.)

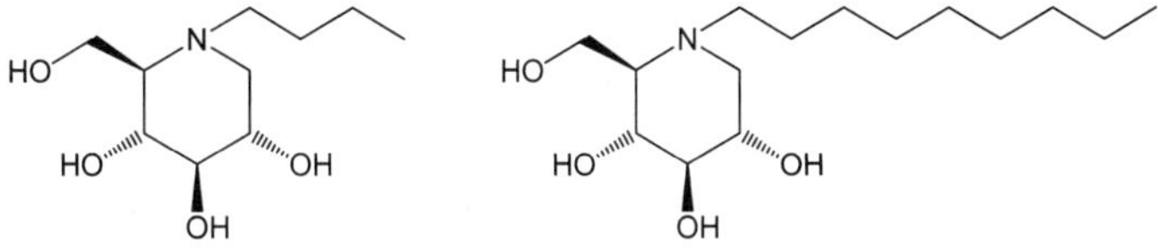

131   **132**

SCHEME 33.

distorted into a skew boat ($^1S_3$) and castanospermine (PDB 2CBU) was bound as $^{1,4}B$ boat (Fig. 6).

Glucotetrazole **79** (Scheme 21) in complex with the broad-spectrum maize (PDB 1V08), GH 1 β-glucosidase, showed the inhibitor in a $^4H_3$ half-chair[267] which was also the case in myrosinase (thioglucoside glucohydrolase, EC 3.2.3.1) from *Sinapsis alba* (PDB 1E6Q).[268] In this interesting study, which also involved glucohydroximolactam **124** (Scheme 31; PDB 1E6S), it was concluded that ascorbate functioning as a co-enzyme is a substitute for the catalytic base absent in the active site of this enzyme.

(ii) Lysosomal Enzymes Related to Storage Diseases. The structures of complexes of *N*-butyl-**131** (Scheme 33; K_i 116 μM; PDB 2V3D) and *N*-nonyl-1-deoxynojirimycin **132** (K_i 300 nM; PDB 2V3E) with recombinant human acid β-

glucosidase (E.C. 3.2.1.45, GH 30), the enzyme responsible for the degradation of glucosylceramide, were compared to the corresponding adduct with isofagomine (**61**, Scheme 17; PDB 2NSX).[269] Mutants of this particular enzyme are responsible for Gaucher's disease, the most common lysosomal storage disorder. The three inhibitors probed were found to bind in very similar orientations, with the alkyl chains of compounds **131** and **132** oriented toward the entrance to the active site, with stabilizing interactions found with a tyrosine residue (Tyr 313). The tertiary amines were located in positions close to the ring oxygen atom of the natural substrate. The ring nitrogen atom of isofagomine was positioned between the catalytic glutamic acid residues Glu 325 and Glu 340. The N-nonyl residue was also in contact with Leu 314, a residue close to the entrance to the catalytic site. Notably, the difference of inhibitory power of more than two orders of magnitude between the two 1-deoxynojirimycin derivatives probed cannot be accounted for by this small difference in interactions, and it was proposed to be a consequence of the increased general hydrophobicity of compound **132** and its interaction with the hydrophobic surface at the entrance to the active site.

Despite the fact that no global structural changes were observed upon inhibitor binding with conduritol B epoxide **56** (Scheme 16; PDB 1Y7V),[270] N-alkyl-1-deoxynojirimycins[269] as well as isofagomine (PDB 3GXF)[271,272] triggered conformational changes of loops at the entrance to the active site of acid β-glucosidase, which serve as "gatekeepers" upon substrate or inhibitor binding. Inhibitor complexing induces stability to these partial structures, as is evident upon comparison of free and inhibitor-bound enzyme.

(iii) Castanospermine Versus 1-Deoxynojirimycin. The seemingly rigid castanospermine structure[28] was found in a $^{1,4}B$ boat (carbohydrate numbering) when complexed with the exo-1,3-β-glucanase from *Candida albicans*, a family GH 5 enzyme (PDB 1EQC).[273]

The complexes of trehalulose synthase from *P. mesoacidophila* MX-45, a representative of the sucrose isomerases (GH 13, retaining), with the glucosidase inhibitors 1-deoxynojirimycin **2** (K_i 40 µM; PDB 2PWD) and castanospermine **8** (K_i 15 µM; 2PWG), respectively, have provided an interesting view of inhibitor recognition by glycosidases (Fig. 7).[242] This enzyme binds 1-deoxynojirimycin in a relaxed $^4C_1$ chair conformation, whereas the closely related but structurally rigid castanospermine is found to be in a slightly twisted boat conformation. Moreover, superposition of the two structures reveals a laterally inverted binding mode for castanospermine with 6-OH (equivalent to 2-OH of 1-deoxynojirimycin) in the position of 6-OH of compound **2** and 1-OH (representing 6-OH of **2**) in the location of 2-OH of 1-deoxynojirimycin.

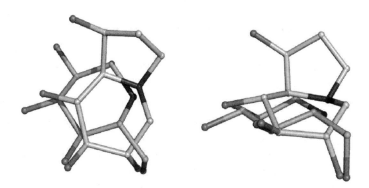

FIG. 7. Superposition from two different viewpoints of the ligands castanospermine, **8** (gray, PDB 2PWG) and 1-deoxynojirimycin **2** (green, PDB 2PWD) bound to trehalulose synthase from *Pseudomonas mesoacidophila* MX-45. The lateral inversion of **2** relative to the pyranoid ring of **8**, with O-6 of **2** matching O-6 (corresponding to O-2 in carbohydrate numbering) of **8** is clearly visible. (See Color Insert.)

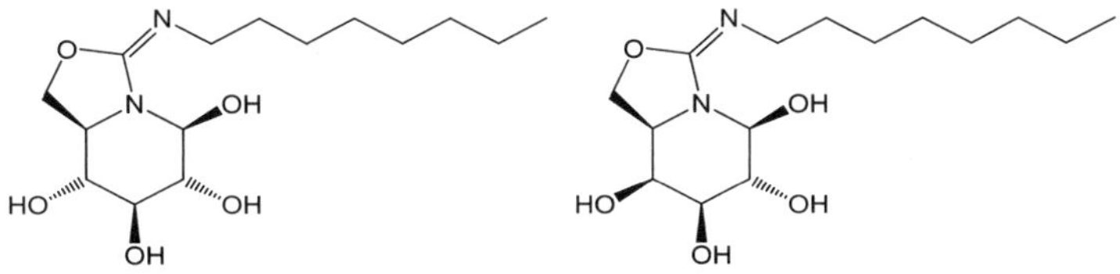

133   **134**

SCHEME 34.

This finding constitutes the first unambiguously observed example and "proof of concept" of a hypothesis,[241] put forward by one of the authors of this account.

Another example of an unusual conformation of castanospermine, which is an inhibitor of the enzyme activity of *Clostridium difficile* toxin B (IC_{50} 400 μM) as well as a of Rac glucosylation by this toxin (IC_{50} 120 μM), was found in its complex with the Rho/Ras-glucosylating toxin from *Clostridium sordellii* (PDB 2VL8).[274] The inhibitor was found as a $^{5,8}B$ boat (equivalent to $^{1,4}B$ of compound **2**). It should also be noted that neither 1-deoxynojirimycin (**2**) nor its *N*-butyl derivative **131** (Scheme 33), which both usually maintain a $^1C_4$ chair in solution as well as in complexes with many carbohydrate-processing enzymes, are inhibitors of this protein. The structures of "designer hybrids" related to both 1-deoxynojirimycin and castanospermine in complex with *T. maritima* β-glucosidase were reported by Ortiz Mellet, Garcia Fernandez,

Davies, and their coworkers.[275] Interestingly, D-*gluco* configured inhibitor **133** (Scheme 34) was found weaker (K_i 35 μM; PDB 2WBG) than the galacto epimer **134** (K_i 0.8 μM; PDB 2WC3). These workers also provided the structure of the complex with lysosomal β-glucosidase (K_i 4 μM; PDB 2WCG) with the ring nitrogen as well as hydroxyl groups matching the positions of the corresponding functional groups in the structure with *N*-butyl-1-deoxynojirimycin **131**. The additional "anomeric" hydroxyl function in compound **133** apparently induced high selectivity for the β-specific enzyme over human lysosomal α-glucosidase.[276]

Structures solved for the inverting, calcium ion-dependent GH 97 α-glucosidase from *Bacteroides thetaiotaomicron* with 1-deoxynojirimycin (**2**, K_i 59 μM; PDB 2JKE) as well as with castanospermine (**8**, K_i 59 μM; PDB 2JKP) showed both inhibitors with the 6-membered ring in the $^4C_1$ conformation and practically superimposably coordinating active-site residues.[277]

In the context of the strong commercial interest in alternative sweeteners, in particular isomaltulose (which is produced industrially by the conversion of sucrose with the sucrose isomerase SmuA from *Protaminobacter rubrum*, a GH 13 enzyme), the complex with 1-deoxynojirimycin (**2**, K_i 10 μM) was investigated (PDB 3GBE)[278] and compared to the already mentioned one of MutB liganded to the same inhibitor.

(iv) Castanospermine Versus Calystegine. Calystegine B$_2$ (**17**, Scheme 6; K_i 1.25 μM), in complex with *T. maritima* β-glucosidase (GH 1), was compared with the corresponding castanospermine adduct (K_i 0.95 μM). Castanospermine (**8**) was, once more, found in a $^{5,8}B$ boat conformation (PDB 2CBU), whereas calystegine B$_2$ **17** was in a relaxed $^4C_1$ chair in the "isofagomine-like" binding mode (PDB 2CBV).[279] The ethylene bridge is located "above" the ring, while 2-OH, 3-OH, as well as 4-OH were superposable on the corresponding hydroxyl groups 6, 7, and 8 in the castanospermine system. Calystegine 5-OH was found in a position comparable to castanospermine 1-OH (or 6-OH of 1-deoxynojirimycin) interacting with Glu 405, the usual partner of 6-OH groups in the monocyclic series of compounds under consideration.[266] Interestingly, a calystegine-related compound **135** (K_i 0.5 μM; Scheme 35)

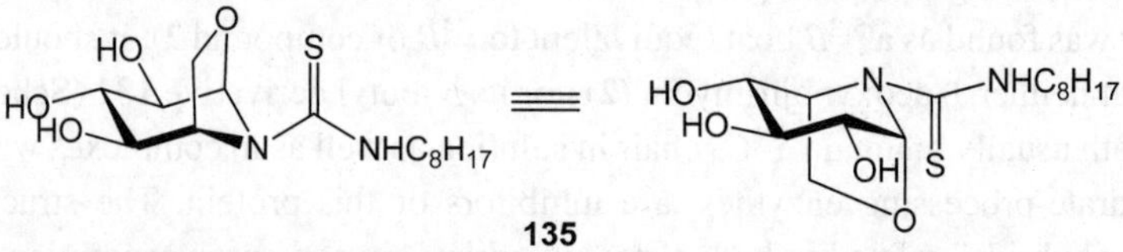

135

SCHEME 35.

lacking a functional group equivalent to 6-OH was found to bind in the "1-deoxynojirimycin-type" mode, with the ethylene bridge "underneath" the ring system (PDB 2VRJ).[280] Again, this example points out how synthetic compounds may offer selective and powerful inhibition by sterically and/or electronically imposing alternative binding modes when compared to the respective natural substrates and standard structures among inhibitors.

b. D-Mannosidase Inhibitors.—Mannosidases, in particular α-mannosidases involved in glycoprotein trimming, are very important enzymes with a large variety of functions. Profound understanding of these enzymes and their modes of action may be of value in the search for new anticancer therapeutic substances.

(i) Endoplasmatic Reticulum α-1,2-Mannosidase I The endoplasmatic reticulum α-1,2-mannosidase I (mannosyl-oligosaccharide 1,2-α-mannosidase, EC 3.3.1.113) is a calcium ion-dependent GH 47 inverting enzyme and is essential for the maturation of N-linked glycoproteins and ER-located degradation trimming of $Man_9GlcNAc_2$ to $Man_8GlcNAc_2$. This modification is a signal for the translocation out of the ER to the proteasome for degradation.[281] Inhibition of α-1,2-mannosidase I results in delayed degradation of various proteins. In the human enzyme, complexed with the inhibitors kifunensine **16** (Scheme 6; IC_{50} 0.2 μM; PDB 1FO3) and 1-deoxymannojirimycin **11** (Scheme 4; IC_{50} 20 μM; PDB 1FO2), the 2-OH and 3-OH groups were discovered to be in strong coordination with calcium ion.[282] This interaction resulted in a conformational change into the $^1C_4$ chair with 3-OH, 4-OH, as well as the hydroxymethyl moiety at C-5 in axial orientations. All hydroxyl groups replace water molecules present in the respective positions in the apo-enzyme. Similar characteristics were found with the 1-deoxymannojirimycin adduct of the corresponding enzyme from yeast (PDB 1G6I) and also confirmed in the complex (PDB 1KRE) with the *Penicillium citrinum* enzyme.[283] From these binding characteristics, in combination of data acquired from the binding of a thiosaccharide, a $^3H_4$ half-chair transition state was invoked for family 47 glycosidases.[244]

(ii) Golgi α-Mannosidase II. Golgi α-mannosidase II (mannosyl-oligosaccharide 1,3-1,6 α-mannosidase, EC 3.2.1.114; GH 38) is a zinc- or cobalt-dependent retaining enzyme, and it catalyzes hydrolysis of the terminal α-(1 → 3)- and α-(1 → 6)-linked mannosyl residues in the high-mannose oligosaccharide $GlcNAcMan_5GlcNAc_2$ to $GlcNAcMan_3GlcNAc_2$, thus providing the substrate for formation of complex N-glycans.[284] Its central role in the trimming process has made it an interesting target for chemical anticancer intervention in certain breast, colon, and skin cancers. Inhibition by swainsonine resulted in decreased tumor growth and metastasis.[285] The crystal

structures of *Drosophila* α-mannosidase II with swainsonine **13** (Scheme 5; IC$_{50}$ 20 nM; PDB 1HWW) as well as with 1-deoxymannojimycin **11** (Scheme 4; IC$_{50}$ 400 μM; PDB 1HXK) revealed a strong coordination to the *cis*-oriented hydroxyl groups in both inhibitors.[284] The superior inhibitory activity of swainsonine was attributed to its structural analogy with the skewed-boat transition-state conformation of the natural substrate.

Whereas kifunensine (**16**, Scheme 6) binds to α-mannosidase I as a strong inhibitor in the $^1C_4$ chair conformation, it is a rather poor inhibitor of Golgi α-mannosidase II (K_i 5.2 mM), where it was found to complex as a $^{1,4}B$ boat (PDB 1PS3).[286] 1-Deoxymannojirimycin (**11**), contrasting its "high energy" conformation in complexation with α-mannosidase I ($^1C_4$), was found in the relaxed $^4C_1$ chair. The authors concluded that family GH 38 glycosidases hydrolyze via a $B_{2,5}$ transition state, as opposed to the $^3H_4$ half-chair utilized by the family 47 ER α-1,2-mannosidases. Thus, kifunensine (**16**) may be considered a "posttransition state"-conformation analogue. It is noteworthy that a powerful mannosidase inhibitor, the cyclopentane-derived mannostatin A **136** (Scheme 36; K_i 36 nM),[287,288] was found (PDB 2F7O) to best fit the covalent intermediate, which was observed to be bound in a distorted $^1S_5$ skew-boat conformation which also lies along the $^4C_1 \leftrightarrow {^OH_5} \leftrightarrow {^OS_2} \leftrightarrow B_{2,5} \leftrightarrow {^1S_5} \leftrightarrow {^{1,4}B}$ pathway.[289]

From a library of 500 compounds, five inhibitors were selected by Withers and coworkers and subsequently employed in structural studies with the *Drosophila* enzyme.[290] Mannoimidazole **137** (K_i 2 μM, PDB 3D4Y) was found to be the most potent in the series. Interestingly, the remaining four inhibitors, glucoimidazole **77** (Scheme 21; K_i 13 μM, PDB 3D4Z), glucohydroxyiminolactam **124** (Scheme 31; 70 μM, PDB 3D51), its extended derivative **138** (520 μM, PDB 3D52), as well as the carba sugar *N*-octyl-1-epivalienamine **65** (Scheme 18; 17 μM, PDB 3D50) all feature the D-*gluco* configuration. Again, in all instances, the strong coordination of 2-OH and 3-OH of the inhibitors with zinc at the active site determines the binding, with

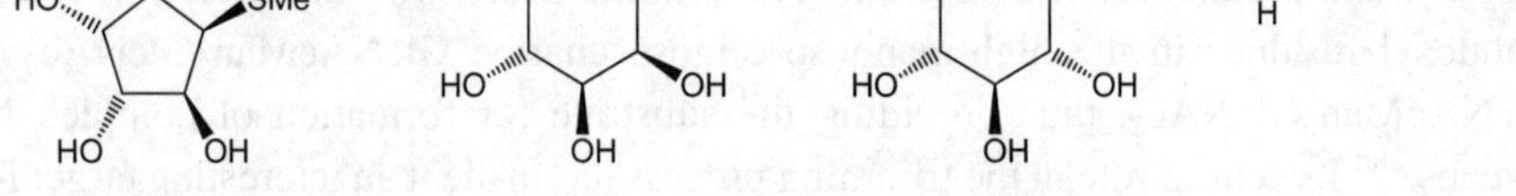

SCHEME 36.

practically all secondary hydroxyl groups colocalized with regard to the active site. Ring systems of swainsonine (PDB 3BLB) as well as of mannose (PDB 3BUP) and a covalently bound substrate analogue, 5-fluoro-β-L-gulopyranosyl fluoride (PDB 1 QWN), were found to be considerably tilted out of the plane of the mannoimidazole. Not unexpectedly, comparison with the complexes of mannoimidazole **137** (PDB 2VMF) as well as noeuromycin **75** (Scheme 20; PDB 2VL4) with the retaining GH 2 β-mannosidase from *B. thetaiotaomicron* also revealed different distortions of the pyranoid rings.[291]

An investigation into modifications of C-5 in swainsonine by introduction of various 4-substituted benzoyl groups at this position provided powerful inhibitors of *Drosophila* Golgi α-mannosidase II, with K_i values (2.7 nM) comparable to that of swainsonine (**13**, Scheme 5; 3 nM).[292] Characteristics of the enzyme–inhibitor complexes are quite similar to those of swainsonine, with the aryl substituents all in a similar environment (PDB 3EJP, 3EJQ, 3EJR, 3EJS, 3EJT, and 3EJU). Interestingly, this location differs from the one found with monocyclic pyrrolidines.[293]

(iii) Other Mannosidases. Many bacteria, frequently pathogenic species, possess retaining α-1,3- and α-1,6 mannosidases of unknown functions and activities, which are putative members of family GH 38. In addition to *Bacteroides*, *Clostridia*, *Listeria*, and *Mycobacteria*, the genus *Streptococcus* shows the highest proportion of bacterial GH 38s. The structure (PDB 2WYI) of the α-mannosidase from pathogenic *Streptococcus pyogenes* in its complex with swainsonine (**13**, Scheme 5; K_i 18 μM)[294] showed strong coordination of the active-site Zn^{2+} and nonpolar interactions of tryptophan (Trp 18) and arginine (Arg 149) residues with the five-membered ring and the unsubstituted ethylene moiety of the six-membered ring, respectively. Hydroxyl groups were found matching the positions of 2-OH, 3-OH, and 4-OH of α-D-mannopyranose in the catalytic subsite − 1. The activity of the enzyme in context with genomic items of information was inferred to play a role in the degradation of host N- or O-glycans.

Structures of the complexes of swainsonine (**13**, K_i 5 and 14 μM, respectively; PDB 2WW0, 2WW2), mannoimidazole **137** (140 μM/93 μM; 2WZS), as well as kifunensine, **16** (Scheme 6; 96 μM/233 μM; PDB 2WVZ) with the Ca^{2+}-dependent inverting exo-α-1,2-mannosidases Bt3990 and Bt2199 (GH 92) from *B. thetaiotaomicron*, a strain of bacteria colonizing the human colon, were determined by Gilbert, Davies, and coworkers[295], and it was shown that, not unexpectedly for metal-dependent mannosidases, the calcium ion at the active site is chelated by O-2 and O-3. Mannoimidazole **137** was found in the $^1S_5/B_{2,5}$ conformation, while kifunensine **16**

(Scheme 6) showed the $^{1,4}B/^1S_5$ conformation, which is opposed to the conformations in solution, which are equivalent to $^3H_4$ in D-mannose.

Two structural examples of a new family, GH 125, of inverting metal-independent exo-α-1,6-mannosidases, from *Clostridium perfringens* and from *S. pneumoniae*, respectively, were recently reported. 1-Deoxymannojirimycin (**11**, Scheme 4) in the *S. pneumoniae* enzyme (PDB 3QRY) was found in the active site in the relaxed $^4C_1$ chair conformation.[296]

c. **D-Galactosidase.**—The complex of retaining human acid α-galactosidase (EC 3.2.1.22; GH 27) with 1-deoxynojirimycin **2** (PDB 3GXT) was reported by Lieberman and coworkers[272]; the inhibitor (K_i 18 μM) adopted a flattened $^4C_1$ chair. Other workers showed that the enzyme catalyzes via a $^1S_3$ skew-boat transition state, suggesting that suitable inhibitors might adopt this or a $^4H_3$ half-chair and the closely related ^4E envelope conformations, respectively.[297]

d. **N-Acetylhexosaminidases.**—The *N*-acetyl-D-galactosamine-related derivative of isofagomine **139** (Scheme 37) was employed for the characterization of GH 20 β-*N*-acetylhexosaminidases.[298] Its complex (K_i 2.7 μM) with *Streptomyces plicatus* β-*N*-acetylhexosaminidase (PDB 1JAK) shows the "typical isofagomine-type" $^4C_1$ chair, with the exocyclic nitrogen as well as 3-OH and the hydroxymethyl group all in equatorial orientations. When compared with the corresponding complex with GlcNAc-thiazoline **140** (PDB 1HP5),[299] the ring nitrogen atom is positioned in "aza-sugar binding mode" close to the location of C-1. A strong interaction of the general acid–base Glu 314 was observed, with the equatorial proton of the ionized ring nitrogen.

The crystal structure of human β-hexosaminidase B (GH 20) in its complex with these inhibitors (GlcNAc-thiazoline **140**: PDB 1NP0; "GalNAc-isofagomine" **139**: PDB 1NOW) revealed a very similar picture, with an activated water molecule above the β face ready to attack the "anomeric center."[300]

Because of the importance of protein O-*N*-acetylglucosaminidases (EC 3.2.1.169) which are β-*N*-acetylglucosaminidases and that catalyze the cleavage of β-linked *N*-acetyl-D-glucosaminides with serine and threonine residues of many eukaryotic

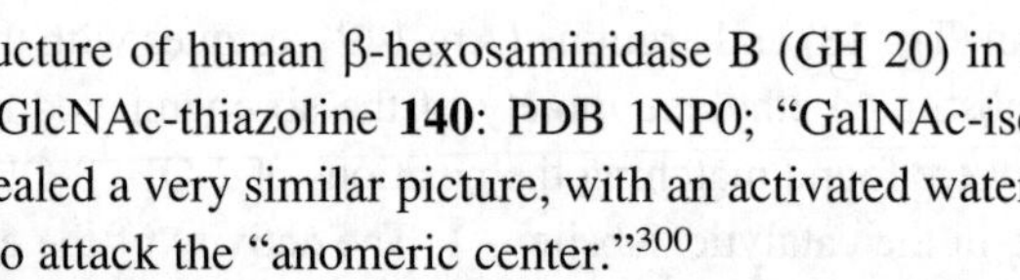

SCHEME 37.

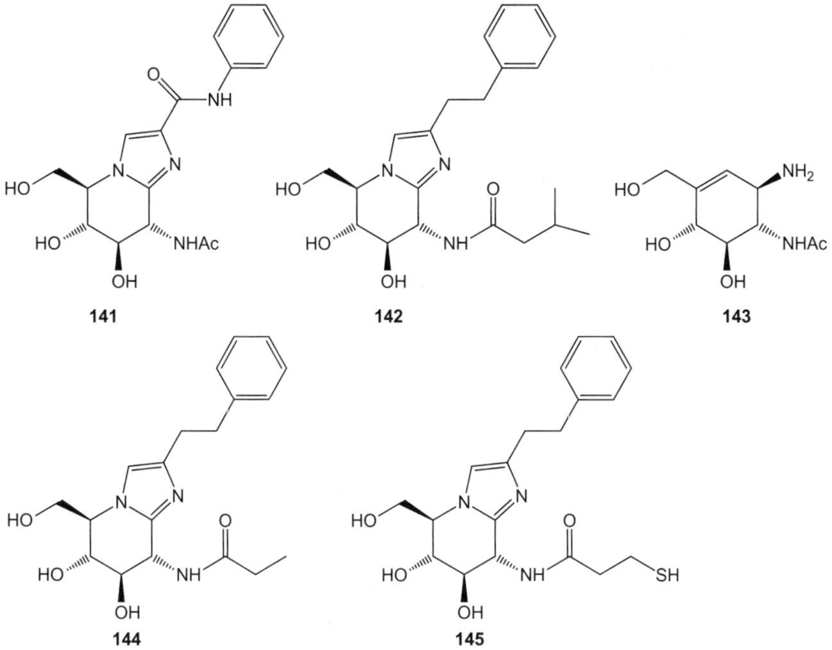

SCHEME 38.

proteins—excellent reviews on O-GlcNAc modification are available[301,302]—these enzymes have been probed with a range of hexosaminidase inhibitors as ligands.

The "PUGNAc"-imidazole hybrid **141** (K_i 3.9 μM; Scheme 38) was found as $^4H_3$ half-chair in its complex with the N-acetylhexosaminidase (PDB 2J47) of B. thetaiotaomicron (GH 84).[303] The powerful inhibitor GlcNAcstatin **142**, a 2-acetamido-2-deoxy analogue of Vasella's excellent lead, the 2-phenylethyl-substituted glucoimidazole **63** (Scheme 17), inhibits bacterial O-N-acetylhexosaminidase highly selectively over lysosomal N-acetylhexosaminidases A and B. For the nagJ (gene symbol) N-acetylhexosaminidase of C. perfringens (family GH 84), the K_i value of 4.6 pM in comparison with the inhibition of human HexA/B (K_i 0.52 μM) indicates a selectivity of five orders of magnitude higher than that of compound **142**.[304] The enzyme–inhibitor complex (PDB 2J62) shows the ligand in a ^4H half-chair similar to the binding mode ($^1S_3$ referring to GlcNAc) of acetamido-epi-valienamine **143** (K_i 6.2 μM with human and 26 μM with bacterial O-GlcNAcase, PDB 2JIW),[305] and displays a nonpolar interaction off the active site with a tryptophan residue, Trp 490. Structural modifications of the acetamido group of compound **142** provided such cell-permeable nanomolar inhibitors as **144** (PDB 2WB5) of human protein *O*-GlcNAcase with even higher selectivities of up to 9×10^5-fold over the lysosomal hexosaminidases.[306,307] The *N*-(3-thiopropanoyl) derivative **145** in complex with recombinant enzyme (PDB 2XPK) showed the acyl residue in a "selectivity pocket," which distinguishes the enzyme from the human lysosomal hexosaminidases, which feature more shallow pockets that cannot accommodate the extended acyl moiety.

The 2-acetamido-2-deoxy derivative **146** (Scheme 39) of glucono-1,5-lactam **126** (Scheme 31) was another ligand (K_i 2 μM) probed with the bacterial *O*-GlcNAcase from *B. thetaiotaomicron* (GH 84).[308] The lactam (PDB 2XM1) was found in the ^4E conformation with its carbonyl group coordinated by the catalytic acid–base Asp 243, Tyr 282 as well as a water molecule.

6-Amino-6-deoxycastanospermine **147** (K_i 0.61 μM; PDB 2X05) and the corresponding aminodeoxy derivative (**148**) of australine (K_i 175 μM; PDB 2X09) in complex with a chitosan-degrading exo-β-D-glucosaminidase from *Amycolatopsis orientalis* (GH 2) showed the amino groups of both inhibitors in virtually identical positions, coordinating a unique acidic pocket provided by glutamates 394 and 591 as well as aspartate 649.[309] Despite its flexibility, the aminomethyl substituent of the australine complex was found to have less contact with the active site than the distinctly more rigid secondary amine of the indolizidine. 6-Acetamido-6-deoxycastanospermine **149** (K_i 0.3 μM) liganding (PDB 2XJ7) with the *B. thetaiotaomicron* *O*-GlcNAcase showed the inhibitor in a boat conformation with an axial N-acetyl substituent.[310]

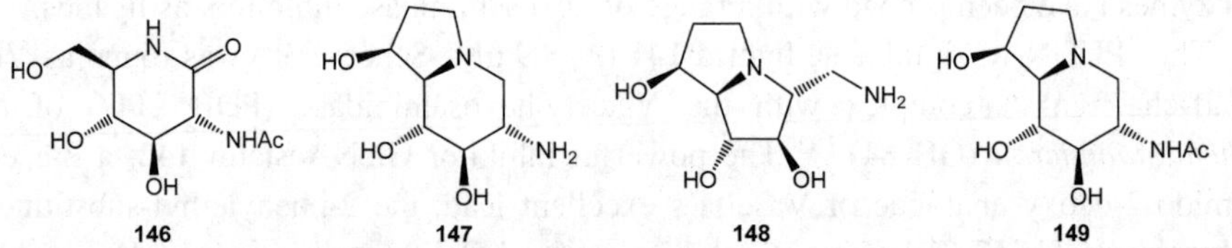

SCHEME 39.

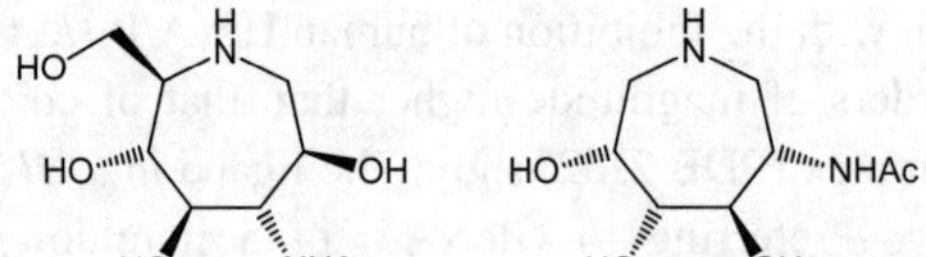

SCHEME 40.

N-Acetyl-D-glucosamine-related azepanes[311] **150** (K_i 89 μM, PDB 2W66; Scheme 40) and **151** (8 μM; 2W67) were probed with *B. thetaiotaomicron* hexosaminidase (GH 84).[311a] In compound **150**, both the pseudo-anomeric hydroxyl group and the acetamido substituent were found in *trans*-diaxial orientations. Both inhibitors feature the pseudo-anomeric carbon on the β face of the ring, corresponding to ^4E and $^{1,4}B$ conformations of the natural substrate.

e. Fucosidases.—α-L-Fucosidase is among the glycosidases most sensitive to inhibition by basic substrate analogues, as was shown for 1-deoxy-L-fuconojirimycin (**20**, Scheme 7), the first iminoalditol found to bind in the submicromolar range. A range of 1-acylamidomethyl derivatives featuring ionizable, lipophilic indole substituents exhibited even more pronounced inhibitory activities with compound **152** (Scheme 41), binding to *T. maritima* GH 29 as a $^1C_4$ chair in the subpicomolar range (K_i 0.5 pM, PDB 2ZX5).[312] This was found to be caused by the nonpolar interactions with two loops of the protein that serve as "gatekeepers" to the active site (akin to observations made with acid β-glucosidase)[269–272] and "close" the access by hydrophobic interactions with the aromatic aglycon moieties (PDB 2ZX6, 2ZX7, 2ZX8, 2ZX9). Furthermore, replacement of water molecules in the active site was also assumed to contribute favorable entropic effects to the powerful binding events.

1-Deoxy-L-fuconojirimycin (**20**) and another lipophilic derivative (**153**) were examined in their complexes with *B. thetaiotaomicron* (GH 29) α-L-fucosidase at different pH values between pH 6.0 and 8.5, with the unsubstituted ring being the better inhibitor between pH 6 (K_d 15 vs. 47 nM) and 7 (K_d 55 vs. 63 nM) and the fluorenone derivative **153** being more potent at pH 8.5 (K_d 372 vs. 157 nM).[313] Both inhibitors were bound in the $^1C_4$ chair conformation (PDB 2XIB, 2XII), thus resembling a hydrophilic 1-hydroxymethyl analogue (PDB 2WVT).[314] As with isofagomine (**61**, Scheme 17)[249] and other examples,[250,266,298] the charged nitrogen atom in the ring is clamped between the carboxylates of the catalytic nucleophile Asp 229 and the general acid–base Glu 288.

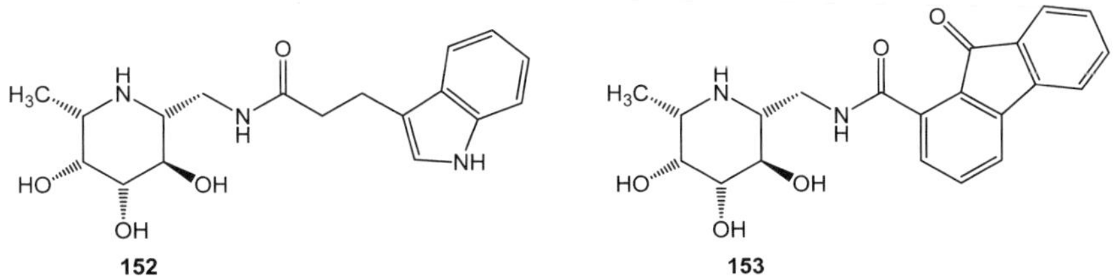

SCHEME 41.

The inverting GH 95 1,2-α-L-fucosidase (EC 3.2.1.63) from *Bifidobacterium bifidum* in complex (PDB 2EAC) with 1-deoxy-L-fuconojirimycin (**20**) showed the inhibitor in $^1C_4$ chair conformation, resembling L-fucose and the entrance to the active site was narrowed by the "gate keeper" loops.[315] A calcium ion associated with the protein was not involved in the catalytic process but was considered to stabilize the enzyme.

f. Pyrrolidine- and Pyrrolizidine-Type Inhibitors.—To date, despite their proven power as isosteric mimics of pyranoid substrates and inhibitors, 1,4- and 2,5-imino sugars have found only a few applications as ligands in the structural analysis of glycosidases and related proteins. The first example, 2,5-dideoxy-2,5-imino-D-glucitol **113** (Scheme 27), in complex with glucose isomerase has already been mentioned.[245]

1,4-Dideoxy-1,4-imino-D-arabinitol (**14**, DAB, Scheme 5; IC$_{50}$ 0.4 μM; PDB 2G9Q) bound to the nonphosphorylated form (GPb) of rabbit-muscle glycogen phosphorylase (EC 2.4.1.1, GT 35) was compared with a range of complexes of pyranoid inhibitors, such as isofagomine **61** (Scheme 17; IC$_{50}$ 1.2 μM; PDB 2G9V), together with N-(3-phenylpropyl)isofagomine **154** (Scheme 42; IC$_{50}$ 0.85 μM; PDB 2G9U).[316] The binding for complex formation was mutually dependent on the presence of high concentrations of phosphate and inhibitor. An ^NE envelope conformation was found for the pyrrolidine component, with very similar positioning of ring nitrogen as well as hydroxyl groups, mimicking 3-OH, 4-OH, and 6-OH as in the $^4C_1$ chairs of the two pyranoid inhibitors probed. A similar conformation was observed with the N-modified derivative ghavamiol (**155**),[317] a nitrogen analogue of the thio sugar salacinol (**60**, Scheme 17), in complex with *Drosophila* Golgi α-mannosidase II (PDB 1TQU).

2,5-Dideoxy-2,5-imino-D-mannitol **7** (Scheme 3) in complex (PDB 2AEY) with chicory fructan 1-exohydrolase (EC 3.2.1.153; GH 32) was found to bind in a similar way as does fructose, as well as the fructofuranosyl subunit of sucrose and

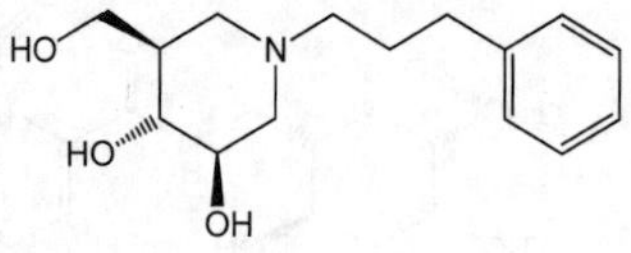

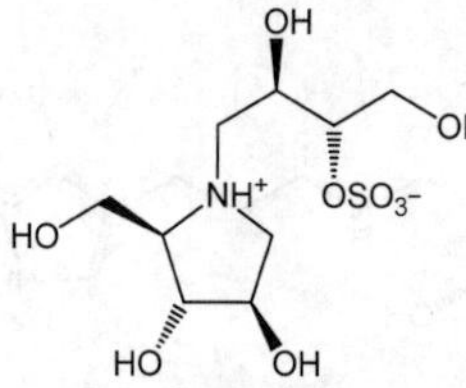

SCHEME 42.

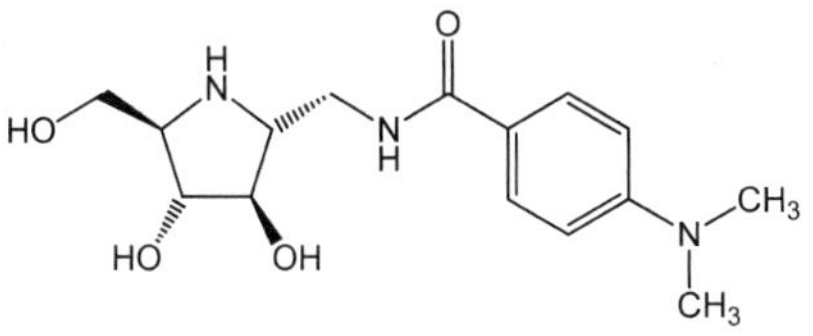

SCHEME 43.

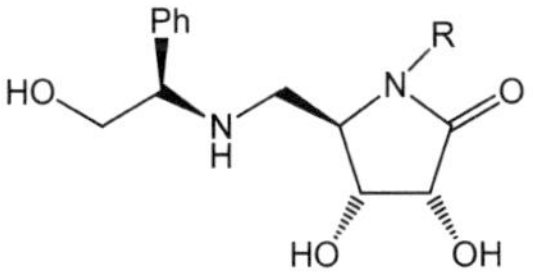

1-kestose.[318] A long-standing hypothesis concerning the isosteric features of 2,5-dideoxy-2,5-imino-D-mannitol when compared with 1 deoxynojirimycin (**2**) was verified by Withers, Strynadka, and their collaborators with the inhibitor's complex with endoglycoceramidase, a retaining GH 5 enzyme that cleaves the β-glycosidic linkage between the sugar and the ceramide moieties of gangliosides.[258] Inhibitor **156** (Scheme 43), featuring a 4-dimethylaminobenzoic amide at C-1, was bound with 6-OH in the position of 2-OH of substrates or pyranoid inhibitors, with additional contacts of the amide moiety with Glu 233, the general acid–base, and a water molecule coordinating the amide carbonyl oxygen as well as 3-OH (PDB 2OYM). This water bridge was also found with cellobio-isofagomine **120** (Scheme 30) between 6-OH of this inhibitor and Glu 233. Adjacent Trp 178 only coordinated the primary hydroxyl groups of compound **120** and inhibitor **123** (Scheme 31) via water bridges, but pyrrolidine **156** (Scheme 43) was not coordinated by this residue.

A series of aminomethyldihydroxypyrrolidine-based inhibitors bearing lipophilic aryl substituents were prepared by Gerber-Lemaire, Rose, Juillerat-Jeanneret, and coworkers as monocyclic structural analogues of swainsonine as potential agents against human glioblastoma, and their structures when complexed with the *Drosophila* α-mannosidase were determined (PDB 3DDF and 3DDG, respectively).[293] Compounds **157** and **158** inhibited the enzyme by around three orders of magnitude less than did swainsonine. The typical zinc chelation by the two *cis* hydroxyl groups was observed in all instances, but no particular interactions with the aryl moieties were found.

The N-terminal subunit of human maltase-glucoamylase (EC 3.2.1.20, GH 31) in complex (PDB 3CTT) with the pyrrolizidine alkaloid casuarine **159** (K_i 0.45 μM; Scheme 44), as well as *E. coli* periplasmatic trehalase (EC 3.2.1.28; GH 37) with the corresponding 6-*O*-α-D-glucopyranosyl derivative **160** (K_i 12 nM; PDB 2JJB), have been reported.[319] Together with the C-7-branched casuarine derivative **161** (K_i

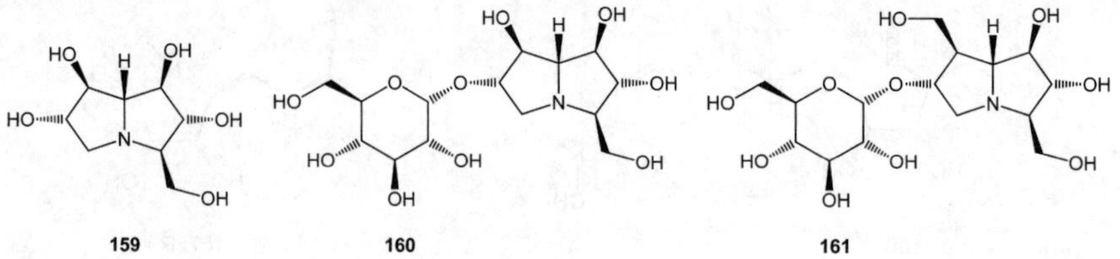

SCHEME 44.

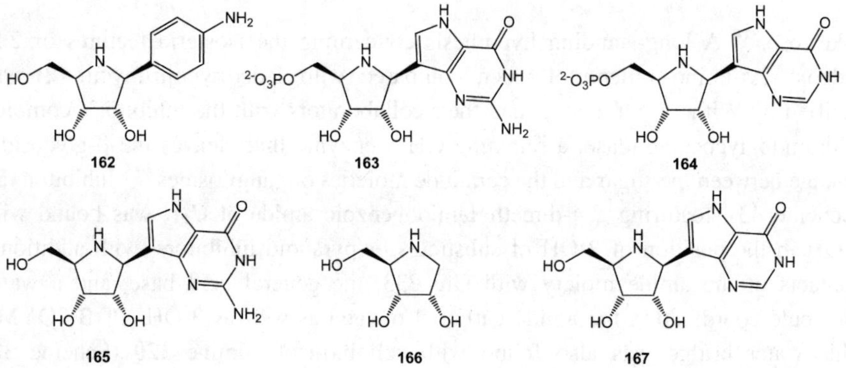

SCHEME 45.

2.8 µM) in complex with the trehalase (PDB 2WYN),[320] these are the only structures of pyrrolizidine complexes with glycosidases thus far reported.

g. Nucleoside Hydrolases.—Imino sugar-related nucleoside analogues are a class of compounds designed as transition-state analogues of nucleoside hydrolase action. The complex of 1-C-(4-aminophenyl)-1,4-dideoxy-1,4-imino-D-ribitol **162** (K_i 30 nM; Scheme 45) with trypanosomal nucleoside hydrolase from *Crithidia fasciculata* (PDB 2MAS) showed the inhibitor strongly bound to Ca^{2+}, which is chelated by 2-OH and 3-OH of the sugar moiety.[321] Invented by rational design by Schramm, Tyler, and their coworkers, the immucillin-type inhibitors of various nucleoside hydrolases and related proteins, including enzymes of pathogenic microorganisms, have turned out to be highly active and have thus attracted considerable attention as biological probes and promising drug candidates.

Immucillin-GP **163** (K_i 4.6 nM), a "C-nucleoside" of iminoribitol in complex (PDB 1BZY) with human hypoxanthine-guanine phosphoribosyltransferase, was resolved to 2.0 Å and showed 2-OH and 3-OH of the imino sugar ring strongly coordinated to Mg^{2+} as well as a catalytic loop that closes the active site of this Mg-dependent enzyme toward bulk water, trapping a limited number of solvent molecules in the catalytic site.[322] Subsequently reported was the structure of the purine phosphoribosyltransferase from the malaria-causing organism *P. falciparum* coordinated by immucillin-HP **164** (K_i 1 nM; PDB 1CJB),[323] the complexes of immucillin-GP (K_i^* 10 nM; PDB 1DQN) and its dephosphorylated parent, immucillin-G **165** (PDB 1DQP) with guanine phosphoribosyltransferase from *Giardia lamblia*, the causative agent of giardiasis,[324] as well as the structures of 1,4-dideoxy-1,4-imino-D-ribitol **166** (PDB 1I80) and immucillin-H **167** (K_i 0.03 nM; PDB 1G2O) with *M. tuberculosis* PNP.[325] Again, catalytic-loop segments close to the active site were identified. Bovine PNP was probed with immucillins G (K_i 30 pM; PDB 1B8N) and H (K_i 23 pM; PDB 1B8O).[326] The authors investigated the binding characteristics of 24 derivatives based on their constitutional structures and concluded that inhibitory activity is a function of a complex pattern of H-bonds as a prerequisite for transition-state resemblance.

Immucillin-H (**167**, K_i^* 72 pM) complexed with human PNP (PDB 1PF7) provided insights into the high selectivity of this inhibitor for the human enzyme.[327] Another structure at 2.3 Å was subsequently deposited by Schramm and coworkers (PDB 1RT9), together with the structure of the enzyme with "DADMe" [4′-deaza-1′-aza-2′-deoxy-1′-(9-methylene)]-immucillin-H **168** (Scheme 46; PDB 1RSZ).

This new group of immucillin derivatives with altered arrangement of functional groups and increased distance between the "anomeric," charge-bearing atom and the purine-type nucleus was further applied in compounds DADMe-immucillin-H (**168**, K_i 42 pM; PDB 1N3I) and G **72** (K_i 24 pM; Scheme 19), which were employed to elaborate structural and catalytic features of the PNP of *M. tuberculosis*.[328] The

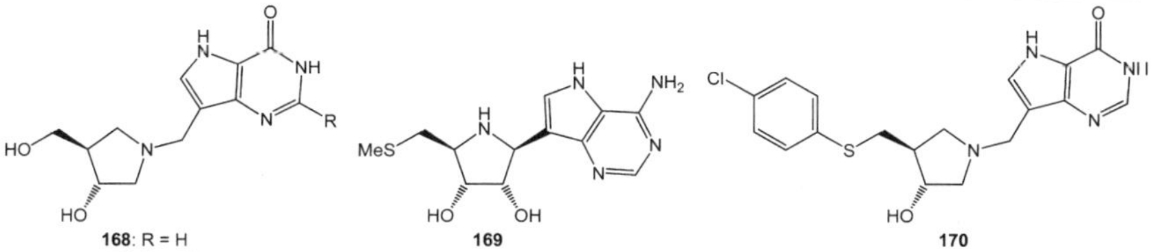

168: R = H
72: R = NH_2

169

170

SCHEME 46.

complex of 5′-deoxy-5′-methylthio-immucillin-H **169** (MT-immucillin-H), a powerful inhibitor of human 5′-methylthioadenosine phosphorylase (K_i* 1 nM) was resolved (PDB 1K27), and some structural analogues, including the corresponding (4-chlorophenylthio)-DADMe relative **170** (Ki* 10 pM), turned out to be among the strongest noncovalent inhibitors of the polyamine pathway known.[329] The PNP of *P. falciparum* in complex with immucillin-H and sulfate (K_d 0.86 nM; PDB 1NW4) as well as with phosphate (PDB 1RR6) and the complex with 5′-deoxy-5′-methylthio-immucillin-H (**169**, K_d 2.7 nM; PDB 1Q1G) were reported in 2004.[330] MT-immucillin-H **73** (Scheme 19) was the first inhibitor found with specificity for the malarial enzyme.

Two interesting structures with immucillin-H (**167**, K_i 6.2 nM) in complex with the purine-specific nucleoside hydrolase from *Trypanosoma vivax* provided additional pieces of information on the "catalytic loop(s)."[331] When the enzyme was soaked with the inhibitor (PDB 2FF1), no conformational changes could be observed when compared to the structure of the free enzyme. In contrast to this observation, in the co-crystallized complex (PDB 2FF2), significant conformational changes were detected resulting from an ordered loop now closing the active site. This structural feature was subsequently further investigated, leading to the conclusion that fast substrate binding and fast aglycon release occurs with the hydrolysis step as the rate-limiting event.[332] The loop closure results in separation of the active site from bulk water and influences the dihedral angle of the anomeric bond. Structures of various mutants of human PNP lacking His 257, a contact point of the primary hydroxyl group 5′-OH of the natural substrates, were investigated by employing immucillin H (**167**, PDB 2OC9, 2ON6, and 2OC4) and DADMe-immucillin H, **168** (PDB 2A0W, 2A0X, and 2A0Y), revealing the importance of this substituent for loosening the lone-pair electrons of the ring oxygen by neighboring-group participation, thus facilitating their interaction with the substrate's scissile *N*-glycosylic bond.[333] The active site of *Anopheles gambiae* PNP was investigated, employing DADMe-immucillin-H, **168** (K_i 3.5 pM). The corresponding adduct (PDB 2P4S) shows that a conserved water molecule was missing when compared to the human enzyme in a complex with the same compound, and this may affect the substrate specificity as well as the tight inhibitor binding.[334] Immucillin A **171** (Scheme 47; K_i* 87 pM; PDB 2I4T) as well as DADMe-Immucillin A **172** (K_i* 30 pM; PDB 2ISC) in its complex with the PNP of *Trichomonas vaginalis* gave insight into the unique purine-salvage pathway of the microorganism.[335] This consists of the bacterial-type PNP and a purine nucleoside kinase. The phosphorylase thus works in the reverse direction relative to the corresponding enzymes of other organisms. Complexes of the L enantiomers of immucillin H **173** (PDB 2Q7O) and DADMe-immucillin H **174**

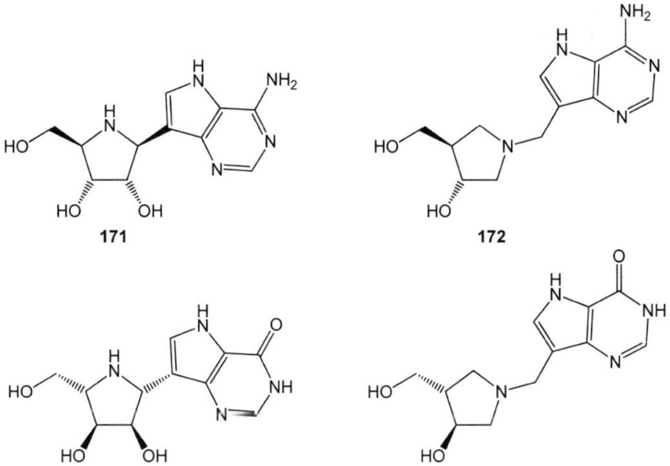

SCHEME 47.

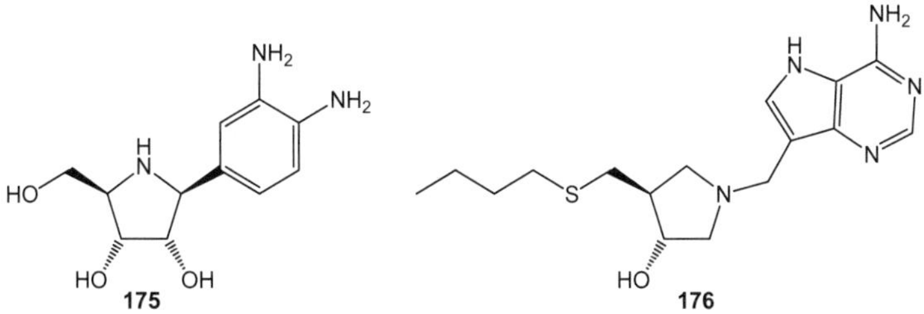

SCHEME 48.

(PDB 3BGS) with human PNP revealed that the pK_a at the NH-7 position as well as the location of the positive charge in the glycon moiety are more relevant for powerful inhibition than the exact placement of the hydroxyl groups.[336]

Pyrimidine nucleoside hydrolase YeiK from *E. coli* in complex (PDB 3MKN) with 1-(3,4-diaminophenyl)-1,4-dideoxy-1,4-imino-D-ribitol (**175**, K_i 76 μM; Scheme 48) was exploited to investigate the conformational changes of two loops adjacent to the active site and their influence on the catalytic pathway of the hydrolysis reaction, introducing a new computational procedure.[337] Recent developments indicate that replacement of ribosyl-simulating systems by open-chain aminoalcohols may allow for even more powerful inhibitors than the furanoid imino sugar structures.[338] Besides the obvious impact on finding leads for drugs against various pathogenic microorganisms employing highly selective enzyme inhibition, quorum-sensing interruption by enzyme inhibition[339] has become a new target for the nucleoside analogs. Butylthio-DADMe-immucillin A (**176**) in complex (PDB 3DP9) with *Vibrio cholerae* 5′-methylthioadenosine/*S*-adenosylhomocysteine nucleosidase (MTAN) served as one of the first probes in this context.[340]

Concluding this section, it should be noted that imino sugars in the widest sense have become most important and versatile tools, in conjunction with X-ray diffraction studies, for the thriving area of structure determination of glycosidases and the exploration of catalytic pathways. In synergy with the powerful tools provided by CAZy, and kinetic considerations employed by many able researchers in the field, tremendous leaps in our understanding of glycosidases and their work have already been made, and more can also be foreseen in the further development of this area, with many fascinating structures being currently "in progress" or still to be determined.

6. "Non-Glycon" Binding Sites and Inhibitor Activity

The tolerance, and frequently even preference, of many glycosidases for glycoside substrates having lipophilic (particularly aromatic) aglycons has been known and exploited during more than 50 years for analytical investigations of enzyme kinetics by employing aromatic residues such as (nitro)phenyl groups or umbelliferyl moieties as aglycons.[341] The general acceptance of a putative nonpolar aglycon binding site adjacent or close to subsite − 1, the catalytic "hot spot," providing cooperative effects for substrate and inhibitor binding, is widely documented in the literature. The finding that phenyl β-D-glucopyranoside and 4-nitrophenyl β-D-glucopyranoside inhibited the action of concanavalin A suggested "that nonspecific binding of the aromatic group occurs probably due to a corresponding hydrophobic region of the protein in close proximity to the carbohydrate-binding region."[342] Mutation of *E. coli* β-galactosidase

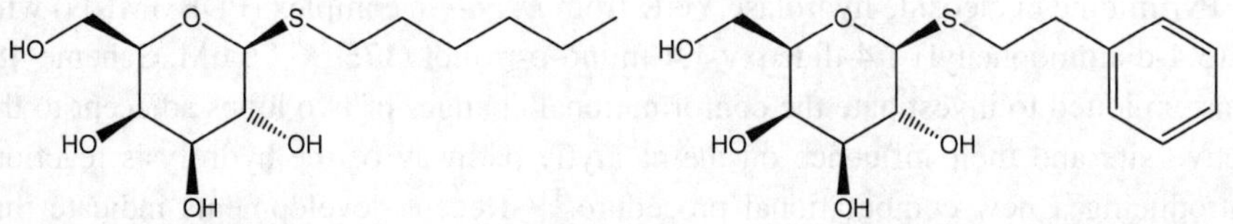

SCHEME 49.

"increased the hydrophobic nature of the the enzyme near the aglycon binding site and facilitated the hydrolysis of more-hydrophobic galactosides."[343]

In another investigation of the aglycon binding site of *E. coli* β-galactosidase, alkylthio (e.g., **177**; Scheme 49) and such ω-phenylalkylthio galactosides as **178** having different lengths of alkyl chain were utilized.[344] The authors reported that "in general, only unspecific hydrophobic forces, rather than directed interactions with amino acid side chains, were involved." Restructuring of the water phase upon substrate binding was inferred, with the nonpolar aglycon "buried in a hydrophobic micro-region which is limited in size."[345] Legler and Liedtke probed glucoceramidase from calf spleen with a set of 4-alkylumbelliferyl β-D-glucopyranosides as well as with *N*-alkyl derivatives of 1-deoxynojirimycin. The kinetic data obtained "could be interpreted in terms of an aglycon binding site that has an extended hydrophobic region starting approximately five carbon atoms from the catalytic site."[346] From the observation of mixed competitive and noncompetitive inhibition of β-D-glucosidase from *S. atra* by alkyl and aryl β-D-glucopyranosides and their 1-thio analogues, it was concluded that the noncompetitive inhibition originates from unspecific binding of the aglycon to the aglycon binding site of the intermediary enzyme–glycosyl complex, with an increase of entropy as the sole driving force.[347] A "large hydrophobic aglycone binding site" was inferred[348] for almond β-glucosidase, and that indeed the aglycon is the main contributor to substrate binding for this enzyme.[349] The powerful inhibition of almond β-glucosidase by 4-phenylimidazole (**50**, Scheme 14; K_i 0.83 μM in unprotonated form) and 4-(3-phenylpropyl)imidazole (K_i 0.07 μM), surpassing 1-deoxynojirimycin **2** as well as D-glucono-1,5-lactam **126** (Scheme 31), was also attributed to the strong binding of the phenyl moiety to the hydrophobic binding site.[142,350] Apparently, the propyl spacer arm was the best fit between the catalytic site with its acid–base catalytic nucelophile pair and the aglycon binding pocket.

For GH 10 xylanases, the ligand binding in the aglycon region of the substrate-binding cleft was found to be dominated by hydrophobic interactions between xylose

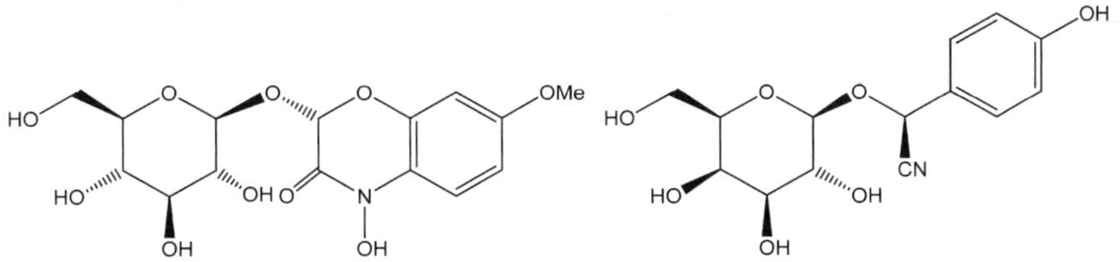

SCHEME 50.

rings and aromatic amino acids.[351] Details on aglycon binding were revealed by the investigation into the aglycon specificity in β-glucosidases from maize and and *Sorghum* (GH 1). Whereas the maize isoenzyme Glu1 is a broad spectrum β-glucosidase accepting a large variety of substrates (**179**, Scheme 50), the *Sorghum* β-glucosidase isoenzyme Dhr1 exclusively hydrolyzes its natural substrate, dhurrin (**180**), which, interestingly, is not a substrate of the maize enzyme. Three phenylalanine residues and one tryptophan were found to form the aglycon binding site of the maize enzyme,[352] whereas in the active site of the *Sorghum* enzyme, Asn 259, Phe 261, and Ser 462, all located in the aglycon binding site, are crucial for aglycon recognition and binding. Tight binding of the aglycon was found important for this glucosidase, forcing the glycon into a $^1S_3$ conformation ready for nucleophilic attack.[353]

a. Imino Sugars with Increased Hydrophobicity

(i) Pyranoid Systems and Glucosidases. As early as 1979, researchers described a wide range of lipophilic *N*-alkyl as well as ω-arylalkyl derivatives of 1-deoxynojirimycin, with chain lengths between C_1 and C_{18}.[354,355]

Glucosidase 1 from calf liver was found to be more strongly inhibited by *N*-alkyl-1-deoxynojirimycins than by the parent compound.[356] This enzyme is the first of a well-ordered cascade trimming the juvenile tetradecasaccharide of N-glycoproteins back to the so-called core, removing the outermost (1→2)-connected glucopyranosyl residue. *N*-Butyl-1-deoxynojirimycin **131** (Scheme 33) and the corresponding *N*-hexyl derivative **181** (Scheme 51) exhibited K_i values of 90 and 130 nM, respectively, with the *N*-methyl analogue **182** being the most potent (K_i 70 nM) when compared to parent compound **2** (K_i 1000 nM). It was concluded that "some hydrophobic interaction of the alkyl chain at the active site of the enzyme" was responsible for this pronounced effect.

A study of anti-HIV replication effects of nearly 50 imino sugars demonstrated that *N*-alkyl derivatives of 1-deoxynojirimycin exhibit markedly enhanced activity when compared to the unsubstituted parent compound. In particular, compound **131**

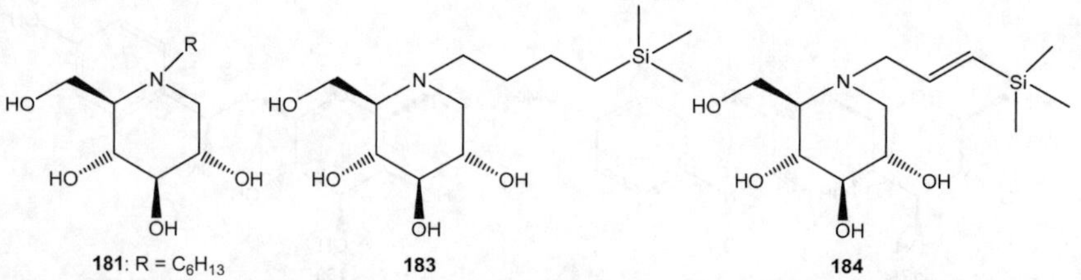

181: R = C_6H_{13}
182: R = CH_3

183

184

SCHEME 51.

(Scheme 33) was highly effective.[357] Nonpolar N-alkyl substituents containing alkylsilyl or arylsilyl moieties in such 1-deoxynojirimycin derivatives as **183** were shown to increase the inhibitory activities on human intestinal α-glucosidases, with short- and medium-length chains containing a double bond (**184**) giving best results.[358]

In contrast to the inactive parent compound **2**, inhibition of glycoprotein-processing α-glucosidases was also found by Butters and coworkers for 1-deoxynojirimycin derivatives having phenylalkyl substituents of medium chain lengths (C_4—C_6) at the ring nitrogen atom.[359]

The increased inhibitory power of N-alkyl derivatives in comparison with compound **2** was observed with human acid β-glucosidase.[360] It was concluded that two hydrophobic regions, the aglycon binding site as well as a "third site," both about four to five carbon bond lengths separated by hydrophilic regions from the active-site pocket, may be accommodating the alkyl chains of the natural substrate or the N-alkyl chain of the inhibitors probed. With cytosolic β-glucosidase from calf liver, the K_i values of parent compound **2**, its N-decyl **185** (Scheme 52), and N-dodecyl derivative **186** were 210, 8.2, and 3.8 μM, respectively.[361] Shorter chains diminished the activity (C_2, 3500 μM; C_4, 850 μM). Interestingly, 5-(butylaminocarbonyl)pentyl-1-deoxynojirimycin **187**, despite its substantially hydrophobic substituent, had a K_i value of only 1700 μM. Comparison of the inhibition by compound **186** and the N-dodecylglycosylamine (**188**) showed unfavorable interaction of the alkyl moiety of the iminoalditol with active-site residues complementary to the ring oxygen and anomeric oxygen of

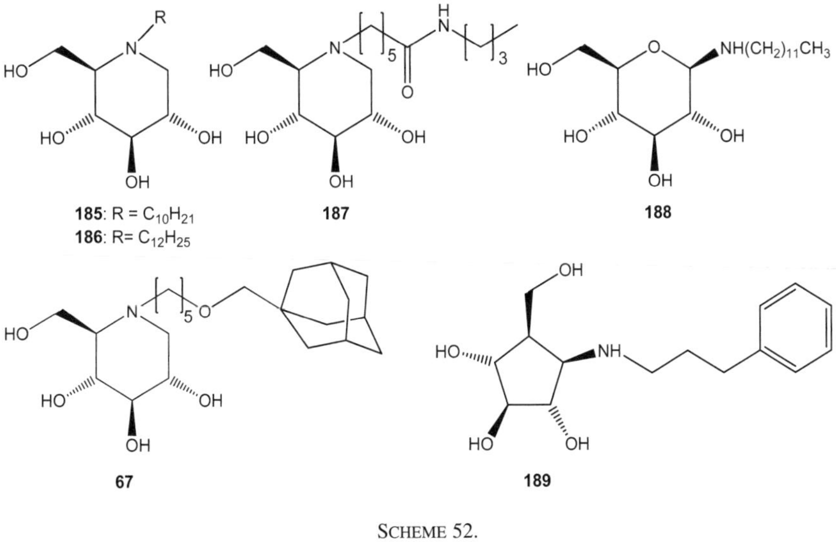

185: R = $C_{10}H_{21}$
186: R = $C_{12}H_{25}$

187

188

67

189

SCHEME 52.

the substrate's glycon. Favorable interactions with the (lipophilic) aglycon binding site required chain lengths longer than C_4. The contribution of the long side chain in compound **186** was found to compensate for unfavorable interactions of the protonated inhibitor (K_i 21 μM) with the catalytic site.

In an important contribution, Overkleeft, Aerts, and coworkers introduced adamantyl-terminated N-alkyl substituents, which provided excellent inhibitors, for example, **67** (Scheme 18), of non-lysosomal glucosylceramidase and glucocerebrosidase.[362]

The "fusion" of a lipophilic 2-(phenyl)ethyl- or (3-phenylpropyl)-imidazole moiety[350] with the D-*gluco*-configured glycon provided two of the most powerful β-glucosidase inhibitors, **63** (Scheme 17) and the corresponding propyl homolog, thus far known.[171,265] Hydrophobic interactions with putative aromatic residues were inferred for this notable effect of the nonpolar substituent. Together with the related structures **128** and **129** (Scheme 32), and their unexpectedly few (or completely absent) directed interactions[265,266] with the active sites probed (PDB 1LQ2, 2CET, 2J7C), it may be inferred that additional or other binding energy-controlling forces account for the significant inhibition-enhancing effects of nonpolar substituents. Similar effects of enhanced inhibition by N-phenylalkyl substitution were found by Reymond and coworkers in such aminocyclopentitol-derived carba sugar glycosidase inhibitors as compound **189** (Scheme 52).[363]

A model for the structural relationship of N-butyl-1-deoxynojirimycin (**131**, Scheme 33) with ceramide, the substrate of ceramide glucosyltransferase, was suggested by Butters and coworkers. They conducted a comparative investigation on a range of 14 derivatives of 1-deoxynojirimycin substituted at the ring nitrogen and/or C-1, as well as 5 pyrrolidine-type inhibitors selected by superpositioning of the iminoalditol with a portion of the ceramide structure.[364]

A range of 2,5-dideoxy-2,5-imino-D-mannitol derivatives (**190–194**, Scheme 53) was prepared as probes for the aglycon binding site of β-glucosidases.[365] The influence on the activity of 1-*O*-alkyl as well as 1-*N*-acyl or -*N*-sulfonyl substituents bearing a variety of lipophilic alkyl- and arylalkyl groups, also in combination with ring nitrogen alkylation,[366] was reported. Several compounds showing considerably enhanced β-glucosidase inhibitory activity were identified among the collection of inhibitors screened.

Following up on previous results, Wong and colleagues found beneficial effects on the activity of some lysosomal glucocerebrosidase mutants related to Gaucher disease. They employed various lipophilic 1-deoxynojirimycin derivatives, along with other iminoalditols having medium-length straight-chain N-substituents, and also with adamantyl-capped ethers and amides **195** (Scheme 54).[367,368]

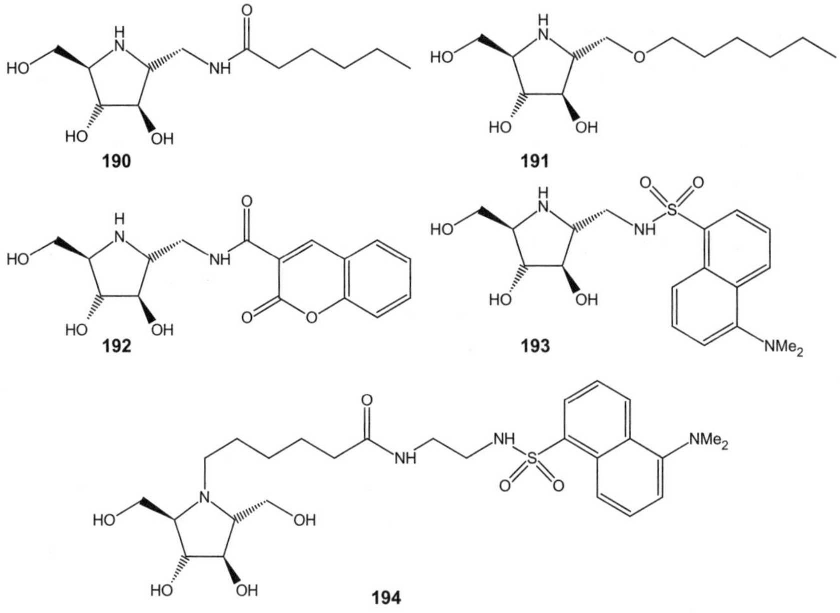

SCHEME 53.

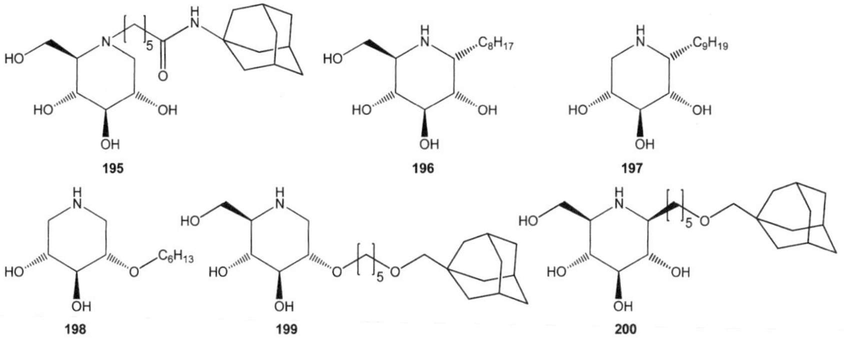

SCHEME 54.

Because of the known detrimental effects of N-alkylation in isofagomine (**61**), Fan and coworkers prepared a range of *C*-alkyl derivatives such as compound **62** (Scheme 17), which constitute the most powerful inhibitors of β-glucosidases known thus far.[170]

Imino sugars bearing extended carbon chains featuring alkyl groups or chains at C-1 have been found as natural products, in both the piperidine and pyrrolidine series. Synthetic derivatives such as α-1-*C*-octyl-1-deoxynojirimycin (**196**, Scheme 54) were introduced by Martin and coworkers as pharmacological chaperones for Gaucher disease, based on their strong interaction with the active site of β-glucocerebrosidase. The C_8 alkyl chain was found to enhance the activity by nearly 500-fold when compared to the unsubstituted parent **2**.[369] Similar efficacy was found with the corresponding C_9 homologue, and even more powerful inhibitors were subsequently introduced. 1-*C*-Alkyl-iminoxylitols, in particular the 1-*C*-nonyl derivative **197**, exhibited a K_i value of 2.2 nM with the human enzyme.[370] Interestingly, despite the counter-intuitive modification at O-2, the recently reported 2-*O*-alkylated iminoxylitol **198** was also a powerful inhibitor (IC_{50} 9 nM).[371] The comparable 2-*O*-alkyl derivative (**199**) of 1-deoxynojirimycin, having an adamantanylmethoxy substituent terminating a C_5 spacer arm, was only active in the micromolar range.[372] 1-*C*-(5-Adamantanylmethoxy)pentyl-1-deoxynojirimycin **200** was a good inhibitor of lysosomal glucocerebrosidase and a potent inhibitor of non-lysosomal glucosylceramidase, albeit not quite as powerful as the corresponding *N*-alkyl derivative **198**. The same groups recently reported an extended range of 1-*C*-substituted D-*gluco*- as well as D-*xylo*-configured lipophilic inhibitors having such additional features as double bonds in the chain, some of which exhibited impressive activities and selectivities.[373]

Notably, only very few directed, nonpolar contacts were found in the comparison of *N*-butyl-1-deoxynojirimycin (**131**, Scheme 33; PDB 2V3D) and its *N*-nonyl analogue (**132**; PDB 2V3E) in the active site of glucocerebrosidase.[269] Neither their numbers nor the difference between the individual contacts for compounds **131** and **132** readily explains the striking difference in biological activity when compared to each other as well as to the parent compound **2**.

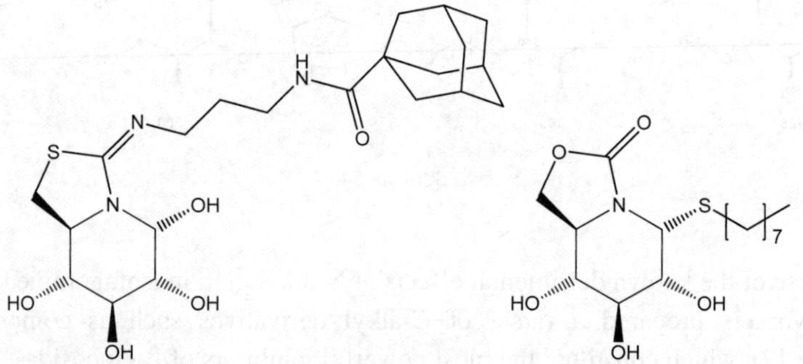

SCHEME 55.

Strongly lipophilic compounds, some closely resembling castanospermine (**8**), were introduced by Ortiz Mellet and Garcia Fernandez. Such compounds as **201** and **202** were found to be potent glycosidase inhibitors featuring interesting selectivities (Scheme 55).[275,276,280,374,375]

(ii) Other Configurations. Medium chain-length N-substituents, for example, in **203** (Scheme 56), were found to augment the inhibitory activity of iminoxylitol with the β-xylosidase from *Thermoanaerobacterium saccharolyticum* by up to one order of magnitude but diminished the effect on β-glucosidase from *Agrobacterium* sp. by up to over 50-fold.[376]

Swainsonine derivatives having nonnatural lipophilic substituents at C-5, such as compound **204**, did not show increased inhibitory power but remained as powerful as the parent compound.[292]

In the series of α-L-fucosidase inhibitors, several potent compounds were generated by N-acylation of 1-aminomethyl-1-deoxy-L-fuconojirimycin with lipophilic acyl moieties. In particular, strongly enhanced activity in the subpicomolar range was observed with compound **152** (Scheme 41).[377] This type of inhibitor could subsequently be exploited in a comparison of human and *T. maritima* α-L-fucosidases and their aglycon binding sites, and as complexing agents for the active site of other GH 29 α-L-fucosidases.[378,312,313]

A typical example demonstrating the advantages of increased lipophilicity for activity enhancement in *N*-acetylhexosaminidase inhibitors is GlcNAcstatin (**142**, Scheme 38).[304] This is a picomolar inhibitor featuring a lipophilic arylalkyl chain reaching toward the aglycon binding site as well as a chain-extended *N*-acyl moiety, both strongly contributing to the superior effects of this compound. Inhibition constants in the nanomolar range were found with *N*-alkyl derivatives of 1-acetamido-1,2,5-trideoxy-2,5-imino-D-mannitol (**205**, Scheme 57).[379]

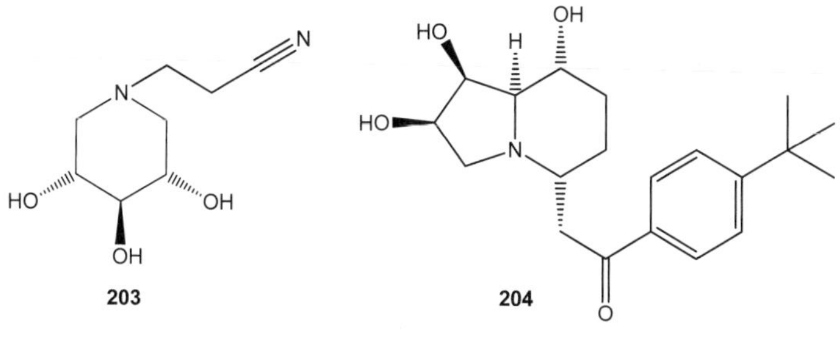

SCHEME 56.

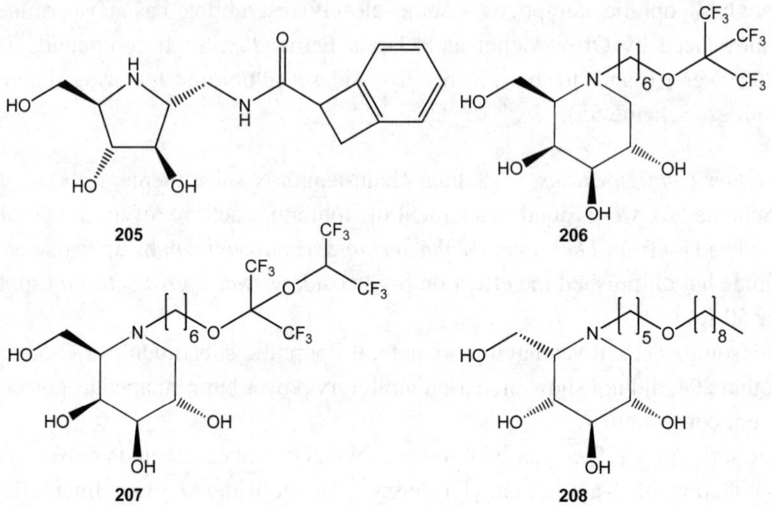

SCHEME 57.

1-Deoxy-D-galactonojirimycin derivatives bearing fluorous N-substituents, for example, compounds **206** and **207**, turned out good inhibitors of β-galactosidases, including human lysosomal β-galactosidase.[380] Interestingly, long-chain N-alkyl-1-deoxy-L-idonojirimycins, in particular the N-nonyloxypentyl derivative **208**, have also recently been found highly potent inhibitors of glucosylceramide synthase.[381]

b. Added-Value Lipophilic Inhibitors.—Emerging from the general interest in the properties and advantages of lipophilic inhibitor derivatives, there is now available an increasingly interesting and versatile range of compounds having structural features that allow novel means of exploitation, as for various analytical purposes of potential utility.

Somewhat unexpectedly, the 1-dansylamino derivative **193** (Scheme 58) of the β-glucosidase inhibitor 2,5-dideoxy-2,5-imino-D-mannitol turned out a nanomolar inhibitor (K_i 2.4 nM) of *Agrobacterium* sp. β-glucosidase.[382] The same was true when the fluorophor and the inhibitor were separated by a C_6 spacer arm. Equally active were the corresponding coumarin-3-carboxylic acid derivative **66** (Scheme 18; 1.2 nM)[175] and the corresponding dapoxyl analogue (4 nM). The autofluorescence of inhibitor **193** (Scheme 58) was exploited for demonstration of the stability of the enzyme–inhibitor complex during migration on native polyacrylamide gel electrophoresis (PAGE), as well as in energy transfer experiments.[382] Subsequently, pyranoid inhibitors of biologically relevant configurations were fluorescently tagged by

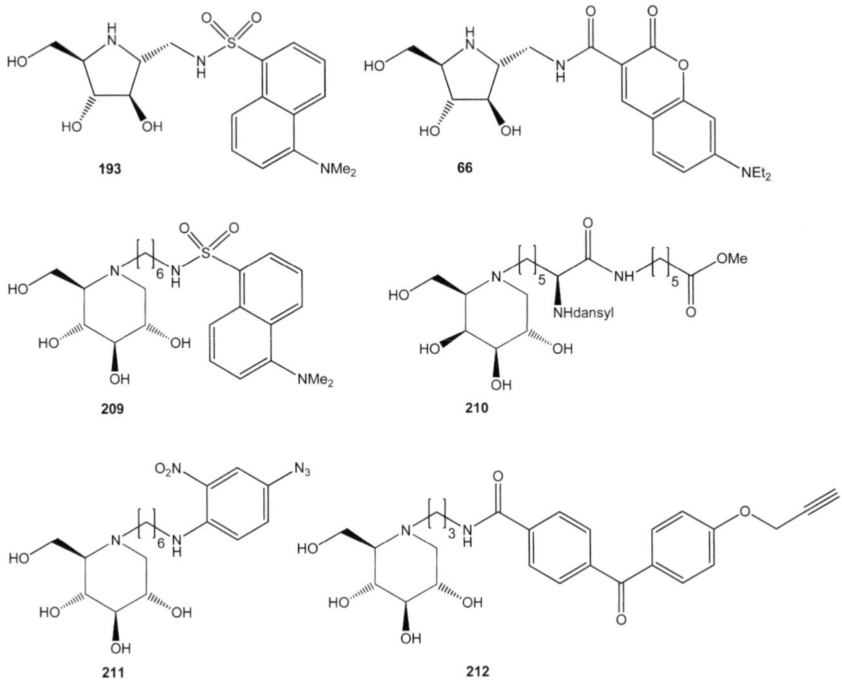

SCHEME 58.

an N-alkylation approach.[383] A lipophilic C_6 spacer arm between the fluorophor and the inhibitor (**209**) turned out to be superior to di- and tri-ethylene glycol spacers in terms of inhibitory activities.[384] Introduction of a bifunctional lysine-derived spacer arm in combination with the dansyl reporter group led to such inhibitors as **210**, which showed enhanced activity in most of the configurational series probed and allowed for additional structural variation with a view to immobilization of the inhibitor.[385,386] In a subsequent study, a very simple glycosidase-binding chip was prepared as a proof of concept.[387] Butters, Fleet, and coworkers prepared novel N-alkyl derivatives of compound **2** as potential photoaffinity probes for α-glucosidases of the endoplasmatic reticulum, with compound **211** being the best inhibitor (IC$_{50}$ 17 nM) of α-glucosidase I known thus far.[388] Other photoaffinity probes, including compound **212** and exhibiting high potency and selectivity toward human β-glucosidases, were developed by Pieters, Overkleeft, Aerts, and coworkers and were applied for labeling non-lysosomal β-glucosidase with high success.[389] Affinity-based profiling of exo-α-glucosidases was also recently achieved by exploiting lipophilic photoaffinity labels.[390]

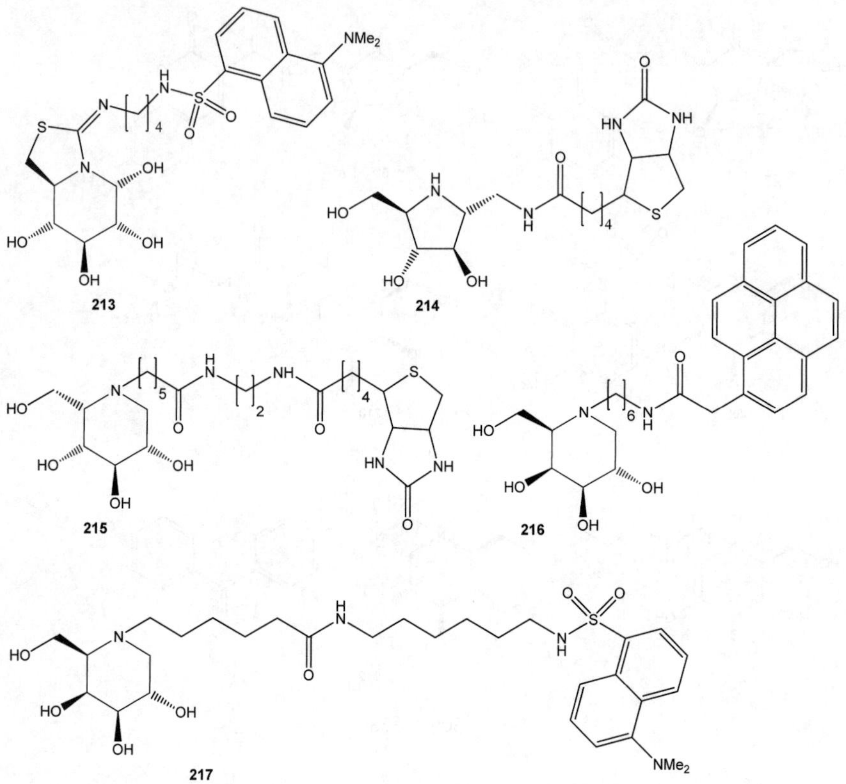

SCHEME 59.

Fluorescent imino sugars **213** (Scheme 59) as inhibitors and probes for lysosomal β-glucocerebrosidase related to Gaucher disease were prepared and screened by Ortiz Mellet and collaborators.[391] Subsequently, these compounds were exploited for lysosome staining in fluorescence-based colocalization studies.[392] Other lipophilic inhibitors, including biotin-tagged compounds **214** and **215**[393] as well as the fluorescent pyrenyl-capped probe **216**,[394] were specifically prepared for the lysosomal β-galactosidase related to Morquio B disease and GM1-gangliosidosis. Intracellular staining of lysosomes was performed with D-galactose inhibitor **217** and related analogues.[178]

VI. Conclusions

From the beginnings of imino sugar research and application, such lipophilic modifications as *N*-alkyl derivatives of 1-deoxynojirimycin and its relevant epimers have been prepared and their properties investigated. These amphiphilic compounds in many cases offer superior inhibitory power when compared to their respective parents. Only a few exceptions, such as isofagomine, whose biological activity is highly sensitive to N-alkylation, have been recorded to date.

In particular, in the glycosidase-inhibitor area, many meaningful structural modifications are actually reduced to N-modification because of the fact that most glycosidases require at least three hydroxyl groups at appropriate distances from each other for recognition and binding. Other options to augment the nonpolar character of potential probes have been the introduction of alkyl branches next to the ring nitrogen atom, as in homo-1-deoxynojirimycins and imino sugar "C-glycosides," or in isofagomine analogues. Pyrrolidine derivatives have been modified by addition of nonpolar substituents via ether- and amide-linked spacer arms.

As to the reasons for the enhanced properties of many lipophilic inhibitors when compared to their unsubstituted parent compounds, only more or less vague hypotheses have thus far been put forward. The general assumption of a "hydrophobic aglycon binding site" has been widely supported across all relevant configurations of inhibitors and their complementary glycosidases. Interestingly, most of the enzyme–inhibitor complexes thus far visualized by X-ray diffraction exhibit only very few directed contacts between lipophilic alkyl substituents and the environment around the active sites; even more surprisingly there may be none at all, with considerable freedom of movement of the nonpolar group under consideration. In quite a few cases, the differences between hydrophilic parent and lipophilic derivative, or between two or more lipophilic derivatives, cannot readily be explained in this way. Displacement of defined structural water molecules in the active site, or decreased access of water by the steric demand of the substituted inhibitor, was also arguments for the advantageous properties of lipophilic compounds. This supposition was supported by the entropic term that beneficially contributes to the binding energy observed for some of the inhibitors under consideration. General "interference with the solvation sphere" covering the protein is a thought that also points into this direction. Influence of the lipophilic partial structure in the aglycon binding site on the positioning and the conformation of the imino sugar moiety of the inhibitor cannot be excluded, either. Clearly, in addition to the requirements concerning basicity, protonation state, and a substitution pattern that complements the requirements of the respective glycosidase on its conformational journey through the catalytic process, the effects of hydrophobic substituents in strategic regions markedly enhance the inhibitory effects. With such substituents, previously uncharted cooperative subsites or regions which obviously influence inhibitor binding may be discovered and characterized.

As biochemical tools, suitable and sensible imino sugar modifications will certainly become helpful "intelligent" probes and reporters addressing some of the future needs of research in carbohydrate-active enzymes.

VII. Paradigm Changes and Emerging Topics

After a long period characterized by an intense search for the most active inhibitors and the ultimate transition-state analogue for each enzyme, one of the powerful driving forces of recent and contemporary glycosidase-inhibitor research has been the search for selectivity or even specificity of inhibition. Being able to select one particular glycosidase in the presence of the entire remaining repertoire of carbohydrate-processing enzymes remains an important challenge and a highly desirable situation. Of the numerous approaches directed toward this ambitious goal, examples of inhibitors with "unusual" or nonnatural configurations,[395–399] ring sizes,[400] or substitution patterns, including the enantiomers of proven inhibitors that are best exemplified by the "looking glass" inhibitors of Fleet and coworkers,[401–406] which all oppose the accepted view on glycon selectivity by glycosidases, constitute noteworthy examples of chemists' contemporary efforts.

The recent finding that powerful glycosidase inhibitors can activate mutant enzymes that are retarded in their efficacy has caused a paradigm change and has considerably shifted the view on the entire area. In particular, a range of lysosomal disorders such as Gaucher disease may soon be treatable with these so-called chemical or pharmacological chaperones providing small-molecule templates for improved protein folding, thus providing a higher proportion and concentration of catalytically competent glycosidase in the lysosomes. Many of the very recent efforts and compounds mentioned in this account have been dedicated to the identification and improvement of such imino sugar-based active-site templates for enzyme mutants that are susceptible to this structural support. This area of imino sugars for chaperone-mediated therapy has recently been excellently reviewed,[407] with important new contributions appearing weekly.

Potentially beneficial roles of imino sugars as chemotherapeutic compounds, based on their complex interactions with such vital and essential metabolic processes as glycosphingolipidoses, tumor development, or the immune system have also been recently reviewed[408] and have remained "hot topics."[409–411]

Considering the huge number of possible target proteins that feature carbohydrate-processing or binding activities, new trends and novel applications[412] appear always to be just around the next corner.

VIII. Table of PDB-Entries: Enzyme–Inhibitor Complexes of Imino Sugars and Selected Other Ligands

Compound	Enzyme	PDB	References
D-Galactose-, D-glucose-, and D-xylose-related ligands (Sections V5a and V5c)			
1-Deoxynojirimycin (**2**) 1,5-Dideoxy-1,5-imino-D-glucitol	Xylose isomerase (EC 5.3.1.5)	1DIE	245
	Glucoamylase *Asp. awamorii*	1DOG	246
	Cyclodextrin glucanotransferase *Bacillus* sp. 1011	1I75	248
	β-Glucosidase	1OIM	249
	Thermotoga maritima (GH 1)	2J77	266
	Trehalulose synthase *Pseudomonas mesoacidophila* MX-45 (GH 13)	2PWD	241
	α-Glucosidase *Bacteroides thetaiotaomicron* (GH 97)	2JKE	277
	Sucrose isomerase SmuA *Protaminobacter rubrum* (GH 13)	3GBE	278
	Acid α-Gal human (GH 27)	3GXT	271
N-Butyl-1-deoxynojirimycin (**131**)	Acid β-glucosidase, human (GH 30)	2V3D	268
N-Nonyl-1-deoxynojirimycin (**132**)	Acid β-glucosidase, human (GH 30)	2V3E	268

Continued

Compound	Enzyme	PDB	References
Castanospermine (**8**) (*1S,6S,7R,8R,8aR*)-1,6,7,8-Tetrahydroxyindolizidine	β-Glucosidase *Thermotoga maritima* (GH 1)	2CBU	266
	exo-1,3-β-Glucanase *Candida albicans* (GH 5)	1EQC	273
	Trehalulose synthase *Pseudomonas mesoacidophila* MX-45 (GH 13)	2PWG	241
	Rho/Ras-glucosylating toxin *Clostridium sordellii*	2VL8	274
	α-Glucosidase *Bacteroides thetaiotaomicron* (GH 97)	2JKP	277
5-Amino-5-deoxy-D-gluconohydroximo-1,5-lactam (**124**)	α-Glucosidase *Sulfolobus solfataricus*	1UWU	259
	Myrosinase (thioglucoside glucohydrolase, EC 3.2.3.1)	1E6S	268
	α-Mannosidase II, *Drosophila* Golgi (GH 38)	3D51	290
O-[2-Acetamido-2-deoxy-D-*gluco*-pyranosylideneamino *N*-(4-chloro)phenyl] carbamate (**138**)	α-Mannosidase II, *Drosophila* Golgi (GH 38)	2D52	290

Compound	Enzyme	PDB	References
(5R,6R,7S,8R,8aR)-5,6,7,8-Tetradeoxy-3-octylimino-2-oxa-indolizidine (133)	β-Glucosidase *Thermotoga maritima* (GH 1)	2WBG	275
	β-Glucosidase lysosomal	2WCG	276
(5R,6R,7S,8S,8aR)-5,6,7,8-Tetradeoxy-3-octylimino-2-oxa-indolizidine (134)	β-Glucosidase *Thermotoga maritima* (GH 1)	2WC3	
Isofagomine (61) 1,4,5-Trideoxy-4-C-hydroxymethyl-1,5-imino-L-xylitol	β-Glucosidase *Thermotoga maritima* (GH 1)	1OIF	249
	Acid β-Glucosidase human (GH 30)	2NSX 3GXF	269 271,272
	Glycogen phosphorylase GPb rabbit muscle (GT 35)	2G9V	316
N-(3-Phenyl)propyl isofagomine (154)	Glycogen phosphorylase GPb rabbit muscle (GT 35)	2G9U	316
	Endocellulase Cel5A *Bacillus agaradhaerens* (GH 5)	1OCQ	250
	Cellulase *Humacola insolens*	1OCN	256
	Cellulase Cel6 *Mycobacterium tuberculosis* (GH 6)	1UP2	257

Continued

Compound	Enzyme	PDB	References
4-O-β-D-Glucopyranosyl isofagomine (**120**)	Endoglycoceramidase II *Rhodococcus* sp. (GH 5)	2OYK	258
Noeuromycin (**75**) 5-Amino-4,5-dideoxy-4-C-hydroxymethyl-L-xylose	β-Glucosidase *Thermotoga maritima* (GH 1)	2J75	266
	β-Mannosidase *Bacteroides thetaiotaomicron* (GH 2)	2VL4	291
Isofagomine lactam (**122**) 5-Amino-4,5-dideoxy-4-C-hydroxymethyl-L-xylonolactam	β-Glucosidase *Thermotoga maritima* (GH 1)	1UZ1	255
	exo-β-Mannanase *Cm*Man5 *Cellvibrio mixtus* (GH 26)	1 UZ4	255
Tetrahydrooxazine (**74**) (4R,5S,6R)-4,5-Dihydroxy-6-(hydroxymethyl)-3,4,5,6-tetrahydro-2H-1,2-oxazine	β-Glucosidase *Thermotoga maritima* (GH 1)	1W3J	251,266

Compound	Enzyme	PDB	References
Azafagomine (**130**) (3*S*,4*S*,5*S*)-4,5-Dihydroxy-3-(hydroxymethyl) hexahydropyridazine	β-Glucosidase *Thermotoga maritima* (GH 1)	2J7H	266
Glucoimidazole (**77**) (5*R*,6*R*,7*S*,8*S*)-Tetrahydro-5-(hydroxymethyl) imidazo[1,2-*a*]pyridine-6,7,8-triol	β-Glucosidase *Thermotoga maritima* (GH 1)	2CES	196
	β-Glucosidase *Sulfolobus solfataricus*	2CEQ	196
	α-Mannosidase II, *Drosophila* Golgi (GH 38)	3D4Z	290
(5*R*,6*R*,7*S*,8*S*)-6-*O*-β-D-Glucopyranosyl-tetrahydro-5-(hydroxymethyl)imidazo [1,2-*a*] pyridine-6,7,8-triol (**123**)	Endoglycoceramidase II *Rhodococcus* sp. (GH 5)	2OYL	259

Continued

Compound	Enzyme	PDB	References
Phenylethyl-glucoimidazole (**63**) (5*R*,6*R*,7*S*,8*S*)-5-(Hydroxymethyl)-2-(2-phenylethyl)-5,6,7,8-tetrahydro-imidazo[1,2-*a*]pyridine-6,7,8-triol	β-Glucosidase *Thermotoga maritima* (GH 1) β-Glucosidase *Sulfolobus solfataricus*	2CEQ 2CET 2CER	196 266 196
Phenyl-glucoimidazole (**128**) (5*R*,6*R*,7*S*,8*S*)-5-(Hydroxymethyl)-2-phenyl-5,6,7,8-tetrahydro-imidazo[1,2-*a*]pyridine-6,7,8-triol	β-D-Glucan glucohydrolase barley (GH 3)	1LQ2	264,265
Anilinomethyl-glucoimidazole (**129**) (5*R*,6*R*,7*S*,8*S*)-5-(Hydroxymethyl)-2-(phenylamino)methyl-5,6,7,8-tetrahydro-imidazo[1,2-*a*]pyridine-6,7,8-triol	β-Glucosidase *Thermotoga maritima* (GH 1)	2J7C	266

Compound	Enzyme	PDB	References
Glucotetrazole (**79**) (5*R*,6*R*,7*S*,8*S*)-5,6,7,8-Tetrahydro-5-(hydroxymethyl)pyrido[1,2-*d*]tetrazole-6,7,8-triol	Maize, β-Glucosidase (GH 1)	1V08	267
	Myrosinase (thioglucoside glucohydrolase EC 3.2.3.1)	1E6Q	268
4-*O*-β-D-Xylopyranosyl-1-deoxy-D-xylonojirimycin (**80**) 4-*O*-β-D-Xylopyranosyl-1,5-dideoxy-1,5-imino-D-xylitol	Xylanase Xyn10A *Streptomyces lividans* (GH 10)	1V0K	252
3-*O*-β-D-Xylopyranosyl-dehydroxymethyl-isofagomine (**81**) 4-*O*-β-D-Xylopyranosyl-1,4,5-dideoxy-1,5-imino-D-*threo*-pentitol	Xylanase Xyn10A *Streptomyces lividans* (GH 10)	1V0N	252

Continued

Compound	Enzyme	PDB	References
4-O-β-D-Xylopyranosyl-dehydroxymethyl-isofagomine lactam (**121**) 3-O-β-D-Xylopyranosyl-5-amino-4,5-dideoxy-D-*threo*-pentonolactam	Xylanase Xyn10A *Streptomyces lividans* (GH 10)	1OD8	253
Calystegine B$_2$ (**17**)	β-Glucosidase *Thermotoga maritima* (GH 1)	2CBV	279
(*1S,2R,3S,4R,5R*)-2,3,4-Trihydroxy-*N*-(*N*-octylthiocarbamoyl)-6-oxa-nor-tropane (**135**)	β-Glucosidase *Thermotoga maritima* (GH 1)	2VRJ	280
5-Amino-5-deoxy-D-galactonohydroximo-1,5-lactam (**125**)	β-Glucosidase *Sulfolobus solfataricus*	1UWT	259
N-Octyl-1-*epi*-valienamine (**65**)	α-Mannosidase II, *Drosophila* Golgi (GH 38)	3D50	290

Compound	Enzyme	PDB	References
Conduritol B epoxide (56)	Acid β-glucosidase, human (GH 30)	1Y7V	270
D-Mannose-related ligands (Section V5b)			
1-Deoxymannojirimycin (11) 1,5-Dideoxy-1,5-imino-D-mannitol	α-1,2-Mannosidase I, human ER (GH 47)	1F02	282
	α-1,2-Mannosidase I, yeast (GH 47)	1G6I	
	α-1,2-Mannosidase I, Penicillium citrinum (GH 47)	1KRE	283
	α-Mannosidase II, Drosophila Golgi (GH 38)	1HXK	284
	exo-α-1,6-mannosidase Streptococcus pneumoniae (GH 125)	3QRY	296
Kifunensine (16)	α-1,2-Mannosidase I, human ER	1F03	282
	α-Mannosidase II, Drosophila Golgi (GH 38)	1PS3	286
	exo-α-mannosidase Bt3990 Bacteroides thetaiotaomicron (GH 92)	2WVZ	295
Mannoimidazole (137) (5R,6R,7S,8R)-5,6,7,8-Tetrahydro-5-(hydroxymethyl)imidazo[1,2-a]pyridine-6,7,8-triol	α-Mannosidase II, Drosophila Golgi (GH 38)	3D4Y	290
	exo-α-Mannosidase Bt3990 Bacteroides thetaiotaomicron (GH 92)	2WZS	295
	β-Mannosidase Bacteroides thetaiotaomicron (GH 2)	2VMF	291

Continued

Compound	Enzyme	PDB	References
Swainsonine (**13**)	α-Mannosidase II, Drosophila Golgi (GH 38)	1HWW 3BLB	284 290
	α-Mannosidase Streptococcus pyogenes (GH 38)	2WYI	294
	exo-α-Mannosidase Bt3990 Bacteroides thetaiotaomicron (GH 92)	2WW0	295
	exo-α-Mannosidase Bt2199 Bacteroides thetaiotaomicron (GH 92)	2WW2	295
(5R)-5-C-(4-t-Butyl)benzoylmethyl swainsonine (**204**)	α-Mannosidase II, Drosophila Golgi (GH 38)	3EJP, 3EJQ, 3EJR, 3EJS, 3EJT, 3EJU	293
Mannostatin A (**136**)	α-Mannosidase II, Drosophila Golgi (GH 38)	2F7O	289

N-Acetyl-D-glucosamine-related ligands (Section V5d)

N-Acetyl-D-glucosaminono lactam (**146**) 2-*N*-Acetyl-5-amino-5-deoxy-D-glucosaminono lactam	*O*-GlcNAcase Bacteroides thetaiotaomicron (GH 84)	2XM1	308

Compound	Enzyme	PDB	References
2-Acetamido-4-epi-isofagomine (**139**) (2R,3R,4S,5R)-2-Acetamido-3,4-dihydroxy-5-hydroxymethylpiperidine	β-NAc-Hexosaminidase *Streptomyces plicatus* (GH 20)	1JAK	298
	β-NAc-Hexosaminidase, human	1NOW	300
NAG-Thiazoline (**140**) 2-Amino-2-deoxy-1-S,2-N-ethylylidene-1-thio-α-D-glucopyranose	β-NAc-Hexosaminidase, *Streptomyces plicatus* (GH 20)	1HP5	299
	β-NAc-Hexosaminidase, human	1NP0	300
PUGNAc-Imidazole hybrid (**141**) (5R,6R,7S,8S)-8-Acetamido-5,6,7,8-tetrahydro-5-hydroxymethyl-2-(phenylamino)carbonyl-imidazo[1,2-*a*]pyridine-6,7-diol	β-NAc-Hexosaminidase *Bacteroides thetaiotaomicron* (GH 84)	2J47	303

Continued

Compound	Enzyme	PDB	References
GlcNAcstatin (**142**) (5*R*,6*R*,7*S*,8*S*)-5,6,7,8-Tetrahydro-5-(hydroxymethyl)-8-(3-methyl)butanoylamino-2-(2-phenyl)ethyl-imidazo[1,2-*a*]pyridine-6,7-diol	*O*-GlcNAcase NagJ *Clostridium perfringens* (GH 84)	2J62	304
(5*R*,6*R*,7*S*,8*S*)-5,6,7,8-Tetrahydro-5-(hydroxymethyl)-2-(2-phenyl)ethyl-8-propanoylamino-imidazo[1,2-*a*]pyridine-6,7-diol (**144**)	*O*-GlcNAcase, human	2WB5	306,307

Compound	Enzyme	PDB	References
N-Thiopropionyl-GlcNAcstatin (145) (5R,6R,7S,8S)-5,6,7,8-Tetrahydro-5-(hydroxymethyl)-2-(2-phenyl)ethyl-8-(3-thio)propanoylamino-imidazo[1,2-a]pyridine-6,7-diol	O-GlcNAcase, human	2XPK	307
Acetamido-epi-valienamine (143) N-[(1S,2R,5R,6R)-2-Amino-5,6-dihydroxy-4-(hydroxymethyl)-cyclohex-3-en-1-yl]acetamide	O-GlcNAcase	2JIW	305
6-Amino-6-deoxycastanospermine (147) (1S,6S,7R,8R,8aR)-6-Amino-1,7,8-trihydroxyindolizidine	exo-β-D-Glucosaminidase Amycolatopsis orientalis (GH 2)	2X05	309

Continued

Compound	Enzyme	PDB	References
6-Acetamido-6-deoxycastanospermine (**149**) (*1S,6S,7R,8R,8aR*)-6-Acetamido-1,7,8-trihydroxyindolizidine	*O*-GlcNAcase *Bacteroides thetaiotaomicron* (GH 84)	2XJ7	310
Aminodeoxyaustraline (**148**) (*1R,2R,3R,7S,7aR*)-3-Aminomethyl-1,2,7-trihydroxy-pyrrolizidine	exo-β-D-Glucosaminidase *Amycolatopsis orientalis* (GH 2)	2X09	309
3-Acetamido-1,3,6-trideoxy-1,6-imino-D-*glycero*-D-*ido*-heptitol (**150**)	*O*-GlcNAcase *Bacteroides thetaiotaomicron* (GH 84)	2W66	311a
2-Acetamido-1,2,6-trideoxy-1,6-imino-D-glucitol (**151**)	*O*-GlcNAcase *Bacteroides thetaiotaomicron* (GH 84)	2W67	311a

Compound	Enzyme	PDB	References
L-*Fucose-related ligands (Section V5e)*			
1-Deoxy-L-fuconojirimycin (**20**) 1,5,6-Trideoxy-1,5-imino-L-galactitol	α-L-Fucosidase *Bacteroides thetaiotaomicron* (GH 29)	2XIB	313
	1,2-α-L-Fucosidase *Bifidobacterium bifidum* (GH 95)	2EAC	315
(*1S*)-1-Hydroxymethyl-1-deoxy-L-fuconojirimycin	α-L-Fucosidase *Bacteroides thetaiotaomicron* (GH 29)	2WVT	314
1-Acylaminomethyl-1-deoxy-L-fuco-nojirimycin (**152**) and analogues *N*-Acyl-1-amino-2,6,7-trideoxy-2,6-imino-L-*glycero*-D-*manno*-heptitols	α-L-Fucosidase *Thermotoga maritima* (GH 29)	2ZX5, 2ZX6, 2ZX7, 2ZX8, 2ZX9	312
1-Acylaminomethyl-1-deoxy-L-fuco-nojirimycin (**153**) *N*-Acyl-1-amino-2,6,7-trideoxy-2,6-imino-L-*glycero*-D-*manno*-heptitol	α-L-Fucosidase *Bacteroides thetaiotaomicron* (GH 29)	2XII	313

Continued

Compound	Enzyme	PDB	References
Pyrrolidines and pyrrolizidines (Section V5f)			
1,4-Dideoxy-1,4-imino-D-arabinitol (**14**)	Glycogen phosphorylase GPb rabbit muscle (GT 35)	2G9Q	316
4,5-Diamino-4,5-deoxy-5-*N*-[(*1R*)-2-hydroxy-1-phenylethyl]-D-ribono-1,4-lactam (**157**)	α-Mannosidase II, *Drosophila* Golgi (GH 38)	3DDF	293
4,5-Diamino-4,5-dideoxy-5-*N*-[(*1R*)-2-hydroxy-1-phenylethyl]-4-*N*-methyl-D-ribono-1,4-lactam (**158**)	α-Mannosidase II, *Drosophila* Golgi (GH 38)	3DDG	293
2,5-Dideoxy-2,5-imino-D-mannitol (**7**)	Fructan 1-exohydrolase Chicory (GH 32)	2AEY	318
1-*N*-(4-Dimethylamino)benzoyl-1-amino-1,2,5-trideoxy-2,5-imino-D-mannitol (**156**)	Endoglycoceramidase (GH 5)	2OYM	258

Compound	Enzyme	PDB	References
2,5-Dideoxy-2,5-imino-D-glucitol (**113**)	Xylose isomerase (EC 5.3.1.5)	1DID	245
Casuarine (**159**) (*1R,2R,3R,6S,7S,7aR*)-3-Hydroxymethyl-1,2,6,7-tetrahydroxy-pyrrolizidine	Maltase-glucoamylase, human (GH 31)	3CTT	319
6-*O*-α-D-Glucopyranosyl casuarine (**160**) (*1R,2R,3R,6S,7S,7aR*)-6-*O*-α-D-Glucopyranosyl-3-(hydroxymethyl)-1,2,6,7-tetrahydroxy-pyrrolizidine	Periplasmatic trehalase *E. coli* (GH 37)	2JJB	319
7-Deoxy-6-*O*-α-D-glucopyranosyl-7-hydroxymethyl casuarine (**161**) (*1R,2R,3R,6S,7R,7aR*)-3,7-Di-(hydroxymethyl)-1,2,6-trihydroxy-pyrrolizidine	Periplasmatic trehalase *E. coli* (GH 37)	2WYN	320

Continued

Compound	Enzyme	PDB	References
Ghavamiol (**155**)	α-Mannosidase II, *Drosophila* Golgi (GH 38)	1TQU	317

Acknowledgment

We thank Martin Thonhofer for his expert assistance during the generation of this article.

References

1. R. S. Cahn, C. Ingold, and V. Prelog, Specificity of molecular chirality, *Angew. Chem. Int. Ed.*, 5 (1966) 385–415.
2. E. J. Corey, Total synthesis of prostaglandins, *Ann. N. Y. Acad. Sci.*, 180 (1971) 24–37.
3. E. J. Corey and W. T. Wipke, Computer-assisted design of complex organic syntheses, *Science*, 166 (1969) 178–192.
4. E. J. Corey and D. Seebach, Synthesis of 1,n-dicarbonyl derivatives using carbanions from 1,3-dithianes, *Angew. Chem. Int. Ed. Engl.*, 4 (1965) 1075–1077.
5. R. Hoffmann and R. B. Woodward, Selection rules for concerted cycloaddition reactions, *J. Am. Chem. Soc.*, 87 (1965) 2046–2048.
6. Y. Kishi, F. Nakatsubo, M. Aratani, T. Goto, and S. Inouye, Synthetic approach towards tetrodotoxin. II. A stereospecific synthesis of a compound having the same six chiral centers on the cyclohexane ring as those of tetrodotoxin, *Tetrahedron Lett.*, 11 (1970) 5129–5132.
7. I. Dyong and N. Jersch, Synthesis of important carbohydrates. 18. The *vic-cis*-oxyamination as the key reaction for amino sugar syntheses: Methyl *N*-acetyl-4-epi-α-garosaminide, *Chem. Ber.*, 112 (1979) 1849–1858.
8. I. Dyong, J. Weigand, and J. Thiem, Syntheses of biologically important carbohydrates. 33. Syntheses of unsaturated amino sugars and aminoalkyl-branched carbohydrates via sigmatropic rearrangement of trichloroacetimidates, *Liebigs Ann. Chem.* (1986) 577–599.
9. D. Horton and W. Weckerle, Synthesis of 3-amino-2,3,6-trideoxy-D-*ribo*-hexose hydrochloride, *Carbohydr. Res.*, 46 (1976) 227–235.
10. T.-M. Cheung, D. Horton, and W. Weckerle, Preparative synthesis of 3-amino-2,3,6-trideoxy-L-*lyxo*-hexopyranose derivatives, *Carbohydr. Res.*, 74 (1979) 93–103.
11. K. Bock, I. Lundt, and C. Pedersen, Reaction of aldonic acids with hydrogen bromide. I. Preparation of some bromodeoxyaldonic acids, *Carbohydr. Res.*, 68 (1979) 313–319.

12. K. Bock, I. Lundt, and C. Pedersen, Amino acids and amino sugars from bromodeoxyaldonolactones, *Acta Chem. Scand. B*, 41 (1987) 435–441.
13. K. Heyns, J. Feldmann, D. Hadamczyk, J. Schwentner, and J. Thiem, Ein Verfahren zur Synthese von α-Glycosiden der 3-Amino-2,3,6-tridesoxyhexopyranosen aus Glycalen, *Chem. Ber.*, 114 (1981) 232–239.
14. J. Thiem and D. Springer, Synthesis and hydrogenation studies of 3-azidohex-2-enopyranosides, precursors of the sugar constituents of anthracycline glycosides, *Carbohydr. Res.*, 136 (1985) 325–334.
15. T. Nishikawa and N. Ishida, A new antibiotic R-468 active against drug-resistant *Shigella*, *J. Antibiot. A*, 18 (1965) 132–133.
16. S. Inouye, T. Tsuruoka, and T. Niida, Structure of nojirimycin, sugar antibiotic with nitrogen in the ring, *J. Antibiot. A*, 19 (1966) 288–292.
17. N. Ishida, K. Kumagai, T. Niida, T. Tsuruoka, and H. Yumoto, Nojirimycin, a new antibiotic. II. Isolation, characterisation, and biological activity, *J. Antibiot. A*, 20 (1967) 66–71.
18. S. Inouye, T. Tsuruoka, T. Ito, and T. Niida, Structure and synthesis of nojirimycin, *Tetrahedron*, 24 (1968) 2125–2144.
19. M. Koyama and S. Sakamura, The structure of a new piperidine derivative from buckwheat seeds (*Fagopyrum esculentum* Moench), *Agric. Biol. Chem.*, 38 (1974) 1111–1112.
20. S. Hanessian and T. H. Haskell, Synthesis of 5-acetamido-5-deoxypentoses. Sugar derivatives containing nitrogen in the ring, *J. Org. Chem.*, 28 (1963) 2604–2610.
21. H. Paulsen, Darstellung von 5-Acetamido-5-deoxy-D-xylopiperidinose, *Angew. Chem.*, 74 (1962) 901.
22. T. J. Adley and L. N. Owen, Thio sugars with sulphur in the ring, *Proc. Chem. Soc.* (1961) 418.
23. J. C. P. Schwarz and K. C. Yule, D-Xylothiapyranose: A sugar with sulfur in the ring, *Proc. Chem. Soc.* (1961) 417.
24. M. S. Feather and R. L. Whistler, Derivatives of 5-thio-D-glucose, *Tetrahedron Lett.*, 3 (1962) 667–668.
25. U. G. Nayak and R. L. Whistler, Synthesis of 5-thio-D-glucose, *J. Org. Chem.*, 34 (1969) 97–100.
26. M. Yagi, T. Kouno, Y. Aoyagi, and H. Murai, The structure of moranoline, a piperidine alkaloid from *Morus* species, *Nippon Nogei Kagaku Kaishi*, 50 (1976) 571–572.
27. A. Welter, J. Jadot, G. Dardenne, M. Marlier, and J. Casimir, 2,5-Dihydroxymethyl 3,4-dihydroxypyrrolidine dans les feuilles de *Derris elliptica*, *Phytochemistry*, 15 (1976) 747–749.
28. L. D. Hohenschutz, E. A. Bell, P. J. Jewess, D. P. Leworthy, R. J. Pryce, E. Arnold, and J. Clardy, Castanospermine, a 1,6,7,8-tetrahydroxyoctahydroindolizine alkaloid, from seeds of *Castanospermum australe*, *Phytochemistry*, 20 (1981) 811–814.
29. R. J. Molyneux, M. Benson, R. Y. Wong, J. E. Tropea, and A. D. Elbein, Australine, a novel pyrrolizidine alkaloid glucosidase inhibitor from *Castanospermum australe*, *J. Nat. Prod.*, 51 (1988) 1198–1206.
30. R. J. Nash, L. E. Fellows, J. V. Dring, G. W. J. Fleet, A. E. Derome, T. A. Hamor, A. M. Scofield, and D. J. Watkin, Isolation from *Alexa leiopetala* and x-ray crystal structure of alexine, ($1R,2R,3R,7S,8S$)-3-hydroxymethyl-1,2,7-trihydroxypyrrolizidine [($2R,3R,4R,5S,6S$)-2-hydroxymethyl-1-azabicyclo [3.3.0]octan-3,4,6-triol], a unique pyrrolizidine alkaloid, *Tetrahedron Lett.*, 29 (1988) 2487–2490.
31. G. E. McCasland, S. Furuta, and L. J. Durham, Alicyclic carbohydrates. XXIX. The synthesis of a pseudo-hexose (2,3,4,5-tetrahydroxycyclohexanemethanol), *J. Org. Chem.*, 31 (1966) 1516–1521.
32. T. Suami and S. Ogawa, Chemistry of carba-sugars (pseudo-sugars) and their derivatives, *Adv. Carbohydr. Chem. Biochem.*, 48 (1990) 21–90.
33. L. E. Fellows, E. A. Bell, D. G. Lynn, F. Pilkiwicz, I. Miura, and K. Nakanishi, Isolation and structure of an unusual cyclic amino alditol from a legume, *J. Chem. Soc., Chem. Commun.* (1979) 977–978.
34. H. Paulsen, Y. Hayauchi, and V. Sinnwell, Monosaccharides containing nitrogen in the ring. XXXVII. Synthesis of 1,5-dideoxy-1,5-imino-D-galactitol, *Chem. Ber.*, 113 (1980) 2601–2608.

35. S. M. Colegate, P. R. Dorling, and C. R. Huxtable, A spectroscopic investigation of swainsonine: An α-mannosidase inhibitor isolated from *Swainsona canescens*, *Aust. J. Chem.*, 32 (1979) 2257–2264.
36. G. Kinast and M. Schedel, Vierstufige 1-Desoxynojirimycin-Synthese mit einer Biotransformation als zentralem Reaktionsschritt, *Angew. Chem.*, 93 (1981) 799–800.
37. K. Leontein, B. Lindberg, and J. Lönngren, Formation of 1,5-dideoxy-1,5-iminohexitols on borohydride reduction of 2-amino-2-deoxyhexofuranurono-6,3-lactones, *Acta Chem. Scand. B*, 36 (1982) 515–518.
38. U. Fuhrmann, E. Bause, G. Legler, and H. Ploegh, Novel mannosidase inhibitor blocking conversion of high mannose to complex oligosaccharides, *Nature*, 307 (1984) 755–758.
39. R. J. Nash, E. A. Bell, and J. M. Williams, 2-Hydroxymethyl-3,4-dihydroxypyrrolidine in fruits of Angylocalyx boutiqueanus, *Phytochemistry*, 24 (1985) 1620–1622.
40. J. Furukawa, S. Okuda, K. Saito, and S. I. Hatanaka, 3,4-Dihydroxy-2-hydroxymethylpyrrolidine from *Arachniodes standishii*, *Phytochemistry*, 24 (1985) 593–594.
41. H. Umezawa, T. Aoyagi, T. Komiyama, H. Morishima, M. Hamada, and T. Takeuchi, Purification and characterization of a sialidase inhibitor, siastatin, produced by *Streptomyces*, *J. Antibiot.*, 27 (1974) 963–969.
42. H. Kayakiri, S. Takase, T. Shibata, M. Okamoto, H. Terano, and M. Hashimoto, Structure of kifunensine, a new immunomodulator isolated from Actinomycete, *J. Org. Chem.*, 54 (1989) 4015–4016.
43. M. Iwami, O. Nakayama, H. Terano, M. Kohsaka, H. Aoki, and H. Imanaka, A new immunomodulator, FR-900494: Taxonomy, fermentation, isolation, and physico-chemical and biological characteristics, *J. Antibiot.*, 40 (1987) 612–622.
44. A. D. Elbein, J. E. Tropea, M. Mitchell, and G. P. Kaushal, Kifunensine, a potent inhibitor of the glycoprotein processing mannosidase I, *J. Biol. Chem.*, 265 (1990) 15599–15605.
45. C. M. Harris, T. M. Harris, R. J. Molyneux, J. E. Tropea, and A. D. Elbein, 1-Epiaustraline, a new pyrrolizidine alkaloid from *Castanospermum australe*, *Tetrahedron Lett.*, 30 (1989) 5685–5688.
46. P.-H. Ducrot and J. L. Lallemand, Structure of the calystegines: New alkaloids of the nortropane family, *Tetrahedron Lett.*, 31 (1990) 3879–3882.
47. M. Shibano, S. Kitagawa, and G. Kusano, Studies on the constituents of *Broussonetia* species. I. Two new pyrrolidine alkaloids, broussonetines C and D, as β-galactosidase and β-mannosidase inhibitors from *Broussonetia kazinoki* SIEB., *Chem. Pharm. Bull.*, 45 (1997) 505–508.
48. M. Shibano, S. Kitagawa, S. Nakamura, N. Akazawa, and G. Kusano, Studies on the constituents of *Broussonetia* species. II. Six new pyrrolidine alkaloids, broussonetine A, B, E, F and broussonetine A and B, as inhibitors of glycosidases, from *Broussonetia kazinoki* SIEB., *Chem. Pharm. Bull.*, 45 (1997) 700–705.
49. G. W. J. Fleet, P. W. Smith, S. V. Evans, and L. E. Fellows, Design synthesis and preliminary evaluation of a potent α-mannosidase inhibitor: 1,4-Dideoxy-1,4-imino-D-mannitol, *J. Chem. Soc., Chem. Commun.* (1984) 1240–1241.
50. G. W. J. Fleet, A. N. Shaw, S. V. Evans, and L. E. Fellows, Synthesis from D-glucose of 1,5-dideoxy-1,5-imino-L-fucitol, a potent α-L-fucosidase inhibitor, *J. Chem. Soc., Chem. Commun.* (1985) 841–842.
51. G. W. J. Fleet, S. J. Nicholas, P. W. Smith, S. V. Evans, L. E. Fellows, and R. J. Nash, Potent competitive inhibition of α-galactosidase and α-glucosidase activity by 1,4-dideoxy-1,4-iminopentitols: Syntheses of 1,4-dideoxy-1,4-imino-D-lyxitol and of both enantiomers of 1,4-dideoxy-1,4-iminoarabinitol, *Tetrahedron Lett.*, 26 (1985) 3127–3130.
52. G. W. J. Fleet, P. W. Smith, R. J. Nash, L. E. Fellows, R. B. Parekh, and T. W. Rademacher, Synthesis of 2-acetamido-1,5-imino-1,2,5-trideoxy-D-mannitol and of 2-acetamido-1,5-imino-1,2,5-trideoxy-D-glucitol, a potent and specific inhibitor of a number of β-N-acetylglucosaminidases, *Chem. Lett.* (1986) 1051–1054.

53. G. Legler and E. Jülich, Synthesis of 5-amino-5-deoxy-D-mannopyranose and 1,5-dideoxy-1,5-imino-D-mannitol, and inhibition of α- and β-D-mannosidases, *Carbohydr. Res.*, 128 (1984) 61–72.
54. J. Schweden, G. Legler, and E. Bause, Purification and characterization of a neutral processing mannosidase from calf liver acting on (Man)$_9$(GlcNAc)$_2$ oligosaccharides, *Eur. J. Biochem.*, 157 (1986) 563–570.
55. K. M. Osiecki-Newman, D. Fabbro, T. Dinur, S. Boas, S. Gatt, G. Legler, R. J. Desnick, and G. A. Grabowski, Human acid β-glucosidase: Affinity purification of the normal placental and Gaucher disease splenic enzymes on *N*-alkyl-deoxynojirimycin-sepharose, *Enzyme*, 35 (1986) 147–153.
56. A. de Raadt, C. Ekhart, G. Legler, and A. E. Stütz, Iminoalditols as affinity ligands for the purification of glycosidases, in A. E. Stütz, (Ed.), *Iminosugars as Glycosidase Inhibitors,* Wiley-VCH, Weinheim, New York, 1999, pp. 207–215.
57. Y. Kishi, Natural products synthesis: Palytoxin, *Pure Appl. Chem.*, 61 (1989) 313–324.
58. R. H. Grubbs, S. J. Miller, and G. C. Fu, Ring closing metathesis and related processes in organic synthesis, *Acc. Chem. Res.*, 28 (1995) 446–452.
59. H. W. Kroto, I. R. Heath, S. C. O'Brien, R. F. Curl, and R. E. Smalley, C$_{60}$: Buckminsterfullerene, *Nature*, 318 (1985) 162–163.
60. B. L. Stocker, E. M. Dangerfield, A. L. Win-Mason, G. W. Haslett, and M. S. M. Timmer, Recent developments in the synthesis of pyrrolidine-containing iminosugars, *Eur. J. Org. Chem.* (2010) 1615–1637.
61. B. G. Davis, A silver-lined anniversary of Fleet iminosugars: 1984–2009, from DIM to DRAM to LABNAc, *Tetrahedron: Asymmetry*, 20 (2009) 652–671.
62. B. G. Winchester, Iminosugars: From botanical curiosities to licensed drugs, *Tetrahedron: Asymmetry*, 20 (2009) 645–651.
63. N. Asano, Naturally occurring iminosugars and related alkaloids: Structure, activity and applications, in P. Compain and O. R. Martin, (Eds.), *Iminosugars: From Synthesis to Therapeutic Applications,* Wiley, Chichester, 2007, pp. 7–24.
64. B. La Ferla, L. Cipolla, and F. Nicotra, General strategies for the synthesis of iminosugars and new approaches towards iminosugar libraries, in P. Compain and O. R. Martin, (Eds.), *Iminosugars: From Synthesis to Therapeutic Applications,* Wiley, Chichester, 2007, pp. 25–61.
65. M. S. J. Simmonds, G. C. Kite, and E. A. Porter, Taxonomic distribution of iminosugars in plants and their biological activities, in A. E. Stütz, (Ed.), *Iminosugars as Glycosidase Inhibitors,* Wiley-VCH, Weinheim, New York, 1999, pp. 8–30.
66. B. La Ferla and F. Nicotra, Synthetic methods for the preparation of iminosugars, in A. E. Stütz, (Ed.), *Iminosugars as Glycosidase Inhibitors,* Wiley-VCH, Weinheim, New York, 1999, pp. 68–92.
67. I. Lundt and R. Madsen, Iminosugars as powerful glycosidase inhibitors—Synthetic approaches from aldonolactones, in A. E. Stütz, (Ed.), *Iminosugars as Glycosidase Inhibitors,* Wiley-VCH, Weinheim, New York, 1999, pp. 93–111.
68. I. Lundt and R. Madsen, Isoiminosugars: Glycosidase inhibitors with nitrogen at the anomeric position, in A. E. Stütz, (Ed.), *Iminosugars as Glycosidase Inhibitors,* Wiley-VCH, Weinheim, New York, 1999, pp. 112–124.
69. B. W. Matthews, The structure of *E. coli* β-galactosidase, *C. R. Biol.*, 328 (2005) 549–556.
70. T. Narikawa, H. Shinoyama, and T. Fujii, A β-rutinosidase from *Penicillium rugulosum* IFO 7242 that is a peculiar flavonoid glycosidase, *Biosci. Biotechnol. Biochem.*, 64 (2009) 1317–1319.
71. V. Patel and A. L. Tappel, Identity of β-glucosidase and β-xylosidase activities in rat liver lysosomes, *Biochim. Biophys. Acta Enzymol.*, 191 (1969) 653–662.
72. R. Kuhn, Über die spezifische Natur und den Wirkmechanismus kohlehydrat- und glykosidspaltender Enzyme (On the specific nature and the mechanism of action of carbohydrate and glycoside cleaving enzymes), *Naturwissenschaften*, 11 (1923) 732–742.

73. S. Veibel, Enzymatic and acid hydrolysis of glucosides, *Kem. Maanedsblad*, 20 (1939) 253–258.
74. W. W. Pigman, Classification of carbohydrases, *J. Res. Natl. Bur. Stand.*, 30 (1943) 257–265.
75. W. W. Pigman, Specificity, classification and mechanism of action of the glycosidases, *Adv. Enzymol. Relat. Subj. Biochem.*, 4 (1944) 41–74.
76. F. Shafizadeh and A. Thompson, An evaluation of the factors influencing the hydrolysis of the aldosides, *J. Org. Chem.*, 21 (1956) 1059–1062.
77. D. E. Koshland, Stereochemistry and the mechanism of enzymatic reactions, *Biol. Rev.*, 28 (1953) 416–426.
78. D. E. Koshland and S. S. Stein, Correlation of bond breaking with enzyme specificity. Cleavage point of invertase, *J. Biol. Chem.*, 208 (1954) 139–148.
79. D. E. Koshland, Physical organic approach to enzymatic mechanisms, *Trans. N. Y. Acad. Sci.*, 16 (1954) 110–113.
80. C. C. F. Blake, D. F. König, G. A. Mair, A. C. T. North, D. C. Phillips, and V. R. Sarma, Structure of hen egg-white lysozyme—A three dimensional Fourier synthesis at 2 A resolution, *Nature*, 206 (1965) 757–761.
81. C. A. Vernon and B. Banks, The enzymatic hydrolysis of glycosides, *Biochem. J.*, 86 (1963) Proc. Biochem. Soc. 7P.
82. C. A. Vernon, Mechanisms of hydrolysis of glycosides and their relevance to enzyme-catalyzed reactions, *Proc. R. Soc. Lond. B Biol. Sci.*, 167 (1967) 389–401.
83. G. Legler, Glycoside hydrolases: Mechanistic information from studies with reversible and irreversible inhibitors, *Adv. Carbohydr. Chem. Biochem.*, 48 (1990) 319–384.
84. J. D. McCarter and S. G. Withers, Mechanisms of enzymatic glycoside hydrolysis, *Curr. Opin. Struct. Biol.*, 4 (1994) 885–892.
85. V. L. Y. Yip and S. G. Withers, Nature's many mechanisms for the degradation of oligosaccharides, *Org. Biomol. Chem.*, 2 (2004) 2707–2713.
86. C. S. Rye and S. G. Withers, Glycosidase mechanisms, *Curr. Opin. Chem. Biol.*, 4 (2000) 573–580.
87. A. White and D. R. Rose, Mechanism of catalysis by retaining β-glycosyl hydrolases, *Curr. Opin. Struct. Biol.*, 7 (1997) 645–651.
88. S. G. Withers, Enzymatic cleavage of glycosides: How does it happen, *Pure Appl. Chem.*, 67 (1995) 1673–1682.
89. A. L. Fink and N. E. Good, Evidence for a glucosyl-enzyme intermediate in the β-glucosidase-catalyzed hydrolysis of p-nitrophenyl-β-D-glucoside, *Biochem. Biophys. Res. Commun.*, 58 (1974) 126–131.
90. D. L. Zechel and S. G. Withers, Glycosidase mechanisms: Anatomy of a finely tuned catalyst, *Acc. Chem. Res.*, 33 (2000) 11–18.
91. E. J. Hehre, A fresh understanding of the stereochemical behaviour of glycosylases: Structural distinction of "inverting" (2-MCO-type) versus "retaining" (1-MCO-type) enzymes, *Adv. Carbohydr. Chem. Biochem.*, 55 (2000) 265–310.
92. C. B. Post and M. Karplus, Does lysozyme follow the lysozyme pathway? An alternative based on dynamics, structural, and stereoelectronic considerations, *J. Am. Chem. Soc.*, 108 (1986) 1317–1319.
93. G. W. J. Fleet, An alternative proposal for the mode of inhibition of glycosidase activity by polyhydroxylated piperidines, pyrrolidines and indolizidines: Implications for the mechanism of action of some glycosidases, *Tetrahedron Lett.*, 26 (1985) 5073–5076.
94. R. W. Franck, The mechanism of β-glycosidases: A reassessment of some seminal papers, *Bioorg. Chem.*, 20 (1992) 77–88.
95. S. Knapp, D. Vocadlo, Z. Gao, B. Kirk, J. Lou, and S. G. Withers, NAG-thiazoline, an N-acetyl-β-hexosaminidase inhibitor that implicates acetamido participation, *J. Am. Chem. Soc.*, 118 (1996) 6804–6805.

96. D. Piszkiewicz and T. C. Bruise, Glycoside hydrolysis. I. Acetamido and hydroxyl group catalysis in glycoside hydrolysis, *J. Am. Chem. Soc.*, 89 (1967) 6237–6243.
97. G. Lowe, G. Sheppard, M. L. Sinnott, and A. Williams, Lysozyme-catalysed hydrolysis of some β-aryl di-*N*-acetylchitobiosides, *Biochem. J.*, 104 (1967) 893–899.
98. http://www.cazy.org/.
99. M. A. Jermyn, The action of the β-glucosidase of *Stachybotrys atra* p-nitrophenyl quinovoside, *Aust. J. Biol. Sci.*, 16 (1963) 926.
100. G. Legler, Action mechanism of glucosidase-splitting enzymes. II. Isolation and enzymatic properties of two β-glucosidases from *Aspergillus wentii*, *Hoppe Seyler's Zeitschr. Physiol. Chem.*, 348 (1967) 1359–1366.
101. G. Legler and L. M. O. Osama, Mechanism of action of glycosidase splitting enzymes. IV. Purification and properties of a β-glucosidase from *Aspergillus oryzae*, *Hoppe Seyler's Zeitschr. Physiol. Chem.*, 349 (1968) 1488–1492.
102. G. Legler, The mechanism of action of glycosidases, *Acta Microbiol. Acad. Sci. Hung.*, 22 (1975) 403–409.
103. D. F. Walker and B. Axelrod, Evidence for a single catalytic site on the "β-D-glucosidase–β-D-galactosidase" of almond emulsin, *Arch. Biochem. Biophys.*, 187 (1978) 102–107.
104. K.-R. Roeser and G. Legler, Role of sugar hydroxyl groups in glycoside hydrolysis. Cleavage mechanism of deoxyglucosides and related substrates by β-glucosidase A_3 from *Aspergillus wentii*, *Biochim. Biophys. Acta Enzymol.*, 657 (1981) 321–333.
105. M. N. Namchuk and S. G. Withers, Mechanism of *Agrobacterium* β-glucosidase: Kinetic studies of the role of noncovalent enzyme/substrate interactions, *Biochemistry*, 34 (1995) 16194–16202.
106. T. Mega and Y. Matsushima, Energy of binding of *Aspergillus oryzae* β-glucosidase with the substrate, and the mechanism of its enzymatic action, *J. Biochem.*, 94 (1983) 1637–1647.
107. B. W. Sigurdskjold, B. Duus, and K. Bock, Hydrolysis of substrate analogues catalysed by β-D-glucosidase from *Aspergillus niger*. Part II: Deoxy and deoxyhalo derivatives of cellobiose, *Acta Chem. Scand.*, 45 (1991) 1032–1041.
108. M. N. Namchuk, J. D. McCarter, A. Becalski, T. Andrews, and S. G. Withers, The role of sugar substituents in glycoside hydrolysis, *J. Am. Chem. Soc.*, 122 (2000) 1270–1277.
109. T. Nishimura and M. Ishihara, Action of fungal β-glucosidase on the mono-*O*-methylated *p*-nitrophenyl β-D-glucopyranoside, *Holzforschung*, 63 (2009) 47–51.
110. K. Bock and K. Adelhorst, Derivatives of methyl β-lactoside as substrates for and inhibitors of β-D-galactosidase from *E. coli*, *Carbohydr. Res.*, 202 (1990) 131–149.
111. J. D. McCarter, M. J. Adam, and S. G. Withers, Binding energy and catalysis. Fluorinated and deoxygenated glycosides as mechanistic probes of *Escherichia coli (lac Z)* β-galactosidase, *Biochem. J.*, 286 (1992) 721–727.
112. A. Rivera-Sagredo, F. J. Canada, O. Nieto, J. Jimenez-Barbero, and M. Martin-Lomas, Substrate specificity of small-intestinal lactase: Assessment of the role of the substrate hydroxyl groups, *Eur. J. Biochem.*, 209 (1992) 415–422.
113. K. Bock and H. Pedersen, The substrate specificity of the enzyme amyloglucosidase (AMG). Part I. Deoxy derivatives, *Acta Chem. Scand. B*, 41 (1987) 617–628.
114. K. Bock and H. Pedersen, The substrate specificity of the enzyme amyloglucosidase (AMG). Part II. 6-Substituted maltose derivatives, *Acta Chem. Scand. B*, 42 (1988) 75–85.
115. M. R. Sierks, K. Bock, S. Refn, and B. Svensson, Active site similarities of glucose dehydrogenase, glucose oxidase, and glucoamylase probed by deoxygenated substrates, *Biochemistry*, 31 (1992) 8972–8977.
116. M. R. Sierks and B. Svensson, Energetic and mechanistic studies of glucoamylase using molecular recognition of maltose OH groups coupled with site-directed mutagenesis, *Biochemistry*, 39 (2000) 8585–8592.

117. R. U. Lemieux, U. Spohr, M. Bach, D. R. Cameron, T. P. Frandsen, B. B. Stoffer, B. Svensson, and M. M. Palcic, Chemical mapping of the active site of the glucoamylase of *Aspergillus niger*, *Can. J. Chem.*, 74 (1996) 319–335.
118. T. P. Frandsen, B. B. Stoffer, M. M. Palcic, S. Hof, and B. Svensson, Structure and energetics of the glucoamylase-isomaltose transition state complex probed by using modeling and deoxygenated substrates coupled with site-directed mutagenesis, *J. Mol. Biol.*, 263 (1996) 79–89.
119. T. B. Frandsen, F. Lok, E. Mirgorodskaya, P. Roepstorff, and B. Svensson, Purification, enzymatic characterisation and nucleotide sequence of a high-isoelectric-point α-glucosidase from barley malt, *Plant Physiol.*, 123 (2000) 275–286.
120. M. Ogawa, T. Nishio, W. Hakamata, Y. Matsuishi, S. Hoshino, A. Kondo, M. Kitagawa, R. Kawachi, and T. Oku, Substrate hydroxyl groups are involved in the ionization of catalytic carboxyl groups of *Aspergillus niger* α-glucosidase, *J. Appl. Glycosci.*, 51 (2004) 9–14.
121. W. Hakamata, T. Nishio, and T. Oku, Synthesis of *p*-nitrophenyl 3- and 6-deoxy-α-D-glucopyranosides and their specificity to rice α-glucosidase, *J. Appl. Glycosci.*, 46 (1999) 459–463.
122. T. Nishio, W. Hakamata, A. Kimura, S. Chiba, A. Takatsuki, R. Kawachi, and T. Oku, Glycon specificity profiling of α-glucosidases using monodeoxy and mono-*O*-methyl derivatives of *p*-nitrophenyl α-D-glucopyranoside, *Carbohydr. Res.*, 337 (2002) 629–634.
123. W. Hakamata, M. Muroi, T. Nishio, T. Oku, and A. Takatsuki, Recognition properties of processing α-glucosidase I and α-glucosidase II, *J. Carbohydr. Chem.*, 23 (2004) 27–39.
124. W. Hakamata, T. Nishio, and T. Oku, Hydrolytic activity of α-galactosidases against deoxy derivatives of *p*-nitrophenyl α-D-galactopyranoside, *Carbohydr. Res.*, 324 (2000) 107–115.
125. K. J. Dean and C. C. Sweeley, Studies on human liver α-galactosidases. III. Partial characterization of carbohydrate-binding specificities, *J. Biol. Chem.*, 254 (1979) 10006–10010.
126. T. Nishio, Y. Miyake, H. Tsujii, W. Hakamata, K. Kadokura, and T. Oku, Hydrolytic activity of α-mannosidase against deoxy derivatives of *p*-nitrophenyl α-D-mannopyranoside, *Biosci. Biotechnol. Biochem.*, 60 (1996) 2038–2042.
127. T. Chrzaszcz and J. Janicki, "Sistoamylase", a natural inhibitor of amylase, *Biochem. Zeitsch*, 260 (1933) 354–368.
128. E. Truscheit, W. Frommer, B. Junge, L. Müller, D. D. Schmidt, and W. Wingender, Chemistry and biochemistry of microbial α-glucosidase inhibitors, *Angew. Chem. Int. Ed. Engl.*, 20 (1981) 744–761.
129. J. J. Marshall and C. M. Lauda, Purification and properties of phaseolamin, an inhibitor of α-amylase, from the kidney bean, *Phaseolus vulgaris*, *J. Biol. Chem.*, 250 (1975) 8030–8037.
130. S. Ueda, Y. Koba, and H. Chaen, Action of amylase inhibitors produced by *Streptomyces* sp. on some carbohydrate hydrolases and phosphorylases, *Carbohydr. Res.*, 61 (1978) 253–264.
131. S. Murao, A. Goto, Y. Matsui, and K. Ohyama, New proteinous inhibitor (Haim) of animal α-amylase from *Streptomyces griseosporeus* YM-25, *Agric. Biol. Chem.*, 44 (1980) 1679–1681.
132. K. Yokose, K. Ogawa, Y. Suzuki, I. Umeda, and Y. Suhara, New α-amylase inhibitor, trestatins. II. Structure of trestatins A, B and C, *J. Antibiot.*, 36 (1983) 1166–1175.
133. G. Limberg, I. Lundt, and J. Zavilla, Deoxyiminoalditols from aldonic acids. VI. Preparation of the four stereoisomeric 4-amino-3-hydroxypyrrolidines from bromodeoxytetronic acids. Discovery of a new α-mannosidase inhibitor, *Synthesis* (1999) 178–183.
134. G.-F. Dai, H.-W. Xu, J.-F. Wang, F.-W. Liu, and H.-M. Liu, Studies on the novel α-glucosidase inhibitory activity and structure-activity relationships for andrographolide analogues, *Bioorg. Med. Chem. Lett.*, 16 (2006) 2710–2713.
135. Atta-ur-Rahman, S. Zareen, M. I. Choudary, M. N. Akhtar, and S. N. Khan, α-Glucosidase inhibitory activity of triterpenoids from *Cichorium intybus*, *J. Nat. Prod.*, 71 (2008) 910–913.
136. J. D. Wansi, J. Wandji, L. Mbaze Meva'a, A. F. Kamdem Waffo, R. Ranjit, S. N. Khan, A. Asma, C. M. Iqbal, M.-C. Lallemand, F. Tillequin, and Z. Fomum Tanee, α-Glucosidase inhibitory and

antioxidant acridone alkaloids from the stem bark of *Oriciopsis glaberrima* ENGL. (Rutaceae), *Chem. Pharm. Bull.*, 54 (2006) 292–296.
137. Y. Nakao, T. Maki, S. Matsunaga, R. W. M. van Soest, and N. Fusetani, Penarolide sulfates A_1 and A_2, new α-glucosidase inhibitors from a marine sponge *Penares* sp, *Tetrahedron*, 56 (2000) 8977–8987.
138. J. H. Kim, Y. B. Ryu, N. S. Kang, B. W. Lee, J. S. Heo, I.-Y. Jeong, and K. H. Park, Glycosidase inhibitory flavonoids from Sophora flavescens, *Biol. Pharm. Bull.*, 29 (2006) 302–305.
139. W. D. Seo, J. H. Kim, J. E. Kang, H. W. Ryu, M. J. Curtis-Long, H. S. Lee, M. S. Yang, and K. H. Park, Sulfonamide chalcone as a new class of α-glucosidase inhibitors, *Bioorg. Med. Chem. Lett.*, 15 (2005) 5514–5516.
140. I. Miwa, J. Okuda, T. Horie, and M. Nakayama, Inhibition of intestinal α-glucosidases and sugar absorbtion by flavones, *Chem. Pharm. Bull.*, 34 (1986) 838–844.
141. J. Y. Kim, J. W. Lee, Y. S. Kim, Y. Lee, Y. B. Ryu, S. Kim, H. W. Ryu, M. J. Curtis-Long, K. W. Lee, W. S. Lee, and K. H. Park, A novel competitive class of α-glucosidase inhibitors: (E)-1-Phenyl-3-(4-styrylphenyl)urea derivatives, *ChemBioChem*, 11 (2010) 2125–2131.
142. Y.-K. Li and L. D. Byers, Inhibition of β-glucosidase by imidazoles, *Biochim. Biophys. Acta*, 999 (1989) 227–232.
143. R. A. Field, A. H. Haines, E. J. T. Chrystal, and M. C. Luszniak, Histidines, histamines and imidazoles as glycosidase inhibitors, *Biochem. J.*, 274 (1991) 885–889.
144. P. Magdolen and A. Vasella, Monocyclic, substituted imidazoles as glycosidase inhibitors, *Helv. Chim. Acta*, 88 (2005) 2454–2469.
145. J. L. Goldstein and T. Swain, The inhibition of enzymes by tannins, *Phytochemistry*, 4 (1965) 185–192.
146. G. H. B. Maegawa, M. Tropak, J. Buttner, T. Stockley, F. Kok, J. T. R. Clarke, and D. J. Mahuran, Pyrimethamine as a potential pharmacological chaperone for late-onset forms of GM2 gangliosidosis, *J. Biol. Chem.*, 282 (2007) 9150–9161.
147. G. H. B. Maegawa, M. B. Tropak, J. D. Buttner, B. A. Rigat, M. Fuller, D. Pandit, L. Tang, G. J. Kornhaber, Y. Hamuro, J. T. R. Clarke, and D. J. Mahuran, Identification and characterization of ambroxol as an enzyme enhancement agent for Gaucher disease, *J. Biol. Chem.*, 284 (2009) 23502–23516.
148. A. D. Abell, M. J. Ratcliffe, and J. Gerrard, Ascorbic acid-based inhibitors of α-amylases, *Bioorg. Med. Chem. Lett.*, 8 (1998) 1703–1706.
149. G. Legler and W. Lotz, Mechanism of action of glycoside-splitting enzymes. VII. Functional groups at the active site of an α-glucosidase from *Saccharomyces cerevisiae*, *Hoppe-Seyler's Z. Physiol. Chem.*, 354 (1973) 243–254.
150. S. Atsumi, K. Umezawa, H. Iinuma, H. Naganawa, H. Nakamura, Y. Iitaka, and T. Takeuchi, Production, isolation and structure determination of a novel β-glucosidase inhibitor, cyclophellitol, from *Phellinus* sp, *J. Antibiot.*, 43 (1990) 49–53.
151. S. Atsumi, H. Iinuma, C. Nosaka, and K. Umezawa, Biological activities of cyclophellitol, *J. Antibiot.*, 43 (1990) 1579–1985.
152. J. Marco-Contelles, Cyclohexane epoxides—Chemistry and biochemistry of (+)-cyclophellitol, *Eur. J. Chem.* (2001) 1607–1618.
153. S. G. Withers and K. Umezawa, Cyclophellitol: A naturally occurring mechanism-based inactivator of β-glucosidases, *Biochem. Biophys. Res. Commun.*, 177 (1991) 532–537.
154. T. M. Gloster, R. Madsen, and G. J. Davies, Structural basis for cyclophellitol inhibition of β-glucosidase, *Org. Biomol. Chem.*, 5 (2007) 444–446.
155. K. Tatsuta, Y. Niwata, K. Umezawa, K. Toshima, and M. Nakata, Syntheses and enzyme inhibiting activities of cyclophellitol analogs, *J. Antibiot.*, 44 (1991) 912–914.
156. M. Nakata, C. Chong, Y. Niwata, K. Toshima, and K. Tatsuta, A family of cyclophellitol analogs: Synthesis and evaluation, *J. Antibiot.*, 46 (1993) 1919–1922.

157. M. K. Tong and B. Ganem, A potent new class of active-site-directed glycosidase inactivators, *J. Am. Chem. Soc.*, 110 (1988) 312–313.
158. M. L. Sinnot and P. J. Smith, Affinity labelling with deaminatively generated carbonium ion. Kinetics and stoicheiometry of the alkylation of methionine-500 of the lacZ β-galactosidase of *Escherichia coli* by β-D-galactopyranosylmethyl-*p*-nitrophenyltriazene, *Biochem. J.*, 175 (1978) 525–538.
159. F. Naider, Z. Bohak, and J. Yariv, Reversible alkylation of a methionyl residue near the active site of β-galactosidase, *Biochemistry*, 11 (1972) 3202–3207.
160. M. L. Shulman, S. D. Shiyan, and A. Y. Khorlin, Specific irreversible inhibition of sweet-almond β-glucosidase by some β-glycopyranosylepoxyalkanes and β-D-glucopyranosyl isothiocyanate, *Biochim. Biophys. Acta*, 445 (1976) 169–181.
161. B. P. Rempel and S. G. Withers, Covalent inhibitors of glycosidases and their applications in biochemistry and biology, *Glycobiology*, 18 (2008) 570–586.
162. G. Legler, Glycosidase inhibition by basic sugar analogs and the transition state of enzymatic glycoside hydrolysis, in A. E. Stütz, (Ed.), *Iminosugars as Glycosidase Inhibitors,* Wiley-VCH, Weinheim, New York, 1999, pp. 31–67.
163. A. Berecibar, C. Grandjean, and A. Siriwardena, Synthesis and biological activity of natural aminocyclitol glycosidase inhibitors: Mannostatins, trehazolin, allosamidines, and their analogues, *Chem. Rev.*, 99 (1999) 779–844.
164. L. Sim, K. Jayakanthan, S. Mohan, R. Nasi, B. D. Johnston, B. M. Pinto, and D. R. Rose, New glucosidase inhibitors from ayurvedic herbal treatment for type 2 diabetes: Structures and inhibition of human intestinal maltase-glucoamylase with compounds from *Salacia reticulata*, *Biochemistry*, 49 (2010) 443–451.
165. S. Mohan and B. M. Pinto, Sulfonium-ion glycosidase inhibitors isolated from *Salacia* species used in traditional medicine, and related compounds, *Coll. Czech. Chem. Commun.*, 74 (2009) 1117–1136.
166. A. Steiner, A. Stütz, and T. Wrodnigg, Sulfur-containing glycomimetics, in B. O. Fraser-Reid, K. Tatsuta, and J. Thiem, (Eds.), *Glycoscience Chemistry and Chemical Biology,* 2nd ed. Springer, Berlin, 2008, pp. 1999–2020; Part 9.
167. S. G. Withers, M. Namchuk, and R. Mosi, Potent glycosidase inhibitors: Transition state mimics or simply fortuitous binders? in A. E. Stütz (Ed.), *Iminosugars as Glycosidase Inhibitors,* Wiley-VCH, Weinheim, New York, 1999, pp. 188–206.
168. R. Wolfenden, X. Lu, and G. Young, Spontaneous hydrolysis of glycosides, *J. Am. Chem. Soc.*, 120 (1998) 6814–6815.
169. T. M. Jespersen, W. Dong, T. Skrydstrup, M. R. Sierks, I. Lundt, and M. Bols, Isofagomine, a potent, new glycosidase inhibitor, *Angew. Chem. Int. Ed.*, 33 (1994) 1778–1779.
170. X. Zhu, K. A. Sheth, S. Li, H.-H. Chang, and J.-Q. Fan, Rational design and synthesis of highly potent β-glucocerebrosidase inhibitors, *Angew. Chem. Int. Ed.*, 44 (2005) 7450–7453.
171. N. Panday, Y. Canac, and A. Vasella, Very strong inhibition of glucosidases by *C(2)*-substituted tetrahydroimidazopyridines, *Helv. Chim. Acta*, 83 (2000) 58–79.
172. R. Saul, R. J. Molyneux, and A. D. Elbein, Studies on the mechanism of castanospermine inhibition of α- and β-glucosidases, *Arch. Biochem. Biophys.*, 230 (1984) 668–675.
173. J. N. Greul, M. Kleban, B. Schneider, S. Picasso, and V. Jäger, Amino(hydroxymethyl)cyclopentane-triols, an emerging class of potent glycosidase inhibitors—Part II: Synthesis and optimisazion of β-D-galactopyranoside analogues, *ChemBioChem*, 2 (2001) 368–370.
174. S. Ogawa, M. Ashiura, C. Uchida, S. Watanabe, C. Yamazaki, K. Yamagishi, and J. Inokuchi, Synthesis of potent β-D-glucocerebrosidase inhibitors: *N*-Alkyl-β-valienamines, *Bioorg. Med. Chem. Lett.*, 6 (1996) 929–932.
175. T. M. Wrodnigg, S. G. Withers, and A. E. Stütz, Novel, lipophilic derivatives of 2,5-dideoxy-2,5-imino-D-mannitol (DMDP) are powerful β-glucosidase inhibitors, *Bioorg. Med. Chem. Lett.*, 11 (2001) 1063–1064.

176. T. Wennekes, R. J. B. H. N. van den Berg, K. M. Bonger, W. E. Donker-Koopman, A. Ghisaidoobe, G. A. van der Marel, A. Strijland, J. M. F. G. Aerts, and H. S. Overkleeft, Synthesis and evaluation of dimeric lipophilic iminosugars as inhibitors of glucosylceramide metabolism, *Tetrahedron: Asymmetry*, 20 (2009) 836–846.
177. Y. Ichikawa and Y. Igarashi, An extremely potent inhibitor for β-galactosidase, *Tetrahedron Lett.*, 36 (1995) 4585–4586.
178. R. F. G. Fröhlich, R. H. Furneaux, D. J. Mahuran, R. Saf, A. E. Stütz, M. B. Tropak, J. Wicki, S. G. Withers, and T. M. Wrodnigg, 1-Deoxy-D-galactonojirimycins with dansyl capped N-substituents as β-galactosidase inhibitors and potential probes for G_{M1} gangliosidosis affected cell lines, *Carbohydr. Res.*, 346 (2011) 1592–1598.
179. S. Ogawa, Y. K. Matsunaga, and Y. Suzuki, Chemical modification of the β-glucocerebrosidase inhibitor N-octyl-β-valienamine: Synthesis and biological evaluation of 4-epimeric and 4-O-(β-D-galactopyranosyl) derivatives, *Bioorg. Med. Chem.*, 10 (2002) 1967–1972.
180. V. L. Schramm and P. C. Tyler, Transition state analogue inhibitors of N-ribosyltransferases, in P. Compain and O. R. Martin, (Eds.), *Iminosugars: From Synthesis to Therapeutic Applications*, Wiley, Chichester, 2007, pp. 177–208.
181. R. V. Stick, The synthesis of novel enzyme inhibitors and their use in defining the active sites of glycan hydrolases, *Top. Curr. Chem.*, 187 (1997) 187–213.
182. P. Bach and M. Bols, Synthesis of an 1-azaglucose analog with ring oxygen retained, *Tetrahedron Lett.*, 40 (1999) 3461–3464.
183. W. M. Best, J. M. MacDonald, B. W. Skelton, R. V. Stick, D. M. G. Tilbrook, and A. H. White, The synthesis of a carbohydrate-like dihydrooxazine and tetrahydrooxazine as putative inhibitors of glycoside hydrolases: A direct synthesis of isofagomine, *Can. J. Chem.*, 80 (2002) 857–865.
184. H. Liu, X. Liang, H. Sohoel, A. Bülow, and M. Bols, Noeuromycin, a glycosyl cation mimic that strongly inhibits glycosidases, *J. Am. Chem. Soc.*, 123 (2001) 5116–5117.
185. G. Legler, M. L. Sinnott, and S. G. Withers, Catalysis by β-glucosidase A_3 of *Aspergillus wentii*, *J. Chem. Soc., Perkin Trans. 2* (1980) 1376–1383.
186. H. H. Jensen and M. Bols, Synthesis of 1-azagalactofagomine, a potent galactosidase inhibitor, *J. Chem. Soc., Perkin Trans. 1* (2001) 905–909.
187. H. H. Jensen, L. Lyngbye, and M. Bols, A free-energy relationship between the rate of acidic hydrolysis of glycosides and the pK_a of isofagomines, *Angew. Chem. Int. Ed.*, 40 (2001) 3447–3449.
188. H. H. Jensen, L. Lyngbye, A. Jensen, and M. Bols, Stereoelectronic substituent effects in polyhydroxylated piperidines and hexahydropyridazines, *Chem. Eur. J.*, 8 (2002) 1218–1226.
189. J. P. Snyder, N. S. Chandrakumar, H. Sato, and D. C. Lankin, The unexpected diaxial orientation of cis-3,5-difluoropiperidine in water: A potent CF-NH charge dipole effect, *J. Am. Chem. Soc.*, 122 (2000) 544–545.
190. D. C. Lankin, N. S. Chandrakumar, S. N. Rao, D. P. Spangler, and J. P. Snyder, Protonated 3-fluoropiperidines. An unusual fluoro directing effect and a test for quantitative theories of solvation, *J. Am. Chem. Soc.*, 115 (1993) 3356–3357.
191. M. Bols, X. Liang, and H. H. Jensen, Equatorial contra axial polar substituents. The relation of a chemical reaction to stereochemical substituent constants, *J. Org. Chem.*, 67 (2002) 8970–8974.
192. C. McDonnell, O. Lopez, P. Murphy, J. G. Fernandez Bolanos, R. Hazell, and M. Bols, Conformational effects on glycoside reactivity: Study of the high reactive conformer of glucose, *J. Am. Chem. Soc.*, 126 (2004) 12374–12385.
193. T. D. Heightman and A. T. Vasella, Recent insights into inhibition, structure, and mechanism of configuration-retaining glycosidases, *Angew. Chem. Int. Ed.*, 38 (1999) 750–770.
194. S. Vonhoff, K. Piens, M. Pipelier, C. Braet, M. Claeyssens, and A. Vasella, Inhibition of cellobiohydrolases from *Trichoderma reesei*. Synthesis and evaluation of some glucose-, cellobiose-, and cellotriose-derived hydroximolactams and imidazoles, *Helv. Chim. Acta*, 82 (1999) 963–980.

195. A. Varrot, M. Schülein, M. Pipelier, A. Vasella, and G. J. Davies, Lateral protonation of a glycosidase inhibitor. Structure of the *Bacillus agaradhaerens* Cel5A in complex with cellobiose-derived imidazole at 0.97 Å resolution, *J. Am. Chem. Soc.*, 121 (1999) 2621–2622.
196. T. M. Gloster, S. Roberts, G. Perugino, M. Rossi, M. Moracci, N. Panday, M. Terinek, A. Vasella, and G. J. Davies, Structural, kinetic and thermodynamic analysis of glucoimidazole-derived glycosidase inhibitors, *Biochemistry*, 45 (2006) 11879–11884.
197. B. Shanmugasundaram and A. Vasella, Synthesis of new *C(2)*-substituted *gluco*-configured tetrahydroimidazopyridines and their evaluation as glucosidase inhibitors, *Helv. Chim. Acta*, 88 (2005) 2593–2602.
198. V. Notenboom, S. J. Williams, R. Hoos, S. G. Withers, and D. R. Rose, Detailed structural analysis of glycosidase/inhibitor interactions: Complexes of Cex from *Cellulomonas fimi* with xylobiose-derived aza-sugars, *Biochemistry*, 39 (2000) 11553–11563.
199. S. Williams, R. Hoos, and S. G. Withers, Nanomolar versus millimolar inhibition by xylobiose-derived azasugars: Significant differences between structurally distinct xylanases, *J. Am. Chem. Soc.*, 122 (2000) 2223–2235.
200. W. Nerinckx, T. Desmet, K. Piens, and M. Claeyssens, An elaboration on the *syn-anti* proton donor concept of glycoside hydrolases. Electrostatic stabilisation of the transition state as a general strategy, *FEBS Lett.*, 579 (2005) 302–312.
201. N. K. Vyas, Atomic features of protein-carbohydrate interactions, *Curr. Opin. Struct. Biol.*, 1 (1991) 732–740.
202. T. Collins, D. De Vos, A. Hoyoux, S. N. Savvides, C. Gerday, J. Van Beeumen, and G. Feller, Study of the active site residues of a glycoside hydrolase family 8 xylanase, *J. Mol. Biol.*, 354 (2005) 425–435.
203. C. W. Ekhart, M. H. Fechter, P. Hadwiger, E. Mlaker, A. E. Stütz, A. Tauss, and T. M. Wrodnigg, Tables of glycosidase inhibitors with nitrogen in the sugar ring and their activities, in A. E. Stütz, (Ed.), *Iminosugars as Glycosidase Inhibitors*, Wiley-VCH, Weinheim, New York, 1999, pp. 254–390.
204. N. Asano, K. Oseki, E. Kaneko, and K. Matsui, Enzymic synthesis of α- and β-D-glucosides of 1-deoxynojirimycin and their glycosidase inhibitory activities, *Carbohydr. Res.*, 258 (1994) 255–266.
205. A. Kato, N. Asano, H. Kizu, and K. Matsui, Fagomine isomers and glycosides from *Xanthocercis zambesiaca*, *J. Nat. Prod.*, 60 (1997) 312–314.
206. J.-Y. Goujon, D. Gueyrard, P. Compain, O. R. Martin, K. Ikeda, A. Kato, and N. Asano, General synthesis and biological evaluation of α-1-*C*-substituted derivatives of fagomine (2-deoxynojirimycin-α-*C*-glycosides), *Bioorg. Med. Chem.*, 13 (2005) 2313–2324.
207. N. Asano, K. Oseki, H. Kizu, and K. Matsui, Nitrogen-in-the-ring pyranoses and furanoses: Structural basis of inhibition of mammalian glycosidases, *J. Med. Chem.*, 37 (1994) 3701–3706.
208. T. Kajimoto, K. K.-C. Liu, R. L. Pederson, Z. Zhong, Y. Ichikawa, J. C. Porco, and C.-H. Wong, Enzyme-catalyzed aldol condensation for asymmetric synthesis of azasugars: Synthesis, evaluation, and modelling of glycosidase inhibitors, *J. Am. Chem. Soc.*, 113 (1991) 6187–6196.
209. G. Legler, unpublished.
210. B. Winchester, S. Al Daher, N. C. Carpenter, I. Cenci di Bello, S. S. Choi, A. J. Fairbanks, and G. W. J. Fleet, The structural basis of the inhibition of human α-mannosidases by azafuranose analogues of mannose, *Biochem. J.*, 290 (1993) 743–749.
211. J. L. Reymond and P. Vogel, Application of new optically pure ketene equivalents derived from tartaric acids to the total, asymmetric syntheses of (+)-6-deoxycastanospermine and (+)-6-deoxy-6-fluorocastanospermine, *J. Chem. Soc., Chem. Commun.* (1990) 1070–1072.
212. J. L. Reymond, A. A. Pinkerton, and P. Vogel, Total asymmetric synthesis of (+)-castanospermine, (+)-6-deoxycastanospermine, and (+)-6-deoxy-6-fluorocastanospermine, *J. Org. Chem.*, 56 (1991) 2128–2135.

213. C.-K. Lee, K. Y. Sim, and J. Zhu, Enantiomeric synthesis of polyhydroxylated indolizidine analogues related to castanospermine: 1-Deoxy-7-epicastanospermine and 1,7-dideoxy-7-fluorocastanospermine, *Tetrahedron*, 48 (1992) 8541–8544.
214. R. H. Furneaux, J. M. Mason, and P. C. Tyler, The chemistry of castanospermine. Part II: Synthesis of deoxyfluoro analogues of castanospermine, *Tetrahedron Lett.*, 35 (1994) 3143–3146.
215. R. H. Furneaux, G. J. Gainsford, J. M. Mason, and P. C. Tyler, The chemistry of castanospermine. Part I: Synthetic modifications at C-6, *Tetrahedron*, 50 (1994) 2131–2160.
216. R. H. Furneaux, G. J. Gainsford, J. M. Mason, P. C. Tyler, O. Hartley, and B. Winchester, The chemistry of castanospermine. Part IV: Synthetic modifications at C-8, *Tetrahedron*, 51 (1995) 12611–12630.
217. R. H. Furneaux, G. J. Gainsford, J. M. Mason, and P. C. Tyler, The chemistry of castanospermine. Part V: Synthetic modifications at C-1 and C-7, *Tetrahedron*, 53 (1997) 245–268.
218. K. Dax, V. Grassberger, and A. E. Stütz, Simple synthesis of 1,5,6-trideoxy-1,5-imino-D-glucitol, the first fluorine-containing derivative of glucosidase inhibitor 1-deoxynojirimycin, *J. Carbohydr. Chem.*, 9 (1990) 903–908.
219. J. Di, B. Rajanikanth, and W. A. Szarek, Fluorinated 1,5-dideoxy-1,5-iminoalditols: Synthesis of 1,5,6-trideoxy-6-fluoro-1,5-imino-D-glucitol (1,6-dideoxy-6-fluoronojirimycin) and 1,4,5-trideoxy-4-fluoro-1,5-imino-D-ribitol (1,2,5-trideoxy-2-fluoro-1,5-imino-L-ribitol), *J. Chem. Soc., Perkin Trans. 1* (1992) 2151–2152.
220. D. P. Getman and D. P. De Crescenzo, *Eur. Pat. Appl.* (1991) EP 410953 A2 19910130.
221. G. A. De Crescenzo and D. P. Getman, *Eur. Pat. Appl.* (1992) EP 481950 A2 19920422.
222. C.-K. Lee, H. Jiang, L. L. Koh, and Y. Xu, Synthesis of 1,3-dideoxy-3-fluoronojirimycin, *Carbohydr. Res.*, 239 (1993) 309–315.
223. C.-K. Lee and H. J. Jiang, Asymmetric synthesis of derivatives of 1-deoxynojirimycin. 3. Asymmetric synthesis of 1,3,6-trideoxy-3,6-difluoronojirimycin, *J. Carbohydr. Chem.*, 14 (1995) 407–416.
224. C.-K. Lee, H. Jiang, A. Linden, and A. Scofield, 1,3,6-Trideoxy-3,6-difluoronojirimycin: Structure and evaluation as a glucosidase inhibitor, *Carbohydr. Lett.*, 1 (1996) 417–423.
225. A. Arnone, P. Bravo, A. Donadelli, and G. Resnati, Fluorinated analogues of nojirimycin and mannojirimycin from a non-carbohydrate precursor, *Tetrahedron*, 52 (1996) 131–142.
226. M. A. Szarek, X. Wu, and W. A. Szarek, Synthesis and evaluation of 1,5,6-trideoxy-6,6-difluoro-1,5-imino-D-glucitol (1,6-dideoxy-6,6-difluoronojirimycin) as a glucosidase inhibitor, *Carbohydr. Res.*, 299 (1997) 165–170.
227. R.-W. Wang, X.-L. Qiu, M. Bols, F. Ortega-Caballero, and F.-L. Qing, Synthesis and biological evaluation of glycosidase inhibitors: *gem*-Difluoromethylenated nojirimycin analogues, *J. Med. Chem.*, 49 (2006) 2989–2997.
228. R. Csuk, E. Prell, C. Korb, R. Kluge, and D. Ströhl, Total synthesis of 3,3-difluorinated 1-deoxynojirimycin analogues, *Tetrahedron*, 66 (2010) 467–472.
229. E. Prell and R. Csuk, Amplification of the inhibitory activity of miglitol by monofluorination, *Bioorg. Med. Chem. Lett.*, 19 (2009) 5673–5674.
230. R. Li, M. Bols, C. Rousseau, X. Zhang, R. Wang, and F.-L. Qing, Synthesis and biological evaluation of potent glycosidase inhibitors: 4-Deoxy-4,4-difluoroisofagomine and analogues, *Tetrahedron*, 65 (2009) 3717–3727.
231. R. Csuk, E. Prell, S. Reissmann, and C. Korb, First total synthesis of a fluorinated calystegin, *Z. Naturforsch.*, 65b (2009) 445–451.
232. S. M. Andersen, M. Ebner, C. W. Ekhart, G. Gradnig, G. Legler, I. Lundt, A. E. Stütz, S. G. Withers, and T. Wrodnigg, Two isosteric fluorinated derivatives of the powerful glucosidase inhibitors, 1-deoxynojirimycin and 2,5-dideoxy-2,5-imino-D-mannitol: Syntheses and glycosidase inhibitory activities of 1,2,5-trideoxy-2-fluoro-1,5-imino-D-glucitol and of 1,2,5-trideoxy-1-fluoro-2,5-imino-D-mannitol, *Carbohydr. Res.*, 301 (1997) 155–166.

233. Y.-X. Li, M.-H. Huang, Y. Yamashita, A. Kato, Y.-M. Jia, W.-B. Wang, G. W. J. Fleet, R. J. Nash, and C.-Y. Yu, L-DMDP, L-homoDMDP ad their C-3 fluorinated derivatives: Synthesis and glycosidase inhibition, *Org. Biomol. Chem.*, 9 (2011) 3405–3414.
234. R.-W. Wang, J. Xu, O. Lopez, M. Bols, and F.-L. Qing, Difluoromethylenated polyhydroxylated pyrrolidines: Facile synthesis, crystal structure and biological evaluation, *Future Med. Chem.*, 1 (2009) 991–997.
235. P. C. Tyler and B. G. Winchester, Synthesis and biological activity of castanospermine and close analogs, in A. E. Stütz, (Ed.), *Iminosugars as Glycosidase Inhibitors,* Wiley-VCH, Weinheim, New York, 1999, pp. 125–156.
236. A. Hentges and E. Bause, Affinity purification and characterization of glucosidase II from pig liver, *Biol. Chem.*, 378 (1997) 1031–1038.
237. A. Lewandowicz, E. A. Taylor Ringia, L.-M. Ting, K. Kim, P. C. Tyler, G. B. Evans, O. V. Zubkova, S. Mee, G. F. Painter, D. H. Lenz, R. H. Furneaux, and V. L. Schramm, Energetic mapping of transition state analogue interactions with human and *Plasmodium falciparum* purine nucleoside phosphorylase, *J. Biol. Chem.*, 280 (2005) 30320–30328.
238. G. B. Evans, R. H. Furneaux, A. Lewandowicz, V. L. Schramm, and P. C. Tyler, Exploring structure-activity relationships of transition state analogues of human purine nucleoside phosphorylase, *J. Med. Chem.*, 46 (2003) 3412–3423.
239. J. M. Mason, A. S. Murkin, L. Li, V. L. Schramm, G. J. Gainsford, and B. W. Skelton, A β-fluoroamine inhibitor of purine nucleoside phosphorylase, *J. Med. Chem.*, 51 (2008) 5880–5884.
240. A. A. Edwards, J. M. Mason, K. Clinch, P. C. Tyler, G. B. Evans, and V. L. Schramm, Altered enthalpy-entropy compensation in picomolar transition state analogues of human nucleoside phosphorylase, *Biochemistry*, 48 (2009) 5226–5238.
241. A. E. Stütz, Some reflections on structure-activity relationships in glycosidase-inhibiting iminoalditols and iminosugars, in A. E. Stütz, (Ed.), *Iminosugars as Glycosidase Inhibitors,* Wiley-VCH, Weinheim, New York, 1999, pp. 157–187.
242. S. Ravaud, X. Robert, H. Watzlawick, R. Haser, R. Mattes, and N. Aghajari, Trehalulose synthase native and carbohydrate complexed structures provide insights into sucrose isomerisation, *J. Biol. Chem.*, 282 (2007) 28126–28136.
243. http://www.pdb.org/pdb/home/home.do.
244. K. Karaveg, A. Siriwardena, W. Tempel, Z.-J. Liu, J. Glushka, B.-C. Wang, and K. W. Moremen, Mechanism of class 1 (glycosylhydrolase family 47) α-mannosidases involved in N-glycan processing and endoplasmatic reticulum quality control, *J. Biol. Chem.*, 280 (2005) 16197–16207.
245. C. A. Collyer, J. D. Goldberg, H. Viehmann, D. M. Blow, N. G. Ramsden, G. W. J. Fleet, F. J. Montgomery, and P. Grice, Anomeric specificity of D-xylose isomerase, *Biochemistry*, 31 (1992) 12211–12218.
246. E. M. S. Harris, A. E. Aleshin, L. M. Firsov, and R. B. Honzatko, Refined structure of the complex of 1-deoxynojirimycin with glucoamylase from *Aspergillus awamori* var. X100 to 2.4-Å resolution, *Biochemistry*, 32 (1993) 1618–1626.
247. A. Hempel, N. Camerman, D. Mastropaolo, and A. Camerman, Glucosidase inhibitors: Structures of deoxynojirimycin and castanospermine, *J. Med. Chem.*, 36 (1993) 4082–4086.
248. R. Kanai, K. Haga, K. Yamane, and K. Harata, Crystal structure of cyclodextrin glucanotransferase from alkalophilic *Bacillus* sp. 1011 complexed with 1-deoxynojirimycin at 2.0 Å resolution, *J. Biochem.*, 129 (2001) 593–598.
249. D. L. Zechel, A. B. Boraston, T. Gloster, C. M. Boraston, J. M. Macdonald, D. M. G. Tilbrook, R. V. Stick, and G. J. Davies, Iminosugar glycosidase inhibitors: Structural and thermodynamic dissection of the binding of isofagomine and 1-deoxynojirimycin to β-glucosidases, *J. Am. Chem. Soc.*, 125 (2003) 14313–14323.

250. A. Varrot, C. A. Tarling, J. M. Macdonald, R. V. Stick, D. L. Zechel, S. G. Withers, and G. J. Davies, Direct observation of the protonation state of an iminosugar glycosidase inhibitor upon binding, *J. Am. Chem. Soc.*, 125 (2003) 7496–7497.
251. T. M. Gloster, J. M. Macdonald, C. A. Tarling, R. V. Stick, S. G. Withers, and G. J. Davies, Structural, thermodynamic, and kinetic analyses of tetrahydrooxazine-derived inhibitors bound to β-glucosidases, *J. Biol. Chem.*, 279 (2004) 49236–49242.
252. T. M. Gloster, S. J. Williams, S. Roberts, C. A. Tarling, J. Wicki, S. G. Withers, and G. J. Davies, Atomic resolution analyses of the binding of xylobiose-derived deoxynojirimycin and isofagomine to xylanase Xyn10A, *Chem. Commun.* (2004) 1794–1795.
253. T. Gloster, S. J. Williams, C. A. Tarling, S. Roberts, C. Dupont, P. Jodoin, F. Shareck, S. G. Withers, and G. J. Davies, A xylobiose-derived isofagomine lactam glycosidase inhibitor binds as its amide tautomer, *Chem. Commun.* (2003) 944–945.
254. V. H. Lillelund, H. Z. Liu, X.-F. Liang, H. Sohoel, and M. Bols, Isofagomine lactams, synthesis and enzyme inhibition, *Org. Biomol. Chem.*, 1 (2003) 282–287.
255. F. Vincent, T. M. Gloster, J. Macdonald, C. Moorland, R. V. Stick, F. M. V. Dias, J. A. M. Prates, C. M. G. A. Fontes, H. J. Gilbert, and G. J. Davies, Common inhibition of both β-glucosidases and β-mannosidases by isofagomine lactam reflects different conformational itineraries for pyranoside hydrolysis, *ChemBioChem*, 5 (2004) 1596–1599.
256. A. Varrot, J. Macdonald, R. V. Stick, G. Pell, H. J. Gilbert, and G. J. Davies, Distortion of a cellobio-derived isofagomine highlights the potential conformational itinerary of inverting β-glucosidases, *Chem. Commun.* (2003) 946–947.
257. A. Varrot, S. Leydier, G. Pell, J. M. Macdonald, R. V. Stick, B. Henrissat, H. J. Gilbert, and G. J. Davies, *Mycobacterium tuberculosis* strains possess functional cellulases, *J. Biol. Chem.*, 280 (2005) 20181–20184.
258. M. E. C. Caines, S. M. Hancock, C. A. Tarling, T. M. Wrodnigg, R. V. Stick, A. E. Stütz, A. Vasella, S. G. Withers, and N. C. J. Strynadka, The structural basis of glycosidase inhibition by five-membered iminocyclitols: The clan A glycoside hydrolase endoglycoceramidase as a model system, *Angew. Chem. Int. Ed.*, 46 (2007) 4474–4476.
259. T. M. Gloster, S. Roberts, V. M-A Ducros, G. Perugino, M. Rossi, R. Hoos, M. Moracci, A Vasella, and G. J. Davies, Structural studies of the β-glucosidase from *Sulfolobus solfataricus* in complex with covalently and noncovalently bound inhibitors, *Biochemistry*, 43 (2004) 6101–6109.
260. A. Frankowski, C. Seliga, and J. Streith, Imidazole analogues of 6-epicastanospermine and of 3,7a-diepialexine, *Helv. Chim. Acta*, 74 (1991) 934–940.
261. K. Tatsuta, S. Miura, S. Ohta, and H. Gunji, Total syntheses of de-branched nagstatin and its analogs having glycosidase inhibiting activities, *Tetrahedron Lett.*, 36 (1995) 1085–1088.
262. T. Granier and A. Vasella, Synthesis and evaluation as glycosidase inhibitors of 1*H*-imidazol-2-yl C-glycopyranosides, *Helv. Chim. Acta*, 78 (1995) 1738–1746.
263. T. Granier, N. Panday, and A. Vasella, Structure-activity relations of imidazo-pyridine-type inhibitors of β-D-glucosidases, *Helv. Chim. Acta*, 80 (1997) 979–987.
264. M. Hrmova, R. de Gori, B. J. Smith, A. Vasella, J. N. Varghese, and G. B. Fincher, Three-dimensional structure of the barley β-D-glucan glucohydrolase in complex with a transition state mimic, *J. Biol. Chem.*, 279 (2004) 4970–4980.
265. M. Hrmova, V. A. Streltsov, B. J. Smith, A. Vasella, J. N. Varghese, and G. B. Fincher, Structural rationale for low-nanomolar binding of transitition state mimics to a family GH3 β-D-glucan glucohydrolase from barley, *Biochemistry*, 44 (2005) 16529–16539.
266. T. M. Gloster, P. Meloncelli, R. V. Stick, D. Zechel, A. Vasella, and G. J. Davies, Glycosidase inhibition: An assessment of the binding of 18 putative transition-state mimics, *J. Am. Chem. Soc.*, 129 (2007) 2345–2354.

267. L. Verdoucq, J. Moriniere, D. R. Bevan, A. Esen, A. Vasella, B. Henrissat, and M. Czjzek, Structural determinants of substrate specificity in family 1 β-glucosidases, *J. Biol. Chem.*, 279 (2004) 31796–31803.
268. W. P. Burmeister, S. Cottaz, P. Rollin, A. Vasella, and B. Henrissat, High resolution X-ray crystallography shows that ascorbate is a cofactor for myrosinase and substitutes for the function of the catalytic base, *J. Biol. Chem.*, 275 (2000) 39385–39393.
269. B. Brumshtein, H. M. Greenblatt, T. D. Butters, Y. Shaaltiel, D. Aviezer, I. Silman, A. H. Futerman, and J. L. Sussman, Crystal structures of complexes of N-butyl- and N-nonyl-deoxynojirimycin bound to acid β-glucosidase. Insights into the mechanism of chemical chaperone action in Gaucher disease, *J. Biol. Chem.*, 282 (2007) 29052–29058.
270. L. Premkumar, A. R. Sawkar, S. Boldin-Adamsky, L. Toker, I. Silman, J. W. Kelly, A. H. Futerman, and J. L. Sussman, X-ray structure of human acid-β-glucosidase covalently bound to conduritol-B-epoxide, *J. Biol. Chem.*, 25 (2005) 23815–23819.
271. R. L. Lieberman, B. A. Wustman, P. Huertas, A. C. Powe, Jr., C. W. Pine, R. Khanna, M. G. Schlossmacher, D. Ringe, and G. A. Petsko, Structure of acid β-glucosidase with pharmacological chaperone provides insight into Gaucher disease, *Nat. Chem. Biol.*, 3 (2007) 101–107.
272. R. L. Lieberman, J. A. D'aquino, D. Ringe, and G. A. Petsko, Effects of pH and iminosugar pharmacological chaperones on lysosomal glycosidase structure and stability, *Biochemistry*, 48 (2009) 4816–4827.
273. S. M. Cutfield, G. J. Davies, G. Murshudov, B. F. Anderson, P. C. E. Moody, P. A. Sullivan, and J. F. Cutfield, The structure of the exo-β-(1,3)-glucanase from *Candida albicans* in native and bound forms: Relationship between a pocket and groove in family 5 glycosyl hydrolases, *J. Mol. Biol.*, 294 (1999) 771–783.
274. T. Jank, M. O. P. Ziegler, G. E. Schulz, and K. Aktories, Inhibition of the glucosyltransferase activity of clostridial Rho/Ras-glucosylating toxins by castanospermine, *FEBS Lett.*, 582 (2008) 2277–2288.
275. M. Aguilar-Moncayo, T. M. Gloster, J. P. Turkenburg, M. I. Garcia-Moreno, C. Ortiz Mellet, G. J. Davies, and J. M. Garcia Fernandez, Glycosidase inhibition by ring modified castanospermine analogues: Tackling enzyme selectivity by inhibitor tailoring, *Org. Biomol. Chem.*, 7 (2009) 2738–2747.
276. B. Brumshtein, M. Aguilar-Moncayo, M. I. Garcia-Moreno, C. Ortiz Mellet, J. M. Garcia Fernandez, I. Silman, Y. Shaaltiel, D. Aviezer, J. L. Sussman, and A. H. Futerman, 6-Amino-6-deoxy-5,6-di-N-(N'-octyliminomethylidene)nojirimycin: Synthesis, biological evaluation, and crystal structure in complex with acid β-glucosidase, *ChemBioChem*, 10 (2009) 1480–1485.
277. T. M. Gloster, J. P. Turkenburg, J. R. Potts, B. Henrissat, and G. J. Davies, Divergence of catalytic mechanism within a glycosidase family provides insight into evolution of carbohydrate metabolism by human gut flora, *Chem. Biol.*, 15 (2008) 1058–1067.
278. S. Ravaud, X. Robert, H. Watzlawick, R. Haser, R. Mattes, and N. Aghajari, Structural determinants of product specificity of sucrose isomerases, *FEBS Lett.*, 583 (2009) 1964–1968.
279. T. M. Gloster, R. Madsen, and G. J. Davies, Dissection of conformationally restricted inhibitors binding to a β-glucosidase, *ChemBioChem*, 7 (2006) 738–742.
280. M. Aguilar, T. M. Gloster, M. I. Garcia-Moreno, C. Ortiz Mellet, G. J. Davies, A. Llebaria, J. Casas, M. Egido-Gabas, and J. M. Garcia Fernandez, Molecular basis for β-glucosidase inhibition by ring-modified calystegine analogues, *ChemBioChem*, 9 (2008) 2612–2618.
281. A. Herscovics, Importance of glycosidases in mammalian glycoprotein biosynthesis, *Biochim. Biophys. Acta*, 1473 (1999) 96–107.
282. F. Vallee, K. Karaveg, A. Herscovics, K. W. Moremen, and P. L. Howell, Structural basis for catalysis and inhibition of N-glycan processing class I α-1,2-mannosidases, *J. Biol. Chem.*, 275 (2000) 41287–41298.

283. Y. D. Lobsanov, F. Vallee, A. Imberty, T. Yoshida, P. Yip, A. Herscovics, and P. L. Howell, Structure of *Penicillium citrinum* α1,2-mannosidase reveals the basis for differences in specificity of the endoplasmatic reticulum and golgi class I enzymes, *J. Biol. Chem.*, 277 (2002) 5620–5630.
284. J. M. H. van den Elsen, D. A. Kuntz, and D. R. Rose, Structure of golgi α-mannosidase II: Target for inhibition of growth and metastasis of cancer cells, *EMBO J.*, 20 (2001) 3008–3017.
285. P. E. Goss, M. A. Baker, J. P. Carver, and J. W. Dennis, Inhibitors of carbohydrate processing: A new class of anticancer agents, *Clin. Cancer Res.*, 1 (1995) 935–944.
286. N. Shah, D. G. Kuntz, and D. R. Rose, Comparison of kifunensine and 1-deoxymannojirimycin binding to class I and II α-mannosidases demonstrates different distortions in inverting and retaining catalytic mechanisms, *Biochemistry*, 42 (2003) 13812–13816.
287. S. P. Kawatkar, D. A. Kuntz, R. J. Woods, D. R. Rose, and G.-J. Boons, Structural basis of the inhibition of golgi α-mannosidase II by mannostatin A and the role of the thiomethyl moiety in ligand-protein interactions, *J. Am. Chem. Soc.*, 128 (2006) 8310–8319.
288. D. A. Kuntz, W. Zhong, J. Guo, D. R. Rose, and G.-J. Boons, The molecular basis of inhibition of golgi α-mannosidase II by mannostatin A, *ChemBioChem*, 10 (2009) 268–277.
289. W. Zhong, D. A. Kuntz, B. Ember, H. Singh, K. W. Moremen, D. R. Rose, and G.-J. Boons, Probing the substrate specificity of golgi α-mannosidase II by use of synthetic oligosaccharides and a catalytic nucleophile mutant, *J. Am. Chem. Soc.*, 130 (2008) 8975–8983.
290. D. A. Kuntz, C. A. Tarling, S. G. Withers, and D. R. Rose, Structural analysis of golgi α-mannosidase II inhibitors identified from a focused glycosidase inhibitor screen, *Biochemistry*, 47 (2008) 10058–10068.
291. L. E. Tailford, W. A. Offen, N. L. Smith, C. Dumon, C. Morland, J. Gratien, M.-P. Heck, R. V. Stick, Y. Bleriot, A. Vasella, H. J. Gilbert, and G. J. Davies, Structural and biochemical evidence for a boat-like transition state in β-mannosidases, *Nat. Chem. Biol.*, 4 (2008) 306–312.
292. D. A. Kuntz, S. Nakayama, K. Shea, H. Hori, Y. Uto, H. Nagasawa, and D. R. Rose, Structural investigation of the binding of 5-substituted swainsonine analogues to golgi α-mannosidase II, *ChemBioChem*, 11 (2010) 673–680.
293. H. Fiaux, D. A. Kuntz, D. Hoffman, R. C. Janzer, S. Gerber-Lemaire, D. R. Rose, and L. Juillerat-Jeanneret, Functionalised pyrrolidine inhibitors of human type II α-mannosidases as anti-cancer agents: Optimizing the fit for the active site, *Bioorg. Med. Chem.*, 16 (2008) 7337–7346.
294. M. D. L. Suits, Y. Zhu, E. J. Taylor, J. Walton, D. L. Zechel, H. J. Gilbert, and G. J. Davies, Structure and kinetic investigation of *Streptococcus pyogenes* family GH38 α-mannosidase, *PLoS One*, 5 (2010) e9006.
295. Y. Zhu, M. D. L. Suits, A. J. Thompson, S. Chavan, Z. Dinev, C. Dumon, N. Smith, K. W. Moremen, Y. Xiang, A. Siriwardena, S. J. Williams, H. J. Gilbert, and G. J. Davies, Mechanistic insights into a Ca^{2+}-dependent family of α-mannosidases in a human gut symbiont, *Nat. Chem. Biol.*, 6 (2010) 125–132.
296. K. J. Gregg, W. F. Zandenberg, J.-H. Hehemann, G. E. Whitworth, L. Deng, D. J. Vocadlo, and A. B. Boraston, Analysis of a new family of widely distributed metal-independent α-mannosidases provides unique insight into the processing of *N*-linked glycans, *J. Biol. Chem.*, 286 (2011) 15586–15596.
297. A. I. Guce, N. E. Clark, E. N. Salgado, D. R. Ivanen, A. A. Kulminskaya, H. Brumer, III,, and S. C. Garman, Catalytic mechanism of human α-galactosidase, *J. Biol. Chem.*, 285 (2010) 3625–3632.
298. B. L. Mark, D. J. Vocadlo, D. Zhao, S. Knapp, S. G. Withers, and M. N. G. James, Biochemical and structural assessment of the 1-*N*-azasugar galNAc-isofagomine as a potent family 20 β-*N*-acetylhexosaminidase inhibitor, *J. Biol. Chem.*, 276 (2001) 42131–42137.
299. B. L. Mark, D. J. Vocadlo, S. Knapp, B. L. Triggs-Raine, S. G. Withers, and M. N. G. James, Crystallographic evidence for substrate-assisted catalysis in a bacterial β-hexosaminidase, *J. Biol. Chem.*, 276 (2001) 10330–10337.

300. B. L. Mark, D. J. Mahuran, M. M. Cherney, D. Zhao, S. Knapp, and M. N. G. James, Crystal structure of human β-hexosaminidase B: Understanding the molecular basis of Sandhoff and Tay-Sachs disease, *J. Mol. Biol.*, 327 (2003) 1093–1109.
301. C. Martinez-Fleites, Y. He, and G. J. Davies, Structural analyses of enzymes involved in the *O*-GlcNAc modification, *Biochim. Biophys. Acta*, 1800 (2010) 122–133.
302. G. J. Davies and C. Martinez-Fleites, The *O*-GlcNAc modification: Three-dimentional structure, enzymology and the development of selective inhibitors to probe disease, *Biochem. Soc. Trans.*, 38 (2010) 1179–1188.
303. B. Shanmugasundaram, A. W. Debowski, R. J. Dennis, G. J. Davies, D. J. Vocatlo, and A. Vasella, Inhibition of *O*-GlcNAcase by a *gluco*-configured nagstatin and a PUGNAc-imidazole hybrid inhibitor, *Chem. Commun.* (2006) 4372–4374.
304. H. C. Dorfmueller, V. S. Borodkin, M. Schimpl, S. M. Shepherd, N. A. Shpiro, and D. M. F. van Aalten, GlcNAcstatin: A picomolar, selective *O*-glcNAcase inhibitor that modulates intracellular *O*-glcNAcylation levels, *J. Am. Chem. Soc.*, 128 (2006) 16484–16485.
305. A. Scaffidi, K. A. Stubbs, R. J. Dennis, E. J. Taylor, G. J. Davies, D. J. Vocadlo, and R. V. Stick, A 1-acetamido derivative of 6-*epi*-valienamine: An inhibitor of a diverse group of β-*N*-acetylglucosaminidases, *Org. Biomol. Chem.*, 5 (2007) 3013–3019.
306. H. C. Dorfmueller, V. S. Borodkin, M. Schimpl, and D. M. F. van Aalten, GlcNAcstatins are nanomolar inhibitors of human *O*-GlcNAcase inducing cellular hyper-*O*-GlcNAcylation, *Biochem. J.*, 420 (2009) 221–227.
307. H. C. Dorfmueller, V. S. Borodkin, M. Schimpl, X. Zheng, R. Kime, K. D. Read, and D. M. F. van Aalten, Cell-penetrant, nanomolar *O*-GlcNAcase inhibitors selective against lysosomal hexosaminidases, *Chem. Biol.*, 17 (2010) 1250–1255.
308. Y. He, A. K. Bubb, K. A. Stubbs, T. M. Gloster, and G. J. Davies, Inhibition of a bacterial *O*-glcNAcase homologue by lactone and lactam derivatives: Structural, kinetic and thermodynamic analyses, *Amino Acids*, 40 (2011) 829–839.
309. B. Pluvinage, M. G. Ghinet, R. Brzezinski, A. D. Boraston, and K. A. Stubbs, Inhibition of the exo-β-D-glucosaminidase CsxA by a glucosamine-configured castanospermine and an amino-australine analogue, *Org. Biomol. Chem.*, 7 (2009) 4169–4172.
310. M. S. Macauley, Y. He, T. M. Gloster, K. A. Stubbs, G. J. Davies, and D. J. Vocadlo, Inhibition of *O*-GlcNAcase using a potent and cell-permeable inhibitor does not induce insulin resistance in 3T3-L1 adipocytes, *Chem. Biol.*, 17 (2010) 937–948.
311. J. Saha and M. W. Peczuh, Synthesis and properties of septanose carbohydrates, *Adv. Carbohydr. Chem. Biochem.*, 66 (2011) 121–186.
311a. F. Marcelo, Y. He, S. A. Yuzwa, L. Nieto, J. Jimenez-Barbero, M. Sollogoup, D. J. Vocadlo, G. D. Davies, and Y. Bleriot, Molecular basis for inhibition of GH84 glycoside hydrolases by substituted azepanes: Conformational flexibility enables probing of substrate distortion, *J. Am. Chem. Soc.*, 131 (2009) 5390–5392.
312. H.-J. Wu, C.-W. Ho, T.-P. Ko, S. D. Popat, C.-H. Lin, and A. H.-J. Wang, Structural basis of α-fucosidase inhibition by iminocyclitols with K_i values in the micro- to picomolar range, *Angew. Chem. Int. Ed.*, 49 (2010) 337–340.
313. A. Lammerts van Bueren, S. D. Popat, C.-H. Lin, and G. J. Davies, Structural and thermodynamic analyses of α-L-fucosidase inhibitors, *ChemBioChem*, 11 (2010) 1971–1974.
314. A. Lammerts van Bueren, A. Ardevol, J. Fayers-Kerr, B. Luo, Y. Zhang, M. Sollogoub, Y. Bleriot, C. Rovira, and G. J. Davies, Analysis of the reaction coordinate of α-L-fucosidases: A combined structural and quantum mechanical approach, *J. Am. Chem. Soc.*, 132 (2010) 1804–1806.
315. M. Nagae, A. Tsuchiya, T. Katayama, K. Yamamoto, S. Wakatsuki, and R. Kato, Structural basis of the catalytic reaction mechanism of novel 1,2-α-L-fucosidase from *Bifidobacterium bifidum*, *J. Biol. Chem.*, 282 (2007) 18497–18509.

316. N. G. Oikonomakos, C. Tiraidis, D. D. Leonidas, S. E. Zographos, M. Kristiansen, C. U. Jessen, L. Nørskov-Lauritsen, and L. Agius, Iminosugars as potential inhibitors of glycogenolysis: Structural insights into the molecular basis of glycogen phosphorylase inhibition, *J. Med. Chem.*, 49 (2006) 5687–5701.
317. D. A. Kuntz, A. Ghavami, B. D. Johnston, B. M. Pinto, and D. R. Rose, Crystallographic analysis of the interactions of *Drosophila melanogaster* golgi α-mannosidase II with the naturally occurring glycomimetic salacinol and its analogues, *Tetrahedron: Asymmetry*, 16 (2005) 25–32.
318. M. Verhaest, W. Lammens, K. Le Roy, C. J. De Ranter, A. Van Laere, A. Rabijns, and W. Van den Ende, Insights into the fine architecture of the active site of chicory fructan 1-exohydrolase: 1-Kestose as substrate vs sucrose as inhibitor, *New Phytol.*, 174 (2007) 90–100.
319. F. Cardona, C. Parmeggiani, E. Faggi, C. Bonaccini, P. Gratteri, L. Sim, T. M. Gloster, S. Roberts, G. J. Davies, D. R. Rose, and A. Goti, Total syntheses of casuarine and its 6-*O*-α-glucoside: Complementary inhibition towards glycoside hydroalses of the GH31 and GH37 families, *Chem. Eur. J.*, 15 (2009) 1627–1636.
320. F. Cardona, A. Goti, C. Parmeggiani, P. Parenti, M. Forcella, P. Fusi, L. Cipolla, S. M. Roberts, G. J. Davies, and T. M. Gloster, Casuarine-6-O-a-D-glucoside and its analogues are tight binding inhibitors of insect and bacterial trehalases, *Chem. Commun.*, 46 (2010) 2629–2631.
321. M. Degano, S. C. Almo, J. C. Sacchettini, and V. L. Schramm, Trypanosomal nucleoside hydrolase. A novel mechanism from the structure with a transition-state inhibitor, *Biochemistry*, 37 (1998) 6277–6285.
322. W. Shi, C. M. Li, P. C. Tyler, R. H. Furneaux, C. Grubmeyer, V. L. Schramm, and S. C. Almo, The 2.0 Å structure of human hypoxanthine-guanine phosphoribosyltransferase in complex with a transition-state analog inhibitor, *Nat. Struct. Biol.*, 6 (1999) 588–593.
323. W. Shi, C. M. Li, P. C. Tyler, R. H. Furneaux, S. M. Cahill, M. E. Girvin, C. Grubmeyer, V. L. Schramm, and S. C. Almo, The 2.0 Å structure of malarial purine phosphoribosyltransferase in complex with a transition-state analogue inhibitor, *Biochemistry*, 38 (1999) 9872–9880.
324. W. Shi, N. R. Munagala, C. C. Wang, C. M. Li, P. C. Tyler, R. H. Furneaux, C. Grubmeyer, V. L. Schramm, and S. C. Almo, Crystal structures of *Giardia lamblia* guanidine phosphoribosyltransferase at 1.75 Å, *Biochemistry*, 39 (2000) 6781–6790.
325. W. Shi, L. A. Basso, D. S. Santos, P. C. Tyler, R. H. Furneaux, J. S. Blanchard, S. C. Almo, and V. L. Schramm, Structures of purine nucleoside phophorylase from *Mycobacterium tuberculosis* in complexes with immucillin-H and its pieces, *Biochemistry*, 40 (2001) 8204–8215.
326. G. A. Kicska, P. C. Tyler, G. B. Evans, R. H. Furneaux, W. Shi, A. Fedorov, A. Lewandowicz, S. M. Cahill, S. C. Almo, and V. L. Schramm, Atomic dissection of the hydrogen bond network for transition-state analogue binding to purine nucleoside phosphorylase, *Biochemistry*, 41 (2002) 14489–14498.
327. W. Filgueira de Azevedo, Jr., F. Canduri, D. Marangoni dos Santos, J. H. Pereira, M. V. B. Dias, R. G. Silva, M. A. Mendes, L. A. Basso, M. S. Palma, and D. S. Santos, Structural basis for inhibition of human PNP by immucillin-H, *Biochem. Biophys. Res. Commun.*, 309 (2003) 917–922.
328. A. Lewandowicz, W. Shi, G. B. Evans, P. C. Tyler, R. H. Furneaux, L. A. Basso, D. S. Santos, S. C. Almo, and V. L. Schramm, Over-the barrier transition state analogues and crystal structure with *Mycobacterium tuberculosis* purine nucleoside phosphorylase, *Biochemistry*, 42 (2003) 6057–6066.
329. V. Singh, W. Shi, G. B. Evans, P. C. Tyler, R. H. Furneaux, S. C. Almo, and V. L. Schramm, Picomolar transition state analogue inhibitors of human 5'-methylthioadenosine phosphorylase and X-ray structure with MT-immucillin-A, *Biochemistry*, 43 (2004) 9–18.
330. W. Shi, L.-M. Ting, G. A. Kicska, A. Lewandowicz, P. C. Tyler, G. B. Evans, R. H. Furneaux, K. Kim, S. C. Almo, and V. L. Schramm, *Plasmodium falciparum* purine nucleoside phosphorylase. Crystal structures, immucillin inhibitors, and dual catalytic function, *J. Biol. Chem.*, 279 (2004) 18103–18106.

331. W. Versées, J. Barlow, and J. Steyaert, Transition-state complex of the purine-specific nucleoside hydroalse of *Trypanosoma vivax*: Enzyme conformational changes and implications on catalysis, *J. Mol. Biol.*, 359 (2006) 331–346.
332. A. Vandemeulebroucke, S. de Vos, E. Van Holsbeke, J. Steyaert, and W. Versées, A flexible loop as a functional element in the catalytic mechanism of nucleoside hydrolase from *Trypanosoma vivax*, *J. Biol. Chem.*, 283 (2008) 22272–22282.
333. A. S. Murkin, M. R. Birck, A. Rinaldo-Matthis, W. Shi, E. A. Taylor, S. C. Almo, and V. L. Schramm, Neighboring group participation in the transition state of human purine nucleoside phosphorylase, *Biochemistry*, 46 (2007) 5038–5049.
334. E. A. Taylor, A. Rinaldo-Matthis, L. Li, M. Ghanem, K. Z. Hazleton, M. B. Cassera, S. C. Almo, and V. L. Schramm, *Anopheles gambiae* purine nucleoside phosphorylase: Catalysis, structure, and inhibition, *Biochemistry*, 46 (2007) 12405–12415.
335. A. Rinaldo-Matthis, C. Wing, M. Ghanem, H. Deng, P. Wu, A. Gupta, P. C. Tyler, G. B. Evans, R. H. Furneaux, S. C. Almo, C. C. Wang, and V. L. Schramm, Inhibition and structure of *Trichomonas vaginalis* purine nucleoside phosphorylase with picomolar transition state analogues, *Biochemistry*, 46 (2007) 659–668.
336. A. Rinaldo-Matthis, A. S. Murkin, U. A. Ramagopal, K. Clinch, S. P. H. Mee, G. B. Evans, P. C. Tyler, R. H. Furneaux, S. C. Almo, and V. L. Schramm, L-Enantiomers of transition state analogue inhibitors bound to human purine nucleoside phosphorylase, *J. Am. Chem. Soc.*, 130 (2008) 842–844.
337. A. Fornili, B. Giabbai, G. Garau, and M. Degano, Energy landscapes associated with macromolecular conformational changes from endpoint structures, *J. Am. Chem. Soc.*, 132 (2010) 17570–17577.
338. M.-C. Ho, W. Shi, A. Rinaldo-Matthis, P. C. Tyler, G. B. Evans, K. Clinch, S. C. Almo, and V. L. Schramm, Four generations of transition-state analogues for human purine nucleoside hydrolase, *Proc. Natl. Acad. Sci. USA*, 107 (2010) 4805–4812.
339. R. Guan, M.-C. Ho, S. C. Almo, and V. L. Schramm, Methylthioinosine phosphorylase from *Pseudomonas aeruginosa*. Structure and annotation of a novel enzyme in quorum sensing, *Biochemistry*, 50 (2011) 1247–1254.
340. J. A. Gutierrez, T. Crowder, A. Rinaldo-Matthis, M.-C. Ho, S. C. Almo, and V. L. Schramm, Transition state analogs of 5'-methylthioadenosine nucleosidase disrupt quorum sensing, *Nat. Chem. Biol.*, 5 (2009) 251–257.
341. J. A. R. Mead, J. W. Smith, and R. T. Williams, Detoxication (LXVII). Biosynthesis of the glucuronides of umbelliferone and 4-methylumbelliferone and their use in fluorimetric determination of β-glucuronidase, *Biochem. J.*, 61 (1955) 569–574.
342. I. J. Goldstein and R. N. Iyer, Interaction of concanavalin A, a phytohemagglutinin, with model substrates, *Biochim. Biophys. Acta*, 121 (1966) 197–200.
343. J. Langridge, Mutations conferring quantitative and qualitative increases in β-galactosidase activity in *Escherichia coli*, *Mol. Gen. Genet.*, 105 (1969) 74–83.
344. M. Yde, C. K. De Bruyne, and F. G. Loontiens, Study of the binding of 1-thio-β-D-galactosides on β-galactosidase from *E. coli*, *Arch. Int. Physiol. Biochim.*, 84 (1976) 210–211.
345. M. Yde and C. K. De Bruyne, Binding of substituted phenyl 1-thio-β-D-galactopyranosides to β-D-galactosidase from *E. coli*, *Carbohydr. Res.*, 60 (1978) 155–165.
346. G. Legler and H. Liedtke, Glucosylceramidase from calf spleen. Characterization of its active site with 4-n-alkylumbelliferyl β-glucosides and N-alkyl derivatives of 1-deoxynojirimycin, *Biol. Chem. Hoppe Seyler*, 366 (1985) 1113–1122.
347. G. M. Aerts, O. Van Opstal, and C. K. De Bruyne, Mixed inhibition of β-D-glucosidase from *Stachybotrys atra* by substrate analogues, *Carbohydr. Res.*, 138 (1985) 127–134.
348. N. Baggett, M. A. Case, P. R. Darby, and C. J. Cray, Action of almond β-D-glucosidase on fluorogenic substrates derived from 4-substituted 7-hydroxycoumarins, *Enzyme Microb. Technol.*, 15 (1993) 742–748.

349. M. P. Dale, H. E. Ensley, K. Kern, K. A. R. Shastry, and L. D. Byers, Reversible inhibitors of β-glucosidase, *Biochemistry*, 24 (1985) 3530–3539.
350. Y.-K. Li, H.-S. Hsu, L.-F. Chang, and G. Chen, New imidazoles as probes of the active site topology and potent inhibitors of β-glucosidase, *J. Biochem.*, 123 (1998) 416–422.
351. S. J. Charnock, T. D. Spurway, H. Xie, M.-H. Beylot, R. Virden, R. A. J. Warren, G. P. Hazlewood, and H. J. Gilbert, The topology of the substrate binding clefts of glycosyl hydrolase family 10 xylanases are not conserved, *J. Biol. Chem.*, 273 (1998) 32187–32199.
352. M. Czjzek, M. Cicek, V. Zamboni, D. R. Bevan, B. Henrissat, and A. Esim, The mechanism of substrate (aglycon) specificity in β-glucosidases is revealed by crystal structures of mutant maize β-glucosidase-DIMBOA, -DIMBOAGlc, and -dhurrin complexes, *Proc. Natl. Acad. Sci. USA*, 97 (2000) 13555–13560.
353. L. Verdoucq, J. Moriniere, D. R. Bevan, A. Esen, A. Vasella, B. Henrissat, and M. Czjzek, Structural determinants for substrate specificity in family 1 β-glucosidases, *J. Biol. Chem.*, 279 (2004) 31796–31803.
354. H. Murai, H. Enomoto, Y. Aoyagi, Y. Yoshikuni, M. Yagi, and I. Shirahase, *N*-Alkylpiperidine derivatives. *German Pat.* DE 2824781 A1 19790104.
355. Stoltcfuss, J. 1-Deoxynojirimycin and *N*-substituted derivatives. *Eur. Pat.* 1979–102174 19790629.
356. J. Schweden, C. Borgmann, G. Legler, and E. Bause, Characterization of calf liver glucosidase I and its inhibition by basic sugar analogs, *Arch. Biochem. Biophys.*, 248 (1986) 335–340.
357. G. W. J. Fleet, A. Karpas, R. A. Dwek, L. E. Fellows, A. S. Tyms, S. Petursson, S. K. Namgoong, N. G. Ramsden, P. W. Smith, J. C. Son, F. Wilson, D. R. Witty, G. S. Jacob, and T. W. Rademacher, Inhibition of HIV replication by amino-sugar derivatives, *FEBS Lett.*, 237 (1988) 128–132.
358. B. Lesur, J.-B. Ducep, M.-N. Lalloz, A. Ehrhard, and C. Danzin, New deoxynojirimycin derivatives as potent inhibitors of intestinal α-glucohydrolases, *Bioorg. Med. Chem. Lett.*, 3 (1997) 355–360.
359. D. S. Alonzi, R. A. Dwek, and T. D. Butters, Improved cellular inhibitors for glycoprotein processing α-glucosidases: Biological characterisation of alkyl- and arylalkyl-N-substituted deoxynojirimycins, *Tetrahedron: Asymmetry*, 20 (2009) 897–901.
360. K. Osiecki-Newman, D. Fabbro, G. Legler, R. J. Desnick, and G. A. Grabowski, Human acid β-glucosidase: Use of inhibitors, alternative substrates and amphiphiles to investigate the properties of the normal and Gaucher disease active sites, *Biochim. Biophys. Acta*, 915 (1987) 87–100.
361. G. Legler and E. Bieberich, Isolation of cytosolic β-glucosidase from calf liver and characterization of its active site with alkyl glucosides and basic glycosyl derivatives, *Arch. Biochem. Biophys.*, 260 (1988) 427–436.
362. H. S. Overkleeft, G. H. Renkema, J. Neele, P. Vianello, I. O. Hung, A. Strijland, A. M. van der Burg, G.-J. Koomen, U. K. Pandit, and J. M. F. G. Aerts, Generation of specific deoxynojirimycin-type inhibitors of the non-lysosomal glucoceramidase, *J. Biol. Chem.*, 273 (1998) 26522–26527.
363. L. G. Dickson, E. Leroy, and J.-L. Reymond, Structure-activity relationships in aminocyclopentitol glycosidase inhibitors, *Org. Biomol. Chem.*, 2 (2004) 1217–1226.
364. T. D. Butters, L. A. G. M. Van den Broek, G. W. J. Fleet, T. M. Krulle, M. R. Wormald, R. A. Dwek, and F. M. Platt, Molecular requirements of iminosugars for the selective control of *N*-linked glycosylation and glycosphingolipid biosynthesis, *Tetrahedron: Asymmetry*, 11 (2000) 113–124.
365. T. M. Wrodnigg, F. Diness, C. Gruber, H. Häusler, I. Lundt, K. Rupitz, A. J. Steiner, A. E. Stütz, C. A. Tarling, S. G. Withers, and H. Wölfler, Probing the aglycon binding site of a β-glucosidase: A collection of C-1-modified 2,5-dideoxy-2,5-imino-D-mannitol derivatives and their structure-activity relationships as competitive inhibitors, *Bioorg. Med. Chem.*, 12 (2004) 3485–3495.
366. T. M. Wrodnigg, A. E. Stütz, C. A. Tarling, and S. G. Withers, Fine tuning of β-glucosidase inhibitory activity in the 2,5-dideoxy-2,5-imino-D-mannitol (DMDP) system, *Carbohydr. Res.*, 341 (2006) 1717–1722.

367. A. R. Sawkar, S. L. Adamski-Werner, W.-C. Cheng, C.-H. Wong, E. Beutler, K.-P. Zimmer, and J. W. Kelly, Gaucher disease-associated glucocerebrosidases show mutation-dependent chemical chaperoning profiles, *Chem. Biol.*, 12 (2005) 1235–1244.
368. Z. Yu, A. R. Sawkar, L. J. Whalen, C.-H. Wong, and J. W. Kelly, Isofagomine- and 2,5-anhydro-2,5-imino-D-glucitol based glucocerebrosidase pharmacological chaperones for Gaucher disease prevention, *J. Med. Chem.*, 50 (2007) 94–100.
369. L. Yu, K. Ikeda, A. Kato, I. Adachi, G. Godin, P. Compain, O. Martin, and N. Asano, α-1-C-Octyl-1-deoxynojirimycin as a pharmacological chaperone for Gaucher disease, *Bioorg. Med. Chem.*, 14 (2006) 7736–7744.
370. P. Compain, O. R. Martin, C. Boucheron, G. Godin, L. Yu, K. Ikeda, and N. Asano, Design and synthesis of highly potent and selective pharmacological chaperones for the treatment of Gaucher disease, *Chembiochem*, 7 (2006) 1356–1359.
371. F. Oulaidi, S. Front-Deschamps, E. Gallienne, E. Lesellier, K. Ikeda, N. Asano, P. Compain, and O. R. Martin, Second-generation iminoxylitol-based pharmacological chaperones for the treatment of Gaucher disease, *ChemMedChem*, 6 (2011) 353–361.
372. T. Wennekes, R. J. B. H. N. van den Berg, W. Donker, G. A. van der Marel, A. Strijland, J. M. F. G. Aerts, and H. S. Overkleeft, Development of adamantan-1-yl-methoxy-functionalised 1-deoxynojirimycin derivatives as selective inhibitors of glucosylceramide metabolism in man, *J. Org. Chem.*, 72 (2007) 1088–1097.
373. T. Wennekes, R. J. B. H. N. van den Berg, T. J. Boltje, W. E. Donker-Koopman, B. Kuijper, G. J. A. van der Marel, A. Strijland, C. P. Verhagen, J. M. F. G. Aerts, and H. S. Overkleeft, Synthesis and evaluation of lipophilic aza-C-glycosides as inhibitors of glucosylceramide metabolism, *Eur. J. Org. Chem.* (2010) 1258–1283.
374. E. M. Sanchez-Fernandez, R. Risquez-Cuadro, M. Chasseraud, A. Ahidouch, C. Ortiz Mellet, H. Ouadid-Ahidouch, and J. M. Garcia Fernandez, Synthesis of N-, S-, and C-glycoside castanospermine analogues with selective neutral α-glucosidase inhibitory activity as antitumour agents, *Chem. Commun.*, 46 (2010) 5328–5330.
375. M. Aguilar-Moncayo, M. I. Garcia-Moreno, A. Trapero, M. Egido-Gabas, A. Llebaria, J. M. Garcia Fernandez, and C. Ortiz Mellet, Bicyclic (*galacto*)nojiriymcin analogues as glycosidase inhibitors: Effect of structural modifications in their pharmacological chaperone potential towards β-glucocerebrosidase, *Org. Biomol. Chem.*, 9 (2011) 3698–3713.
376. H. Häusler, K. Rupitz, A. E. Stütz, and S. G. Withers, N-Alkylated derivatives of 1,5-dideoxy-1,5-iminoxylitol as β-xylosidase and β-glucosidase inhibitors, *Chem. Monthly*, 133 (2002) 555–560.
377. C.-F. Chang, C.-W. Ho, C.-Y. Wu, T.-A. Chao, C.-H. Wong, and C.-H. Lin, Discovery of picomolar slow tight-binding inhibitors of α-fucosidase, *Chem. Biol.*, 11 (2004) 1301–1306.
378. C.-W. Ho, Y.-N. Lin, C.-F. Chang, S.-T. Li, Y.-T. Wu, C.-Y. Wu, C.-F. Chang, S.-W. Liu, Y.-K. Li, and C.-H. Lin, Discovery of different types of inhibition between the human and *Thermotoga maritima* α-fucosidases by fuconojirimycin-based derivatives, *Biochemistry*, 45 (2006) 5695–5702.
379. P.-H. Liang, W.-C. Cheng, Y.-L. Lee, H.-P. Yu, Y.-T. Wu, Y.-L. Lin, and C.-H. Wong, Novel five-membered iminocyclitol derivatives as selective and potent glycosidase inhibitors: New structures for antivirals and osteoarthritis, *ChemBioChem*, 7 (2006) 165–167.
380. G. Schitter, A. J. Steiner, G. Pototschnig, E. Scheucher, M. Thonhofer, C. A. Tarling, S. G. Withers, K. Fantur, E. Paschke, D. J. Mahuran, B. A. Rigat, M. B. Tropak, C. Illaszewicz, R. Saf, A. E. Stütz, and T. M. Wrodnigg, Fluorous iminoalditols: A new family of glycosidase inhibitors and pharmacological chaperones, *ChemBioChem*, 11 (2010) 2026–2033.
381. A. Ghisaidoobe, P. Bikker, A. C. J. de Bruijn, F. D. Godschalk, E. Rogaar, M. C. Guijt, P. Hagens, J. M. Halma, S. M. van't Hart, S. B. Luitjens, V. H. S. van Rixel, M. Wijzenbroek, T. Zweegers, W. E. Donker-Koopman, A. Strijland, R. Boot, G. van der Marel, H. S. Overkleeft, J. M. F. G. Aerts, and R. J. B. H. N. van den Berg, *ACS Med. Chem. Lett.*, 2 (2011) 119–123.

382. A. Hermetter, H. Scholze, A. E. Stütz, S. G. Withers, and T. M. Wrodnigg, Powerful probes for glycosidases: Novel, fluorescently tagged glycosidase inhibitors, *Bioorg. Med. Chem. Lett.*, 11 (2001) 1339–1342.
383. P. Greimel, H. Häusler, I. Lundt, K. Rupitz, A. E. Stütz, C. A. Tarling, S. G. Withers, and T. M. Wrodnigg, Fluorescent glycosidase inhibiting 1,5-dideoxy-1,5-iminoalditols, *Bioorg. Med. Chem. Lett.*, 16 (2006) 2067–2070.
384. I. Lundt, A. J. Steiner, A. E. Stütz, C. A. Tarling, S. Ully, S. G. Withers, and T. M. Wrodnigg, Fluorescently tagged iminoalditol glycosidase inhibitors as novel biological probes and diagnostics, *Bioorg. Med. Chem.*, 14 (2006) 1737–1742.
385. A. J. Steiner, A. E. Stütz, C. A. Tarling, S. G. Withers, and T. M. Wrodnigg, Iminoalditol-amino acid hybrids: Synthesis and evaluation as glycosidase inhibitors, *Carbohydr. Res.*, 342 (2007) 1850–1858.
386. A. J. Steiner, G. Schitter, A. E. Stütz, T. M. Wrodnigg, C. A. Tarling, S. G. Withers, K. Fantur, D. Mahuran, E. Paschke, and M. Tropak, 1-Deoxygalactonojirimycin-lysine hybrids as potent D-galactosidase inhibitors, *Bioorg. Med. Chem.*, 16 (2008) 10216–10220.
387. A. J. Steiner, A. E. Stütz, T. M. Wrodnigg, C. A. Tarling, S. G. Withers, A. Hermetter, and H. Schmidinger, Glycosidase profiling with immobilized glycosidase-inhibiting iminoalditols—A proof-of-concept study, *Bioorg. Med. Chem. Lett.*, 18 (2008) 1922–1925.
388. A. J. Rawlings, H. Lomas, A. W. Pilling, M. J.-R. Lee, D. S. Alonzi, J. S. S. Rountree, S. F. Jenkinson, G. W. J. Fleet, R. A. Dwek, J. H. Jones, and T. D. Butters, Synthesis and biological characterisation of novel *N*-alkyl-deoxynojirimycin α-glucosidase inhibitors, *ChemBioChem*, 10 (2009) 1101–1105.
389. M. Van Scherpenzeel, R. J. B. H. N. van den Berg, W. E. Donker-Koopman, R. M. J. Liskamp, J. M. F. G. Aerts, H. S. Overkleeft, and R. J. Pieters, Nanomolar affinity, aminosugar based chemical probes for specific labeling of lysosomal glucocerebrosidase, *Bioorg. Med. Chem.*, 18 (2010) 267–273.
390. M. N. Gandy, A. W. Debowski, and K. A. Stubbs, A general method for affinity-based proteomic profiling of exo-α-glucosidases, *Chem. Commun.*, 47 (2011) 5037–5039.
391. M. Aguilar-Moncayo, M. I. Garcia-Moreno, A. E. Stütz, J. M. Garcia Fernandez, T. M. Wrodnigg, and C. Ortiz Mellet, Fluorescent-tagged sp^2-iminosugars with potent β-glucosidase inhibitory activity, *Bioorg. Med. Chem.*, 18 (2010) 7439–7445.
392. Z. Luan, K. Higaki, M. Aguilar-Moncayo, L. Li, H. Ninomiya, E. Nanba, K. Ohno, M. I. Garcia-Moreno, C. Ortiz Mellet, J. M. Garcia Fernandez, and Y. Suzuki, A fluorescent sp^2-iminosugar with pharmacological chaperone activity for Gaucher disease: Synthesis and intracellular distribution studies, *ChemBioChem*, 11 (2010) 2453–2464.
393. G. Pototschnig, C. Morales de Csaky, J. R. Montenegro Burke, G. Schitter, A. E. Stütz, C. A. Tarling, S. G. Withers, and T. M. Wrodnigg, Synthesis and biological evaluation of novel biotin-iminoalditol conjugates, *Bioorg. Med. Chem. Lett.*, 20 (2010) 4077–4079.
394. G. Schitter, E. Scheucher, A. J. Steiner, A. E. Stütz, M. Thonhofer, C. A. Tarling, S. G. Withers, J. Wicki, K. Fantur, E. Paschke, D. J. Mahuran, B. A. Rigat, M. Tropak, and T. M. Wrodnigg, Synthesis of lipophilic 1-deoxygalactonojirimycin derivatives as D-galactose inhibitors, *Beilstein J. Org. Chem.*, 6(2010) No. 21.
395. A. M. C. H. van den Nieuwendijk, M. Ruben, S. E. Engelsma, M. D. P. Risseeuw, R. J. B. H. N. van den Berg, R. G. Boot, J. M. Aerts, J. Brussee, G. A. van der Marel, and H. Overkleeft, Synthesis of L-altro-1-deoxynojirimycin, D-allo-1-deoxynojirimycin, and D-galacto-1-deoxynojirimycin from a single chiral cyanohydrin, *Org. Lett.*, 12 (2010) 3957–3959.
396. A. Guaragna, D. D'Alonzo, C. Paolella, and G. Palumbo, Synthesis of 1-deoxy-L-gulonojirimycin and 1-deoxy-L-talonojirimycin, *Tetrahedron Lett.*, 50 (2009) 2045–2047.
397. M. Ganesan and N. G. Ramesh, A new and short synthesis of naturally occurring 1-deoxy-L-gulonojirimycin from tri-*O*-benzyl-D-glucal, *Tetrahedron Lett.*, 51 (2010) 5574–5576.

398. M. Ruiz, T. M. Ruanova, O. Blanco, F. Nunez, C. Pato, and V. Ojea, Diastereoselective synthesis of piperidine imino sugars using aldol additions of metalated bislactim ethers to threose and erythrose acetonides, *J. Org. Chem.*, 73 (2008) 2240–2255.
399. B. La Ferla, P. Bugada, and F. Nicotra, Synthesis of the dimethyl ester of 1-deoxy-L-idonojirimycin-1-methylenephosphonate: A new approach to iminosugar phosphonates, *J. Carbohydr. Chem.*, 25 (2006) 151–162.
400. A. Ak, S. Prudent, D. LeNouen, A. Defoin, and C. Tarnus, Synthesis of all-cis 2,5-imino-2,5-dideoxyfucitol and its evaluation as a potent fucosidase and galactosidase inhibitor, *Bioorg. Med. Chem. Lett.*, 20 (2010) 7410–7413.
401. C.-Y. Yu, N. Asano, K. Ikeda, M.-X. Wang, T. D. Butters, M. R. Wormald, R. A. Dwek, A. L. Winters, R. J. Nash, and G. W. J. Fleet, Looking glass inhibitors: L-DMDP, a more potent and specific inhibitor of α-glucosidase than the enantiomeric natural product DMDP, *Chem. Commun.* (2004) 1936–1937.
402. Y. Berliot, D. Gretzke, T. M. Krulle, T. D. Butters, R. A. Dwek, R. J. Nash, N. Asano, and G. W. J. Fleet, Looking glass inhibitors: Efficient synthesis and biological evaluation of D-fuconojirimycin, *Carbohydr. Res.*, 340 (2005) 2713–2718.
403. A. E. Hakansson, J. van Ameijde, L. Guglielmini, G. Horne, R. J. Nash, E. L. Evinson, A. Kato, and G. W. J. Fleet, Looking glass inhibitors: Synthesis of a potent naringinase inhibitor L-DIM (1,4-dideoxy-1,4-imino-L-mannitol), the enantiomer of DIM (1,4-dideoxy-1,4-imino-D-mannitol) a potent α-D-mannosidase inhibitor, *Tetrahedron: Asymmetry*, 18 (2007) 282–289.
404. T. B. Mercer, S. F. Jenkinson, B. Bartholomew, R. J. Nash, S. Miyauchi, A. Kato, and G. W. J. Fleet, Looking glass inhibitors: Bote enantiomeric N-benzyl derivatives of 1,4-dideoxy-1,4-imino-D-lyxitol (a potent competitive inhibitor of α-D-galactosidase) and of 1,4-dideoxy-1,4-imino-L-lyxitol (a weak competitive inhibitor of α-D-galactosidase) inhibit naringinase, an α-L-rhamnosidase competitively, *Tetrahedron: Asymmetry*, 20 (2009) 2368–2373.
405. D. D'Alonzo, A. Guaragna, and G. Palumbo, Glycomimetics at the mirror: Medicinal chemistry of L-iminosugars, *Curr. Med. Chem.*, 16 (2009) 473–505.
406. D. Best, C. Wang, A. C. Weymouth-Wilson, R. A. Clarkson, F. X. Wilson, R. J. Nash, S. Miyauchi, A. Kato, and G. W. J. Fleet, Looking glass inhibitors: Scalable syntheses of DNJ, DMDP, and (3R)-3-hydroxy-L-bulgecinine from D-glucuronolactone and of L-DNJ, L-DMDP, and (3S)-3-hydroxy-D-bulecinine from L-glucuronolactone. DMDP inhibits β-glucosidases and β-galactosidases whereas L-DMDP is a potent and specific inhibitor of α-glucosidases, *Tetrahedron: Asymmetry*, 21 (2010) 311–319.
407. J.-Q. Fan, Iminosugars as active-site-specific chaperones for the treatment of lysosomal storage disorders, in P. Compain and O. R. Martin, (Eds.), *Iminosugars: From Synthesis to Therapeutic Applications,* Wiley, Chichester, 2007, pp. 225–247.
408. P. Compain, V. Desvergnes, V. Liautard, C. Pillard, and S. Toumieux, Tables of iminosugars, their biological activities and their potential as therapeutic agents, in P. Compain and O. R. Martin, (Eds.), *Iminosugars: From Synthesis to Therapeutic Applications,* Wiley, Chichester, 2007, pp. 327–455.
409. O. Martin, Iminosugars: Current and future therapeutic applications, *Ann. Pharm. Fr.*, 65 (2007) 5–13.
410. T. M. Wrodnigg, A. J. Steiner, and B. J. Überbacher, Natural and synthetic iminosugars as carbohydrate processing enzyme inhibitors for cancer therapy, *Anticancer Agents Med. Chem.*, 8 (2008) 77–85.
411. G. Horne, F. X. Wilson, J. Tinsley, D. H. Williams, and R. Storer, Iminosugars past, present and future: Medicines for tomorrow, *Drug Discov. Today*, 16 (2011) 107–118.
412. R. Zhang, W. Zhang, and T. Hu, Dextran glucosidase: A potential target of iminosugars in caries prevention, *Med. Hypotheses*, 76 (2011) 574–575.

AUTHOR INDEX

Page numbers in roman type indicate that the listed author is cited on that page; page numbers in italic denote the page where the literature citation is given.

A

Aamlid, K.H., 53, *66–68*
Abell, A.D., 206, *283*
Achari, B., 164, *184, 185*
Adachi, I., 252, *295*
Adam, M.J., 199, 217–218, *281*
Adamski-Werner, S.L., 250, *295*
Adelhorst, K., 199, *281*
Adley, T.J., 189–190, *276*
Aerts, G.M., 247, *294*
Aerts, J.M.F.G., 250, 252, 254–255, 258, *284, 295–297*
Aghajari, N., 223–224, 230, 232, *288, 290*
Agius, L., 240, *292*
Aguilar-Moncayo, M., 231–233, 253, 256, *290, 296, 297*
Ahidouch, A., 253, *295*
Ahmard, R., 70, *115*
Ak, A., 258, *297*
Akazawa, N., 192, *278*
Akhtar, F., *67*
Akhtar, M.N., 204, *282*
Aktories, K., 231–232, *290*
Alberch, L., 147–148, *181*
Alcazar, E., 145–146, *181*
Al Daher, S., 217, 220, 222, *286*
Aleshin, A.E., 224–227, *288*
Ali, M.A., 48, 51–53, *67*
Ali, M.H., 48, 51–53, *66*
Ali, Y., 20–23, 51–57, *62–64*
Allwein, S.P., 144–145, *181*

Almo, S.C., 242–246, *292–294*
Alonzi, D.S., 249, 254–255, *295, 296*
Altona, C., 156, *183*
Andersen, S.M., 220, 221, *287*
Anderson, B.F., 230, *289*
Anderson, C.B., 159, *183*
Andrews, T., 198–199, *281*
Anet, E.F.L.J., 123, 128–129, *179*
Antholine, W.E., 70–71, 74–75, 77, 87, *115, 118*
Aoki, H., 192, *277*
Aoyagi, T., 192, *277*
Aoyagi, Y., 190, 248, *277, 294*
Arakawa, H., 70, *115*
Aratani, M., 188, *276*
Ardevol, A., 239, *292*
Arion, D., 171, *186*
Arndt, H.-D., 177, *186*
Arnold, E., 190–191, *277*
Arnone, A., 219, *287*
Asakawa, M., 109–110, *120*
Asakura, T., 164, *184*
Asano, N., 193, 216, 221, 252, 258, *279, 286, 295, 297*
Ashiura, M., 209–210, *284*
Asma, A., 204, *282*
Atherton, N.M., 108, *120*
Atria, A.M., 71, 72, 74–75, 104–107, 110–113, *116, 117, 120*
Atsumi, S., 206, *283*
Atta-ur-Rahman, 204, *282*

Aviezer, D., 229–232, 239, 252, 253, *289*, *290*
Axelrod, B., 198, *280*

B

Babu, B.R., 164, *184*
Bach, M., 200–201, *281*
Bach, P., 211, *284*
Baggett, N., 247, *294*
Bainbridge, B.W., *67*
Bakac, A., 71, *116*
Baker, M.A., 233–234, *290*
Baldisseri, D.M., 161, 172–174, *184*
Ballard, J.M., 37, 38, *63–65*
Banks, B., 194–195, *279*
Barberousse, V., 176–177, *186*
Barford, A.D., 20–21, *62*, *63*
Barlow, J., 244–245, *293*
Barr-David, G., 72, 73, 87–89, 99, 100, 110–111, *117*, *120*
Bartholomew, B., 258, *297*
Basso, L.A., 243–244, *293*
Basu, S., 177, *186*
Batchelor, R., 139–140, 152–153, *180*, *182*
Bause, E., 191, 193, 221, 248, *277*, *278*, *287*, *294*
Becalski, A., 198–199, *281*
Bellamy, F., 176–177, *186*
Bell, E.A., 190–191, *277*
Bell, S., *116*, *117*, *120*
Bellú, S., 71, 72, 74–75, 86–87, 93–94, 105–107, 110–113
Bennuaskalmowski, B., 167, *185*
Benson, M., 190–191, *277*
Berecibar, A., 208–209, *283*
Berger, D., 177, *186*
Berliot, Y., 258, *297*
Berman, H.M., 169, *186*
Best, D., 258, *298*
Best, W.M., 211, *285*
Beutler, E., 250, *295*
Bevan, D.R., 229, 247–248, *289*, *294*
Beylot, M.-H., 247–248, *294*
Bhattacharjee, A., 165, *185*
Bhattacharjya, A., 165, *185*

Bhatt, R.S., 6, 26–28, 30, 31, 49, 50, *64*, *65*
Bieberich, E., 249–250, *295*
Bikker, P., 254, *296*
Birch, G.G., 21–23, *62–64*
Birck, M.R., 244–245, *293*
Bird, C.W., *64*
Blake, C.C.F., 194–195, *279*
Blanchard, J.S., 243, *293*
Blanco, O., 258, *297*
Blasiak, J., 70, 109–110, *115*
Blériot, Y., 125, 126, 174–175, *179*, 234–235, 239, 272, *290*, *292*
Blow, D.M., 224–227, *288*
Boas, S., 193, *278*
Bocian, D.F., 156, *182*, *183*
Bock, K., 188–189, 198–201, *276*, *281*
Boggio, J.C., 86–87, 90, 92, *119*
Bohak, Z., 207, *283*
Bolanos, J.G.F., 213–214, *285*
Boldin-Adamsky, S., 230, 239, *289*
Bolli, M., 164, *185*
Bols, M., 209–211, 213–214, 219–222, 224–227, *284*, *285*, *287*, *288*
Boltje, T.J., 252, *295*
Bolton, J.R., 72–75, *117*
Bonaccini, C., 241–242, *292*
Bonger, K.M., 209–210, *284*
Bontchev, P.R., 71, *116*
Boone, M.A., 148–149, *182*
Boons, G.-J., 234, *290*
Booth, H., 159, *183*
Boot, R.G., 258, *297*
Boraston, A.B., 224–227, 236, 239, *288*, *291*
Boraston, A.D., 238, *292*
Boraston, C.M., 224–227, 239, *288*
Borgmann, C., 248, *294*
Borkow, G., 171, *186*
Borodkin, V.S., 237–238, 253, *291*
Boros, S., 176–177, *186*
Borthiry, G.R., 70–71, 74–75, 77, 87, *115*, *118*
Boucheron, C., 252, *295*
Boulineau, F.P., 147–148, *181*, *182*
Bousecksou, A., 98, *119*

Boyko, S., 70–71, 74–75, *116*
Bozell, J.J., 139, *180*
Bozó, E., 176–177, *186*
Braet, C., 214, *285*
Brambley, R., 78, 87–88, 110–111, *118*
Branca, M., 77, 78, 86–88, 101, *118*, *119*
Bravo, P., 219, *287*
Brondino, C., 92–93, 98, *119*
Brouard, I., 126, *179*
Bruise, T.C., 196, *280*
Brukner, I., 171, *186*
Brumby, S., 72, 73, 87–89, 99, 110–111, *117*
Brumer, H. III, 236, *291*
Brumley, R., 72, 73, 89, 99, *117*
Brumshtein, B., 229–232, 239, 252, 253, *289*, *290*
Brunel, L.-C., 77, *118*
Brussee, J., 258, *297*
Brustolon, M., 72–75, *117*
Brzezinski, R., 238, *292*
Bubb, A.K., 238, *292*
Bugada, P., 258, *297*
Bukhari, S.T.K., *63*
Bullock, C., *63*, *66*, *67*
Burke, J.R.M., 256, *297*
Burmeister, W.P., 229, *289*
Buss, D.H., 20–21, *61*
Butcher, M.E., 149–150, *182*
Butters, T.D., 25, 229–230, 239, 249, 250, 252, 254–255, 258, *289*, *295–297*
Buttner, J.D., 206, *283*
Bülow, A., 211, *285*
Byers, L.D., 206, 247, *282*, *294*

C

Caffaratti, E., 78–79, 85–86, *118*
Cahill, S.M., 243, *292*, *293*
Cahn, R.S., 188, 231, 234–236, *276*
Caines, M.E.C., 227, 240–241, *289*
Caldeira, M.M., 105–107, *120*
Calisto, N., 78–79, 109–110, *118*
Camacho, E., 154, *183*
Camerman, A., 224–227, *288*
Camerman, N., 224–227, *288*

Cameron, D.R., 200–201, *281*
Campbell, C.S., 151, *182*
Campos, J., 154, *183*
Canac, Y., 209–210, 250, *284*
Canada, F.J., 199, *281*
Candenas, M.L., 126, *179*
Canduri, F., 243, *293*
Cao, R., 148–149, *182*
Cardona, F., 241–242, *292*
Cargnello, R., 86–88, 92–93, *118*
Carmichael, I., 159–160, *184*
Carpenter, N.C., 217, 220, 222, *286*
Carrington, C., 5–6, *67*
Carver, J.P., 233–234, *290*
Casado, N., 86–87, 90, 92, 96–97, *119*
Casas, J., 232–233, 253, *290*
Case, M.A., 247, *294*
Casimir, J., 190, *277*
Cassera, M.B., 244–245, *293*
Castillón, S., 134, *180*
Castro, S., 145–146, 149–150, 162, *181*, *182*, *184*
Chaen, H., 203, *282*
Chaka, G., 71, *116*
Chandrakumar, N.S., 213–214, *285*
Chang, C.-F., 253, *296*
Chang, H.-H., 209–210, 251, *284*
Chang, L.-F., 247, 250, *294*
Chan, J.Y.C., 53, *66*
Chantereau, C., 125, *179*
Chao, T.-A., 253, *296*
Charara, M., 72, 73, 87–89, 99, 110–111, *117*
Charnock, S.J., 247–248, *294*
Chasseraud, M., 253, *295*
Chattagoon, L., *66*
Chattopadhyay, P., 165, *185*
Chavan, S., 235–236, *291*
Chazan, J.B., 176–177, *186*
Cheah, M., 122, *179*
Chen, G., 247, 250, *294*
Cheng, G., 147–148, *181*
Cheng, R.P., 122, *178*
Cheng, W.-C., 250, 253, *295*, *296*
Chen, X., 123, *179*

Cheong, P.P.L., *66*
Cherney, M.M., 236, *291*
Cheung, T.-M., 188–189, *276*
Chiba, S., 201–202, *281*
Chiu, A.K.B., 21–23, *65*, *66*
Choi, S.S., 217, 220, 222, *286*
Chong, C., 207, *283*
Choudary, M.I., 204, *282*
Chowdhary, M.S., 21, 23, 25–28, *64*, *66*
Chowdhury, S., 165, *185*
Chrystal, E.J.T., 206, *282*
Chrzaszcz, T., 203, *282*
Cicek, M., 247–248, *294*
Cintado, C.G., 126, *179*
Cipolla, L., 193, 241–242, *279*, *292*
Ciullo, L., 72, 86–87, 93–94, *117*
Claeyssens, M., 214, 216, *285*
Clardy, J., 190–191, *277*
Clarke, J.T.R., 206, *283*
Clark, N.E., 236, *291*
Clarkson, R.A., 258, *298*
Clinch, K., 222, 244–246, *288*, *293*
Clowney, L., 169, *186*
Codd, R., 70–75, 78, 81, 86–89, 99, 100, 108–111, *115–118*, *120*
Colegate, S.M., 191, *277*
Collins, T., 216, *286*
Collyer, C.A., 224–227, *288*
Compain, P., 216, 221, 252, 258, *286*, *295*, *298*
Cooke, V.M., *68*
Corbett, C.R.R., *65*
Corey, E.J., 144–145, *181*, 188, 224–227, 231, 234–236, *276*
Cortes-Garcia, R., 29, *65*
Cottaz, S., 229, *289*
Craig, D.C., 129, *180*
Cray, C.J., 247, *294*
Cremer, D., 156, *183*
Crowder, T., 245–246, *294*
Csathy, L., *68*
Csuk, R., 219–221, *287*
Cubero, I.I., *68*
Cummings, D.A., 77, *118*

Curl, R.F., 193, *278*
Curtis-Long, M.J., 204, 206, *282*
Cutfield, J.F., 230, *289*
Cutfield, S.M., 230, *289*
Czader, A., 71, *116*
Czernuszewicz, R.S., 71, *116*
Czjzek, M., 229, 247–248, *289*, *294*

D

Daier, V., 72, 78–79, 82, 86–89, 92–93, 96–98, 104–105, 109–110, *117–119*
Dai, G.-F., 204, *282*
Dalal, N.S., 71, 74–75, *116*
Dale, M.P., 247, *294*
D'Alonzo, D., 258, *297*, *298*
Damha, M.J., 167–168, 171, *185*, *186*
Dangerfield, E.M., 193, *278*
Danilova, V.A., 161, 162, *184*
Danzin, C., 248–249, *294*
D'aquino, J.A., 230, 236, 239, *289*
Darby, P.R., 247, *294*
Dardenne, G., 190, *277*
Das, B.C., *64*
Datta, S., 165, *185*
Davies, G.D., 125, 126, 174–175, *179*, 239, 272, *292*
Davies, G.J., 207, 214, 224–239, 241–242, 250, 253, *283*, *285*, *288–292*
Davis, B.G., 193, *278*
Davison, B.E., *66*
Dax, K., 218, *286*
Dean, K.J., 202, *281*
de Azevedo, W.F. Jr., 243, *293*
DeBeer George, S., 71, 72, 77, *116*
de Bello, I.C., *67*
DeBoer, B.G., 72, 100–101, *117*
Debowski, A.W., 237–238, 254–255, *291*, *297*
de Bruijn, A.C.J., 254, *296*
De Bruyne, C.K., 247, *294*
De Crescenzo, D.P., 218, *286*
De Crescenzo, G.A., 218, *286*
de Csaky, C.M., 256, *297*
de Fina, G.M., 141, *181*
Defoin, A., 258, *297*

Degano, M., 242, 245, *292, 293*
de Gori, R., 227–228, *289*
DeGrado, W.F., 122, *178*
de Lederkremer, R.M., 141, *181*
DeMatteo, M.P., 161, *184*
De Mesmaeker, A., 168, *185*
Demeter, Á., 177, *186*
Deng, H., 244–245, *293*
Deng, L., 236, *291*
Dennis, J.W., 233–234, *290*
Dennis, R.J., 237–238, *291*
Depezay, J.-C., 152–153, *182*
de Raadt, A., 193, *278*
Deraeve, C.l., 177, *186*
De Ranter, C.J., 240–241, *292*
Derome, A.E., 190–191, *277*
Desai, M.C., 144–145, *181*
Desilets, S., 159, *183*
Deslongchamps, P., 159, *183*
Desmet, T., 216, *285*
Desnick, R.J., 193, 249–250, *278, 295*
Dessí, A., 70–71, 74–75, 77, 78, 86–88, 101, *116, 118, 119*
Desvergnes, V., 258, *298*
Dethlefsen, J.R., 77, *118*
De Vos, D., 216, *286*
de Vos, S., 244–245, *293*
Dias, F.M.V., 224–227, *288*
Dias, M.V.B., 243, *293*
Diaz, M.T., 126, *179*
Díaz, Y., 133–134, *180*
di Bello, I.C., 51, 217, 220, 222, *286*
Dickson, L.G., 250, *295*
Di, J., 218, *286*
Dillon, C.T., 71, 78, 81, 86–87, 99, 102, *116*
Diness, F., 250, *295*
Dinev, Z., 235–236, *291*
Dinur, T., 193, *278*
Dixon, J.M., 159, *183*
Dixon, N.E., 94–95, 100, *119, 120*
Dobler, M., 164, *185*
Dominguez, J.F., 154, *183*
Donadelli, A., 219, *287*
Dong, S.Z., 164, *184, 185*

Dong, W., 209–210, *284*
Donker-Koopman, W.E., 209–210, 252, 254–255, *284, 295, 296*
Donker, W., 252, *295*
Dorfmueller, H.C., 253, *291*
Dorling, P.R., 191, *277*
Dorta, R., 159–160, *183*
Dosbaa, I., 152–153, *182*
Døssing, A., 77, *118*
Drew, M.G.B., 164, 165, *184, 185*
Drew, M.J., 67
Dring, J.V., 190–191, *277*
Driver, G.E., 129–132, *180*
Drzewoski J-., 70, 109–110, *115*
Ducep, J.-B., 248–249, *294*
Ducros, G., 227, *289*
Ducrot, P.-H., 192, *278*
Duff, M., 162, *184*
Duff, M.R. Jr., 161, *184*
Dukhan, D., 165, *185*
Dumon, C., 234–236, *290, 291*
Dupont, C., 224–227, *288*
Durette, P.L., 22, 40, 43, *64*
Durham, L.J., 191, *277*
Duus, B., 198–199, *281*
Dwek, R.A., 29, 248–250, 254–255, 258, *294–297*
Dyong, I., 188–189, 231–233, *276*
Dziedzic, S.Z., 21–23, *66*

E

Ebert, M.O., 164, *185*
Ebner, M., 220, 221, *287*
Edwards, A.A., 222, *288*
Edwards, R.G., *64*
Egelman, D., 70, 73, *115*
Egido-Gabas, M., 232–233, 253, *290, 296*
Egli, M., 164, *185*
Ehrhard, A., 248–249, *294*
Eichhorn, G.L., 70, 81, *115*
Ekhart, C.W., 193, 216, 220, 221, *278, 286, 287*
Elbein, A.D., 190–192, 209–210, *277, 278, 284*
Ellinger, B., 177, *186*

Elzagheid, M.I., 168, *186*
Ember, B., 234, *290*
Endo, K., 164, *185*
Engelsma, S.E., 258, *297*
Enomoto, H., 248, *294*
Ensley, H.E., 247, *294*
Entrena, A., 154, *183*
Eschenmoser, A., 164, *185*
Esen, A., 229, 247–248, *289*, *294*
Esim, A., 247–248, *294*
Espinosa, A., 154, *183*
Evans, G.B., 222, 243–246, *287*, *288*, *293*
Evans, S.V., 192–193, *278*
Evinson, E.L., 258, *297*

F

Fabbro, D., 193, 249–250, *278*, *295*
Faggi, E., 241–242, *292*
Fahr, C., 109–110, *120*
Fairbanks, A.J., 217, 222, *286*
Fairclough, P.H., 37–39, *64*
Fan, J.-Q., 209–210, 251, 258, *284*, *298*
Fantur, K., 254–256, *296*, *297*
Farrell, R.P., 72, 73, 86–89, 99, 110–111, *117*, *118*
Favini, G., 154, *182*
Fayers-Kerr, J., 239, *292*
Feather, M.S., 189–190, *277*
Fechter, M.H., 216, *286*
Fedorov, A., 243, *293*
Feldmann, J., 188–189, *276*
Feller, G., 216, *286*
Fellows, L.A., *67*
Fellows, L.E., 190–193, 248–249, *277*, *278*, *294*
Fenton, R., 161, *184*
Fernandez, J.M.G., 231–233, 253, 256, *290*, *295–297*
Ferrier, R.J., *66*
Fiaux, H., 235, 241, *291*
Field, R.A., 206, *282*
Fincher, G.B., 227–228, 250, *289*
Fink, A.L., 194–195, *280*
Firsov, L.M., 224–227, *288*

Fischer, H.O.L., 5, 12–14, *61*
Fitch, W.L., 6, 49, 50, *65*
Flaherty, B., 150–151, *182*
Fleet, G.W.J., 51, 67, 190–193, 196, 217, 220, 222, 224–227, 248–250, 254–255, 258, *277*, *278*, *280*, *286–288*, *294–298*
Foglietti, M.-J., 152–153, *182*
Fontes, C.M.G.A., 224–227, *288*
Foran, G.J., 72, *117*
Forcella, M., 241–242, *292*
Fornili, A., 245, *293*
Foster, A.B., *65*
Franck, R.W., 196, *280*
Frandsen, T.B., 200–201, *281*
Frandsen, T.P., 200, *281*
Frankel, A.E., 161, *184*
Frankowski, A., 227–228, *289*
Frascaroli, M.I., 69–115, *118–120*
Freestone, A.J., *63*, *64*
Freestone, J., *64*
Fröhlich, R.F.G., 210, 256, *284*
Frommer, W., 203–204, *282*
Front-Deschamps, S., 252, *295*
Fuchs, M., 72–73, 77, *117*
Fu, G.C., 193, *278*
Fuhrmann, U., 191, *277*
Fujii, H., 73, 77, 93–94, 104–105, *118*
Fujii, T., 194, *279*
Fuller, M., 206, *283*
Furneaux, R.H., 210, 218, 222, 243–245, 256, *284*, *286*, *287*, *292*, *293*
Furukawa, J., 191, *277*
Furukawa, Y., 109–110, *120*
Furuta, S., 191, *277*
Fusetani, N., 204, *282*
Fusi, P., 241–242, *292*
Futerman, A.H., 229–232, 239, 252, 253, *289*, *290*
Fuzier, M., 152, *182*
Fyvie, W.S., 148, 149, 172, *181*, *182*, *184*

G

Gabrys, H., 70–71, 74–75, *116*
Gainsford, G.J., 218, 222, *286*, *287*

AUTHOR INDEX

Galarneau, A., 168–169, *186*
Galezowski, M., 71, *116*
Gallienne, E., 252, *295*
Gallo, M.A., 154, *183*
Gandolfo, F., 87–89, *119*
Gandy, M.N., 254–255, *297*
Ganem, B., 207, *283*
Ganesan, M., 258, *297*
Ganesh, N.V., 141–143, *181*
Gao, J., 122, *178*, *179*
Gao, Z., 196, *280*
Garau, G., 245, *293*
Garcia-Moreno, M.I., 231–233, 253, 256, *290*, *296*, *297*
Garcia, R.C., 24, 29, *67*
García, S.I., 69–*120*
Garman, S.C., 236, *291*
Garrido, R., 154, *183*
Gascón, J.A., 161, *184*
Gaston, J.L., *65*
Gáti, T., 177, *186*
Gatt, S., 193, *278*
Gelbin, A., 169, *186*
Gellman, S.H., 122, *178*
Gemperli, A.C., 122, *179*
Gerber-Lemaire, S., 235, 241, *291*
Gerday, C., 216, *286*
Gerrard, J., 206, *283*
Getman, D.P., 218, *286*
Gez, S., 71, *116*
Ghanem, M., 244–245, *293*
Ghavami, A., 240, *292*
Ghinet, M.G., 238, *292*
Ghisaidoobe, A., 254, *284*, *296*
Ghosh, A.K., 144–145, *181*
Ghoshal, N., 165, *185*
Giabbai, B., 245, *293*
Giamello, E., 72–75, *117*
Gibbs, C.F., 20–21, *62*
Gilbert, H.J., 224–227, 234–236, 247–248, *288–291*, *294*
Gill, P.L., *63*
Gil, V.M.S., 105–107, *120*
Girvin, M.E., 243, *292*

Gloster, T.M., 207, 214, 224–229, 231–233, 238, 239, 241–242, 250, 253, *283*, *285*, *288–290*, *292*
Glushka, J., 224, 225, *288*
Godin, G., 252, *295*
Godschalk, F.D., 254, *296*
Goldberg, J.D., 224–227, *288*
Goldfarb, D., 72–73, *117*
Goldstein, I.J., 246–247, *294*
Goldstein, J.L., 206, *282*
Gómez, J.A., 154, *183*
González, J.C., 69–*120*
Goodgame, D.M.L., 70–71, 74–75, 90, 92–93, 109–110, *116*
Goodman, B.A., 72, 78–79, 104–105, 109–110, *117*, *118*
Good, N.E., 194–195, *280*
Gosselin, G., 165, *185*
Goss, P.E., 233–234, *290*
Goti, A., 241–242, *292*
Goto, A., 203, *282*
Goto, T., 188, *276*
Goujon, J.-Y., 216, 221, *286*
Grabowski, G.A., 193, 249–250, *278*, *295*
Gradnig, G., 220, 221, *287*
Grandjean, C., 208–209, *283*
Granier, T., 227–228, *289*
Grant, B.D., 6, 49, 50, *65*
Grassberger, V., 218, *286*
Gratien, J., 234–235, *290*
Gratteri, P., 241–242, *292*
Greenblatt, H.M., 229–230, 239, 252, *289*
Gregg, K.J., 236, *291*
Greimel, P., 254–255, *296*
Gretzke, D., 258, *297*
Greul, J.N., 209–210, *284*
Grice, P., 224–227, *288*
Griffe, L., 165, *185*
Griffon, J.F., 165, *185*
Grubbs, R.H., 193, *278*
Gruber, C., 250, *295*
Grubmeyer, C., 243, *292*, *293*
Grupp, A., 75, 99, *117*
Gryko, D.T., 71, *116*

Guan, R., 245–246, *294*
Guaragna, A., 258, *297, 298*
Guce, A.I., 236, *291*
Gueyrard, D., 216, 221, *286*
Guglielmini, L., 258, *297*
Guijt, M.C., 254, *296*
Gunji, H., 227–228, *289*
Guntha, S., 164, *185*
Guo, D.-S., 70, *115*
Guo, J., 234, *290*
Gupta, A., 244–245, *293*
Gurjar, M.K., 21–23, 28–32, 34, *65, 66*
Guthrie, R.D., 18–20, *61*
Gutierrez, J.A., 245–246, *294*
Gyurcsik, B., 102, *120*

H

Hadad, C.M., 147, 161, 172, *181, 184*
Hadamczyk, D., 188–189, *276*
Hadfield, A.F., 44, 46, *64, 65*
Hadwiger, P., 216, *286*
Haener, R., 168, *185*
Haga, K., 224–227, *288*
Hagens, P., 254, *296*
Haines, A.H., 206, *282*
Hakamata, W., 200–202, *281, 282*
Hakansson, A.E., 164, *184*, 258, *297*
Hale, K.J., 53, 56, 57, *66–68*
Halma, J.M., 254, *296*
Hamada, M., 192, *277*
Hama, Y., 109–110, *120*
Hambley, T.W., 72, 100–102, *117, 119*
Hammershoi, A., 100, *120*
Hamor, T.A., 190–191, *277*
Hamuro, Y., 206, *283*
Hancock, S.M., 227, 240–241, *289*
Hanessian, S., 189–190, *276*
Hanson, G.R., 72, 73, 87–88, 99, 110–111, *117*
Hanze, A.R., 133, *180*
Harata, K., 224–227, *288*
Hardcastle, K.I., 149, *182*
Harker, E.A., 122, *179*
Harris, C.M., 192, *278*
Harris, E.M.S., 224–227, *288*

Harris, H.H., 72, *117*
Harris, T.M., 192, *278*
Hartley, O., 218, *286*
Hartung, R., 164, *185*
Harvey, J.E., 139–140, 152, *180, 182*
Haser, R., 223–224, 230, 232, *288, 290*
Hashimoto, M., 192, *277*
Haskell, T.H., 189–190, *276*
Haslett, G.W., 193, *278*
Hatanaka, S.I., 191, *277*
Hatcher, S.A., 149, *182*
Hathway, D.E., 71, 92–93, *116*
Häusler, H., 250, 253–255, *295, 296*
Hayauchi, Y., 191, *277*
Hayman, P.B., 71, 92–93, *116*
Hazell, R., 213–214, *285*
Hazleton, K.Z., 244–245, *293*
Hazlewood, G.P., 247–248, *294*
Headlam, H.A., 71, *116*
Heath, J.R., 193, *278*
Heck, M.-P., 234–235, *290*
Hedegård, E.D., 77, *118*
Hehemann, J.-H., 236, *291*
Hehre, E.J., 195, 224–227, *280*
Heightman, T.D., 214, *285*
He, K., 122, *178*
Hempel, A., 224–227, *288*
Hendrickson, J.B., 154, 155, *182, 183*
Hendry, D., 53, *66, 67*
Hendry, P., 100, *120*
Henrissat, B., 227, 229, 232, 247–248, *289, 290, 294*
Hentges, A., 221, *287*
Heo, J.S., 204, *282*
Hermetter, A., 254–255, *296*
Hernández, A.R., 122, *178*
Herscovics, A., 233, *290*
He, Y., 122, 125, 126, *178, 179*, 236–239, 272, *291, 292*
Heyns, K., 188–189, *276*
Higaki, K., 256, *297*
Hildbrand, S., 164, *184*
Hill, J., *61, 62*
Hindsgaul, O., 133–134, *180*

Hoberg, J.O., 139–140, 152, 165, 167, *180*, *182*
Ho, C.-W., 239, 253, *292*, *296*
Hodsdon, M.E., 122, *179*
Höfer, P., 75, 99, *117*
Hoffman, B.M., 72, *117*
Hoffman, D., 235, 241, *291*
Hoffmann, R., 188, 224–227, 231–233, *276*
Hof, S., 200, *281*
Hohenschutz, L.D., 190–191, *277*
Holman, H.-Y.N., 71, 74–75, *116*
Holzner, A., 164, *185*
Ho, M.-C., 245–246, *293*, *294*
Honzatko, R.B., 224–227, *288*
Hoos, M., 227, *289*
Hoos, R., 215–216, *285*
Horie, T., 206, *282*
Hori, H., 235, 253, *291*
Horne, A.P., 109–110, *120*
Horne, G., 258, *297*, *298*
Horner, M.W., *63*
Horner, W., *63*
Horton, D., 159, 176–177, *183*, *186*, 188–189, *276*
Hoshino, S., 200–201, *281*
Hough, L., 3, 20–46, 48, 51–53, 56, 57, *61–68*
Houpis, I.N., 144–145, *181*
Houseknecht, J.B., 172, *184*
Howell, P.L., 233, *290*
Hoyoux, A., 216, *286*
Hrmova, M., 227–228, 250, *289*
Hsieh, S.H., 169, *186*
Hsu, H.-S., 247, 250, *294*
Huang, M.-H., 220, 222, *287*
Huang, Y.L., 71, *116*
Huertas, P., 230, 239, *289*
Hung, I.O., 250, *295*
Hu, T., 258, *298*
Huxtable, C.R., 191, *277*
Huynh, H.K., 164, *185*

I

Ichikawa, Y., 210, 216, 219, *284*, *286*
Igarashi, Y., 210, *284*
Iinuma, H., 206, *283*

Iitaka, Y., 206, *283*
Ikeda, K., 216, 221, 252, 258, *286*, *295*, *297*
Ikehara, M., 171, *186*
Illaszewicz, C., 254, *296*
Imanaka, H., 192, *277*
Imberty, A., 233, *290*
Ingold, C., 188, 231, 234–236, *276*
Inokuchi, J., 209–210, *284*
Inouye, S., 188, 189, *276*
Iqbal, C.M., 204, *282*
Ireson, J.C., 150, *182*
Irwin, J.A., 71–75, 78, 87–89, 99, 109–111, *116*, *117*
Ishida, N., 189, *276*
Ishihara, M., 199, *281*
Ishikawa, M., 171, *186*
Istomina, N.V., 161, *184*
Ito, T., 189, *276*
Ivanen, D.R., 236, *291*
Iwai, S., 171, *186*
Iwami, M., 192, *277*
Iwanienko, T., 70, 109–110, *115*
Iyer, R.N., 246–247, *294*
Izquierdo-Cubero, I., *68*

J

Jacob, G.S., 248–249, *294*
Jadot, J., 190, *277*
Jäger, V., 209–210, *284*
James, M.N.G., 236, 239, *291*
Janicki, J., 203, *282*
Jank, T., 231–232, *290*
Janzer, R.C., 235, 241, *291*
Jarman, M., *65*
Jaun, B., 164, *185*
Jayakanthan, K., 208–209, *283*
Jayaraman, N., 141–143, *181*
Jenkinson, S.F., 254–255, 258, *296*, *297*
Jensen, A., 213–214, *285*
Jensen, H.H., 213–214, *285*
Jeong, I.-Y., 204, *282*
Jermyn, M.A., 197, *280*
Jersch, N., 188–189, *276*

Jeschke, G., 72–73, *117*
Jespersen, T.M., 209–210, *284*
Jessen, C.U., 240, *292*
Jewess, P.J., 190–191, *277*
Jiang, H.J., 218, *287*
Jiang, J., 70–71, 74–75, *116*
Jia, Y.-M., 220, 222, *287*
Ji, J.Y., 72, 73, 87–89, 99, 110–111, *117*
Jiménez-Barbero, J., 125, 126, 174–175, *179*, 199, 239, 272, *281*, *292*
Jodoin, P., 224–227, *288*
Johnson, C.S., 146, *181*
Johnston, B.D., 208–209, 240, *283*, *292*
Jones, G.S., 141, *181*
Jones, J.H., 254–255, *296*
Joy, A.M., 70–71, 74–75, 90, 109–110, *116*
Juan, X.-Y., 70, *115*
Judd, R.J., 100–101, *119*
Juillerat-Jeanneret, L., 235, 241, *291*
Jülich, E., 193, *278*
Junge, B., 203–204, *282*

K

Kabir, A.K.M.S., 44, 45, *66*
Kadokura, K., 202, *282*
Kajimoto, T., 216, 219, *286*
Kalabegishvili, T.L., 71, 74–75, *116*
Kamada, H., 73, 77, 93–94, 104–105, *118*
Kanai, R., 224–227, *288*
Kaneko, E., 216, *286*
Kang, J.E., 204, *282*
Kang, M.C., 144–145, *181*
Kang, N.S., 204, *282*
Kapre, R., 71, 72, 77, *116*
Karaveg, K., 224, 225, 233, *288*, *290*
Karpas, A., 248–249, *294*
Karplus, M., 196, *280*
Katayama, T., 240, *292*
Kato, A., 216, 220–222, 252, 258, *286*, *287*, *295*, *297*, *298*
Kato, R., 240, *292*
Kaushal, G.P., 192, *277*
Kawachi, R., 200–202, *281*
Kawatkar, S.P., 234, *290*

Kayakiri, H., 192, *277*
Kay, M.S., 122, *179*
Keinicke, L., 164, *184*
Kelly, J.W., 230, 239, 250, *289*, *295*
Kelm, S., 108–109, *120*
Kern, K., 247, *294*
Khamri, W., *68*
Khangarot, B.S., 70, *115*
Khanna, R., 230, 239, *289*
Khan, R., 37, *64*
Khan, S.N., 204, *282*
Khedhair, K.A., 159, *183*
Khorlin, A.Y., 207, *283*
Kicska, G.A., 243–244, *293*
Kime, R., 237–238, *291*
Kim, J.H., 204, *282*
Kim, J.Y., 206, *282*
Kim, K., 222, 243–244, *287*, *293*
Kim, S., 122, *179*, 206, *282*
Kimura, A., 201–202, *281*
Kimura, M., 109–110, *120*
Kim, Y.S., 206, *282*
Kinas, R.W., *65*
Kinast, G., 191, *277*
Kind, P.R.N., *65*, *66*
Kirby, A.J., 159, *183*
Kirk, B., 196, *280*
Kirschner, K.N., 135–136, 159–160, *180*
Kishi, Y., 188, 193, *276*, *278*
Kitagawa, M., 200–201, *281*
Kitagawa, S., 192, *278*
Kite, G.C., 193, *279*
Kittaka, A., 164, *184*
Kizu, H., 216, 221, *286*
Klare, J.P., 72, *117*
Kleban, M., 209–210, *284*
Kluge, R., 219–221, *287*
Knapp, S., 196, 236, 239, *280*, *291*
Koba, Y., 203, *282*
Kochi, J.K., 71, *116*
Koh, L.L., 218, *287*
Kohsaka, M., 192, *277*
Kok, F., 206, *283*
Komiyama, T., 192, *277*

Kondo, A., 200–201, *281*
König, B., 146, *181*
König, D.F., 194–195, *279*
Koo, B., 146, *181*
Kool, E.T., 122, *178*, *179*
Koomen, G.-J., 250, *295*
Korb, C., 219–221, *287*
Korecz, L., 72, 78–79, 82, 85–87, 93–94, 110–111, 113, *117*, *118*, *120*
Kornhaber, G.J., 206, *283*
Koshkin, A.A., 164, *184*
Koshland, D.E., 194–196, *279*
Ko, T.-P., 239, 253, *292*
Kouno, T., 190, *277*
Kover, A., 133–134, *180*
Koyama, M., 189, *276*
Kozlowski, H., 78, 86–88, *118*, *119*
Králová, B., 154, *182*
Kraszni, M., 159–160, *183*
Krishnamurthy, R., 164, *185*
Kristiansen, M., 240, *292*
Kritzer, J.A., 122, *179*
Krivdin, L.B., 161, 162, *184*
Krolikiewicz, K., 167, *185*
Kroto, H.W., 193, *278*
Krueger, A.T., 122, *178*, *179*
Krulle, T.M., 250, 258, *295*, *297*
Krumpolc, M., 72, 100–101, *117*, *120*
Krzystek, J., 77, *118*
Kuhn, R., 194–195, *279*
Kuijper, B., 252, *295*
Kulminskaya, A.A., 236, *291*
Kumagai, K., 189, *276*
Kumar, C.V., 161, 162, *184*
Kumar, R., 164, *184*
Kundu, A.P., 165, *185*
Kuntz, D.A., 233–235, 240, 241, 253, *290*–*292*
Kuntz, D.G., 234, *290*
Kuo, H.-W., 70, *115*
Kuppusamy, P., 72, *117*
Kusano, G., 192, *278*
Kuszmann, J., 176–177, *186*
Kuze, T., 164, *184*

L

Lacombe, P., 125, 143–144, *179*
Lafarga, R., 87–89, *119*
La Ferla, B., 193, 258, *279*, *297*
Lai, J.-S., 70, *115*
Lallemand, J.L., 192, *278*
Lallemand, M.-C., 204, *282*
Lalloz, M.-N., 248–249, *294*
Lammens, W., 240–241, *292*
Langridge, J., 246–247, *294*
Lankin, D.C., 213–214, *285*
Larsen, S.C., 108, *120*
Lauda, C.M., 203, *282*
Lay, P.A., 70–75, 78, 81, 86–89, 94–95, 99, 100, 108–111, *115*–*120*
Lee, A.H.F., 122, *178*
Lee, B.W., 204, *282*
Lee, C.-K., 21–23, *63*, *64*, 218, *286*, *287*
Lee, G., *67*
Lee, H.S., 204, *282*
Lee, J.B., 149–150, *182*
Lee, J.W., 206, *282*
Lee, K.W., 206, *282*
Lee, M.J.-R., 254–255, *296*
Leeuwenburgh, M.A., 143, *181*
Lee, W.S., 206, *282*
Lee, Y., 206, *282*
Lee, Y.-L., 206, 253, *296*
Legler, G., 191, 193–195, 197, 198, 206–209, 213, 216, 220, 221, 247–250, *277*–*280*, *283*, *285*, *287*, *294*, *295*
Le Merrer, Y., 152, *182*
Lemieux, R.U., 159, *183*, 200–201, *281*
LeNouen, D., 258, *297*
Lenz, D.H., 222, *287*
Leonidas, D.D., 240, *292*
Leontcin, K., 191, *277*
Leroy, E., 250, *295*
Leroy, F., 164, 165, 169–170, *185*
Le Roy, K., 241, *292*
Lesellier, E., 252, *295*
Lesur, B., 248–249, *294*
Leumann, C.J., 164, *184*, *185*

Levina, A., 70–72, 74–75, 78, 81, 86–87, 94–95, 99, 100, 109–110, *115–117*, *119*, *120*
Lewandowicz, A., 222, 243–244, 287, *293*
Leworthy, D.P., 190–191, *277*
Leydier, S., 227, *289*
Liang, P.-H., 253, *296*
Liang, X.-F., 211, 213–214, 224–227, *285*, *288*
Liautard, V., 258, *298*
Lichter, J., 149, , *182*
Li, C.M., 243, *292*, *293*
Lieberman, R.L., 230, 236, 239, *289*
Liebross, R.H., 70, 72, *115*
Liedtke, H., 247, *294*
Liew, S.-T., 147–148, *181*
Li, H., 125, *179*
Li, J.Y., 78, *118*, 176, *186*
Li, L., 222, 244–245, 256, *287*, *293*, *297*
Lillelund, V.H., 224–227, *288*
Limberg, G., 203–204, *282*
Lin, C.-H., 239, 253, *292*, *296*
Lindberg, B., 191, *277*
Linden, A., 218, *287*
Lin, T.-L., 70, *115*
Lin, V.S.Y., 71, *116*
Lin, Y.-L., 253, *296*
Lin, Y.-N., 253, *296*
Li, R., 220, 222, *287*
Li, S.-C., 109–110, *120*, 209–210, 251, *284*
Liskamp, R.M.J., 254–255, *296*
Li, S.-T., 253, *296*
Liu, F.-W., 204, *282*
Liu, H., 122, *178*, *179*, 211, *285*
Liu, H.-M., 204, *282*
Liu, H.Z., 224–227, *288*
Liu, J., 70–71, 74–75, *116*
Liu, K.K.-C., 216, 219, *286*
Liu, S.-W., 253, *296*
Liu, Z.-J., 224, 225, *288*
Li, X., 147–148, *181*
Li, Y., 176, *186*
Li, Y.-K., 206, 247, 250, 253, *282*, *294*, *296*
Li, Y.-T., 109–110, *120*

Li, Y.-X., 220, 222, *287*
Llebaria, A., 232–233, 253, *290*, *296*
Lobsanov, Y.D., 233, *290*
Lok, F., 200–201, *281*
Lomas, H., 254–255, *296*
Lönngren, J., 191, *277*
Loontiens, F.G., 247, *294*
Lopez-Espinosa, M.T.P., *68*
Lopez, O., 213–214, 220, *285*, *287*
Lotz, W., 206, *283*
Lou, J., 196, *280*
Lowary, T.L., 172, *184*
Lowe, G., 196, *280*
Luan, Z., 256, *297*
Lubini, P., 164, *185*
Lu, H., 122, *178*, *179*
Luitjens, S.B., 254, *296*
Lundt, I., 188–189, 193, 203–204, 209–210, 220, 221, 250, 254–255, *276*, *279*, *282*, *284*, *287*, *295*, *296*
Luo, B., 239, *292*
Luque, F.J., 171, *186*
Luszniak, M.C., 206, *282*
Luther, A., 164, *185*
Lutz, S., 149, *182*
Lu, X., 209, *284*
Luxenhofer, R., 71, *116*
Lynch, S.R., 122, *178*
Lyngbye, L., 213–214, *285*
Lynn, D.G., 191, *277*

M

Macauley, M.S., 238, *292*
MacDonald, J.M., 211, 224–227, 239, *285*, *288*, *289*
Madsen, R., 193, 207, 232–233, *279*, *283*, *290*
Maegawa, G.H.B., 206, *283*
Magdolen, P., 206, *282*
Mahuran, D.J., 206, 210, 236, 254–256, *283*, *284*, *291*, *296*, *297*
Mair, G.A., 194–195, *279*
Maki, T., 204, *282*
Malecka-Panas, E., 70, 109–110, *115*
Mallet, J.M., 125, *179*

Mandal, S.B., 164, 165, *184*, *185*
Mangiameli, M.F., 73, 77, 86, 104–106, *118*
Mangos, M.M., 168, *186*
Maranda, A., 122, *179*
Marangoni dos Santos, D., 243, *293*
Marcelo, F., 125, 126, , *179*, 239, 272, *292*
Marco-Contelles, J., 207, *283*
Markad, S.D., 123, 147, 161, *179*, *181*, *184*
Mark, B.L., 236, 239, *291*
Marlier, M., 190, *277*
Marquez, M., 171, *186*
Marshall, J.J., 203, *282*
Martinez-Fleites, C., 236–237, *291*
Martin, J.C., 159, *183*
Martin, J.D., 126, *179*
Martin-Lomas, M., 199, *281*
Martino, D., 98, *119*
Martin, O.R., 216, 221, 252, 258, *286*, *295*, *298*
Martin, P., 168, *185*
Mason, J.M., 218, 222, *286–288*
Mastropaolo, D., 224–227, *288*
Matheu, M.I., 133–134, *180*
Mathias, N., *67*
Matsuda, A., 164–165, *185*
Matsui, K., 216, 221, *286*
Matsuishi, Y., 200–201, *281*
Matsui, Y., 203, *282*
Matsunaga, S., 204, *282*
Matsunaga, Y.K., 210, *284*
Matsushima, Y., 198–199, *280*
Mattes, R., 223–224, 230, 232, *288*, *290*
Matthews, B.W., 193–194, *279*
M-A, V., 227, *289*
Mayato, C., 159, *183*
Maynard, L., 122, *178*
McAuliffe, J.C., 133–134, *180*
McCarren, P.R., 172, *184*
McCarter, J.D., 194–195, 198–199, 217–218, *279*, *281*
McCarthy, K.C., *67*
McCasland, G.E., 191, *277*
McDonald, F.E., 145–146, 149, 176–177, *181*, *182*
McDonnell, C., 213–214, *285*
McLauchlan, K.A., *61*
McMaster, J., 77, *118*
Mead, J.A.R., 246–247, *294*
Medgyes, A., 176–177, *186*
Mee, S.P.H., 222, 244–245, 287, 293
Mega, T., 198–199, *280*
Mehring, M., 75, 99, *117*
Meillon, J.C., 164–167, *185*
Mei, S., 161, *184*
Meldgaard, M., 164, *184*
Mellet, C.O., 231–233, 253, 256, *290*, *295–297*
Meloncelli, P., 228–229, 232–233, 239, 250, *289*
Mendes, M.A., 243, *293*
Mercer, T.B., 258, *297*
Meva'a, L.M., 204, *282*
Micera, G., 70–71, 74–75, 77, 78, 86–88, 101, *116*, *118*, *119*
Micheel, F., 127–128, *179*
Miculca, C., 164, *185*
Micura, R., 164, *185*
Midgely, G., *68*
Miles, R.J., 5–6, *67*, *68*
Miller, S.J., 193, *278*
Miller, S.M., 123, *179*
Millet, J., 176–177, *186*
Milsmann, C., 72, *117*
Minasov, G., 164, *185*
Min, K.L., 168, *186*
Minton, M., 164, *185*
Mirgorodskaya, E., 200–201, *281*
Mitchell, H.J., 149–150, *182*
Mitchell, M., 192, *277*
Mitewa, M., 71, *116*
Miura, I., 191, *277*
Miura, S., 227–228, *289*
Miwa, I., 206, *282*
Miyake, Y., 202, *282*
Miyasaka, T., 164, *184*
Miyauchi, S., 258, *297*, *298*
Mlaker, E., 216, *286*
Möbius, K., 72–73, 77, *117*

Mody, V., 71, *116*
Moffatt, J.G., 6, 49, 50, *65*
Mohan, S., 208–209, *283*, *284*
Molyneux, R.J., 190–192, 209–210, *277*, *278*, *284*
Montgomery, F.J., 224–227, *288*
Moody, P.C.E., 230, *289*
Moorland, C., 224–227, *288*
Moracci, A.V., 227, *289*
Moracci, M., 214, *285*
Morales de Csaky, C., 256
Morales, E.Q., 126, *179*
Moremen, K.W., 233–236, *288*, *290*, *291*
Moreno, V., 92–93, *119*
Moriniere, J., 229, 247–248, *289*, *294*
Mori, S., 164, *185*
Morishima, H., 192, *277*
Morís-Varas, F., 125, *179*
Morland, C., 234–235, *290*
Morton, M.D., 123, 147, 148, 161, 162, 172–174, *179*, *181*, *184*
Moser, H.E., 168, *185*
Mosi, R., 209, *284*
Mukhopadhyay, R., 165, *185*
Müller, L., 203–204, *282*
Munagala, N.R., 243, *293*
Munroe, P.A., 33, 36, *63*
Murai, H., 190, 248, *277*, *294*
Murali, R., 140, *180*
Murao, S., 203, *282*
Murkin, A.S., 222, 244–245, *287*, *293*
Muroi, M., 202, *281*
Murphy, D., *61*
Murphy, P., 213–214, *285*
Murshudov, G., 230, *289*
Murusidze, I.G., 71, 74–75, *116*
Myers, C.R., 70–71, 74–75, 77, 87, *115*, *118*
Myers, J.M., 70–71, 74–75, 77, 87, *115*, *118*

N

Nagae, M., 240, *292*
Naganawa, H., 206, *283*
Nagarajan, M., 140, *180*
Nagasawa, H., 235, 253, *291*

Nagy, L., 102, *120*
Nahar, S., 150–151, *182*
Naider, F., 207, *283*
Nair, B.U., 109–110, *120*
Nakagawa, H., 109–110, *120*
Nakamura, H., 171–172, *186*, 206, *283*
Nakamura, K.T., 164, *184*
Nakamura, S., 192, *278*
Nakanishi, K., 191, *277*
Nakao, Y., 204, *282*
Nakata, M., 207, *283*
Nakatsubo, F., 188, *276*
Nakayama, M., 206, *282*
Nakayama, O., 192, *277*
Nakayama, S., 235, 253, *291*
Namchuk, M.N., 198–199, 209, 216, *280*, *281*, *284*
Namgoong, S.K., 248–249, *294*
Nanba, E., 256, *297*
Naomi, M., 70, *115*
Narasinga, R.B.N., 109–110, *120*
Narikawa, T., 194, *279*
Nash, R.J., *67*, 190–193, 220, 222, 258, *277*, *278*, *287*, *297*, *298*
Nasi, R., 208–209, *283*
Nayak, U.G., 149–150, *182*, 189–190, *277*
Nebenführ, H., 75, 99, *117*
Neele, J., 250, *295*
Neese, F., 71–73, 77, *116*, *117*
Nerinckx, W., 216, *285*
Ng, C.J., 129, 130, *180*
Nicholas, S.J., 192–193, *278*
Nicolaou, K.C., 150, *182*
Nicotra, F., 193, 258, *279*, *297*
Nielsen, C., 164, *184*
Nielsen, P., 164, 171, *184*, *186*
Nieto, L., 125, 126, 174–175, *179*, 239, 272, *292*
Nieto, O., 199, *281*
Niida, T., 189, *276*
Ninomiya, H., 256, *297*
Nishikawa, T., 189, *276*
Nishimura, T., 199, *281*
Nishio, T., 200–202, *281*, *282*

Niwata, Y., 207, *283*
Nóbrega, C., 160, *184*
Nomura, M., 164–165, *185*
Noronha, A., 171, *186*
Nørskov-Lauritsen, L., 240, *292*
North, A.C.T., 194–195, *279*
Northcote, P.T., 139–140, *180*
Nosaka, C., 206, *283*
Noszál, B., 160, *183*
Notenboom, V., 215–216, *285*
Noy, A., 171, *186*
Nunez, F., 258, *297*

O

OBrian, P., 72, *117*
O'Brien, S.C., 193, *278*
Oda, Y., 171, *186*
Offen, W.A., 234–235, *290*
Ogawa, K., 203, *282*
Ogawa, M., 200–201, *281*
Ogawa, S., 191, 209–210, 277, *284*
Ohno, K., 256, *297*
Ohta, S., 227–228, *289*
Ohtsuka, E., 171, *186*
Ohyama, K., 203, *282*
Oikonomakos, N.G., 240, *292*
Ojea, V., 258, *297*
Okamoto, M., 192, *277*
Okuda, J., 206, *282*
Okuda, S., 191, *277*
Oku, T., 200–202, *281*, *282*
Olah, V.A., *68*
Olivera, M.S., 72, 86–87, 90, 93–94, *117*
Olivera, S., 92, *119*
Olson, W.K., 169, *186*
Orendt, A.M., 147–148, *181*
Orozco, M., 171, *186*
Ortega-Caballero, F., 219–221, *287*
Osama, L.M.O., 197, *280*
Osborn, H.M.I., 150, *182*
Oseki, K., 216, *286*
Osiecki-Newman, K.M., 193, 249–250, 278, *295*
Ouadid-Ahidouch, H., 253, *295*

Oulaidi, F., 252, *295*
Ovaa, H., 143, *181*
Overend, W.G., 150–152, *182*
Overkleeft, H.S., 143, *181*, 250, 252, 254–255, 258, *284*, *295–297*
Overman, L.E., 177, *186*
Owen, L.N., 189–190, *276*

P

Painter, G.F., 222, *287*
Pakulski, Z., 123, *179*
Pal, A., 165, *185*
Palcic, M.M., 200, *281*
Pallan, P.S., 164–167, *185*
Palma, M.S., 243, *293*
Palmer, A.K., 44, *63*
Palopoli, C., 82, 86–87, 92–93, 96–98, 102, *118–120*
Palumbo, G., 258, *297*, *298*
Panday, N., 209–210, 214, 227–228, 250, *284*, *285*, *289*
Pandit, D., 206, *283*
Pandit, U.K., 250, *295*
Paolella, C., 258, *297*
Paquette, L.A., 164, *184*, *185*
Pardi, L.A., 77, *118*
Parekh, R.B., 192–193, *278*
Parenti, P., 241–242, *292*
Párkányi, L., 176–177, *186*
Parkin, A., *65*
Park, K.H., 204, 206, *282*
Parmeggiani, C., 241–242, *292*
Parniak, M.A., 168, 171, *186*
Paschke, E., 254–256, *296*, *297*
Patel, A., *66*
Patel, V., 194, *279*
Pato, C., 258, *297*
Patra, A., 165, *185*
Pattanayek, R., 164–167, *185*
Paulsen, H., 189–191, *276*, *277*
Pavia, A.A., 159, *183*
Peczuh, M.W., 121–*181*, *184*
Pedersen, C., 188–189, *276*
Pedersen, H., 200–201, *281*

Pederson, R.L., 216, 219, *286*
Pell, G., 227, *288*, *289*
Penglis, A.A.E., 44, *65*
Pereira, J.H., 243, *293*
Perez, A., 171, *186*
Perez, R.L., 126, *179*
Perugino, G., 214, *285*
Perugino, M., 227, *289*
Petersen, M., 164, *184*
Petsko, G.A., 230, 236, 239, *289*
Petursson, S., 248–249, *294*
Peyman, A., 168, *186*
Pfundheller, H.M., 171, *186*
Phillips, D.C., 194–195, *279*
Picart, F., 177, *186*
Picasso, S., 209–210, *284*
Pickett, H.M., 156, *183*
Piens, K., 214, 216, *285*
Pieters, R.J., 254–255, *296*
Pigman, W.W., 194–195, *279*
Pilkiwicz, F., 191, *277*
Pillard, C., 258, *298*
Pilling, A.W., 254–255, *296*
Pine, C.W., 230, 239, *289*
Pinkerton, A.A., 218, *286*
Pinto, B.M., 208–209, 240, *283*, *284*, *292*
Pinto, F.M., 126, *179*
Pipelier, M., 214, *285*
Piszkiewicz, D., 196, *280*
Pitsch, S., 164, *184*, *185*
Pitsch, W., 146, *181*
Plato, M., 72–73, 77, *117*
Platt, F.M., 250, *295*
Pletcher, J.M., 145, *181*
Ploegh, H., 191, *277*
Pluvinage, B., 238, *292*
Pocsi, I., *67*, *68*
Popat, S.-D., 239, 253, *292*
Pople, J.A., 156, *183*
Porco, J.C., 216, 219, *286*
Porter, E.A., 193, *279*
Post, C.B., 196, *280*
Pototschnig, G., 254, 256, *296*, *297*
Potts, J.R., 232, *290*

Powe, A.C. Jr., 230, 239, *289*
Praill, P.F.G., 6–7, *65–67*
Praly, J.-P., 159, *183*
Prates, J.A.M., 224–227, *288*
Pratt, J., 72, *117*
Prell, E., 219–221, *287*
Prelog, V., 188, 231, 234–236, *276*
Premkumar, L., 230, 239, *289*
Preut, H., 177, *186*
Price, R.G., 5–7, *65–68*
Prudent, S., 258, *297*
Pruski, M., 71, *116*
Pryce, R.J., 190–191, *277*

Q

Qian, X.-H., 125, *179*
Qing, F.-L., 219–222, *287*
Qiu, X.-L., 219–221, *287*
Quiroz, M., 102, *120*

R

Rabijns, A., 240–241, *292*
Rademacher, T.W., 192–193, 248–249, *278*, *294*
Radford, T., 149, *182*
Raghothama, S., 141–143, *181*
Rainier, J.D., 144–145, 147–148, *181*
Rajanikanth, B., 218, *286*
Rajwanshi, V.K., 164, *184*
Ramagopal, U.A., 244–245, *293*
Ramana, C.V., 140, *180*
Ramesh, N.G., 258, *297*
Ramos, M.L.D., 105–107, *120*
Ramsden, N.G., 67, 224–227, 248–249, *288*, *294*
Ran, F.A., 122, *179*
Ranjit, R., 204, *282*
Rao, S.N., 213–214, *285*
Rapp, J.L., 71, *116*
Ratcliffe, M.J., 206, *283*
Ravaud, S., 223–224, 230, 232, *288*, *290*
Rawlings, A.J., 254–255, *296*
Ray, K., 71, 72, 77, *116*
Read, K.D., 237–238, *291*

Readshaw, S.A., 159, *183*
Refn, S., 200–201, *281*
Reissmann, S., 220, 221, *287*
Rempel, B.P., 207, 208, *283*
Renaut, P., 176, *186*
Renhowe, P.A., 177, *186*
Renkema, G.H., 250, *295*
Resnati, G., 219, *287*
Reuter, G., 108–109, *120*
Reymond, J.-L., 218, 250, *286, 295*
Richardson, A.C., *3–68*
Rico, M., 126, *179*
Rieger, A.L., 77, *118*
Rieger, P.H., 77, *118*
Rigat, B.A., 206, 254, 256, *283, 296, 297*
Rinaldo-Matthis, A., 244–246, *293, 294*
Ringe, D., 230, 236, 239, *289*
Risquez-Cuadro, R., 253, *295*
Risseeuw, M.D.P., 258, *297*
Rivera-Sagredo, A., 199, *281*
Rizzo, S., 177, *186*
Rizzotto, M., 86–89, 92–95, 98, *118, 119*
Roberts, S.M., 214, 224–227, 241–242, *285, 288, 289, 292*
Roberts, S.W., 147–148, *181*
Robert, X., 223–224, 230, 232, *288, 290*
Rockenbauer, A., 72, 78–79, 82, 85–87, 93–94, 110–111, 113, *117, 118, 120*
Rodriguez, E., 126, *179*
Rodríguez-García, E., 125, *179*
Rodríguez, R.M., 126, 149–150, *179, 182*
Roček, J., 72, 100–101, *117, 120*
Roepstorff, P., 200–201, *281*
Roeser, K.-R., 198, 216, 221, *280*
Rogaar, E., 254, *296*
Roggentin, P., 108–109, *120*
Roldán, V., 78–79, 86–87, 90, 92, 109–110, *118, 119*
Rollin, P., 229, *289*
Rose, D.R., 194–195, 208–209, 215–216, 233–235, 240–242, 253, *280, 283, 285, 290–292*
Rosenberg, A., 108–109, *120*
Rossi, M., 214, *285*

Rossi, R., 227, *289*
Rounds, T.C., 154, *183*
Rountree, J.S.S., 254–255, *296*
Rousseau, C., 220, 222, *287*
Rovira, C., 239, *292*
Roy, A., 164, *185*
Roy, B.G., 164, *184, 185*
Ruanova, T.M., 258, *297*
Ruben, M., 258, *297*
Rübsam, F., 149–150, *182*
Ruiz, M., 258, *297*
Rupitz, K., 250, 253–255, *295, 296*
Russel, A., 146, *181*
Rutledge, S.E., 122, *179*
Ryc, C.S., 194–195, *280*
Ryu, H.W., 204, 206, *282*
Ryu, Y.B., 204, 206, *282*

S

Sabatino, D., 164, 167–168, *185*
Sacchettini, J.C., 242, *292*
Saf, R., 210, 254, 256, *284, 296*
Sahabuddin, S., 164, *185*
Saha, J., 121–178, *180*
St. Jacques, M., 159, 161–162, *183*
Saito, K., 191, *277*
Saito, Y.D., 122, *178*
Sakamura, S., 189, *276*
Sala, L.F., 69–*120*
Salam, M.A., 40–42, *65*
Sala, S., 78–79, 109–110, *118*
Salas, P., 73, 77, 104–106, *118*
Salas Peregrin, J.M., 78–79, 85–87, 90, 92–93, 98–102, 105–107, 110–113, *118–120*
Salgado, E.N., 236, *291*
Salzner, U., 159, *183*
Samreth, S., 176–177, *186*
Samuel, E.G., 71, *116*
Sanchez-Fernandez, E.M., 253, *295*
Sandhu, J., *67*
Sanna, D., 77, 101, *118*
Santoro, M., 78–79, 82, 85–87, 90, 92, 94–97, 102, 109–110, *118–120*

Santos, D.S., 243–244, *293*
Sarma, V.R., 194–195, *279*
Sasaki, T., 164, *185*
Sato, H., 213–214, *285*
Sato, K., 150–151, *182*
Saul, R., 209–210, *284*
Savitsky, A., 72–73, 77, *117*
Savvides, S.N., 216, *286*
Sawkar, A.R., 230, 239, 250, *289, 295*
Scaffidi, A., 237–238, *291*
Schauer, R., 108–110, *120*, 123, *179*
Schedel, M., 191, *277*
Schepartz, A., 122, *179*
Scheucher, E., 254, 256, *296, 297*
Schimpl, M., 237–238, 253, *291*
Schitter, G., 254–256, *296, 297*
Schleyer, P.v.R., 159, *183*
Schlossmacher, M.G., 230, 239, *289*
Schmidinger, H., 254–255, *296*
Schmidt, D.D., 203–204, *282*
Schnegg, A., 72–73, 77, *117*
Schneider, B., 168–169, *186*, 209–210, *284*
Scholze, H., 254–255, *296*
Scholz, P., 164, *185*
Schoning, K.U., 164, *185*
Schramm, V.L., 210–211, 222, 242–245, *284, 287, 288, 292–294*
Schreiber, S.L., 122, *178*
Schülein, M., 214, *285*
Schulz, G.E., 231–232, *290*
Schürmann, M., 177, *186*
Schwarz, J.C.P., 189–190, *277*
Schweden, J., 193, 248, *278, 294*
Schweiger, A., 72–73, *117*
Schwentner, J., 188–189, *276*
Scofield, A.M., 190–191, 218, *277, 287*
Scott, W.J., 141, *181*
Seebach, D., 188, 231, 234–236, *276*
Segre, U., 77, 78, *118*
Seliga, C., 227–228, *289*
Seo, S.-K., 147–148, *181*
Seo, W.D., 204, *282*
Sepp, D.T., 159, *183*
Serianni, A.S., 159–160, *184*

Serra, M.V., 87–88, *119*
Shaaltiel, Y., 229–232, 239, 252, 253, *289, 290*
Shackleton, J.F., 108, *120*
Shafizadeh, F., 194–195, *279*
Shah, N., 234, *290*
Shanmugasundaram, B., 214, 237–238, *285, 291*
Shareck, F., 224–227, *288*
Shastry, K.A.R., 247, *294*
Shaw, A.N., 192–193, *278*
Shaw, L., 108–109, *120*
Shea, K., 235, 253, *291*
Shepherd, S.M., 237–238, 253, *291*
Sheppard, G., 196, *280*
Sheth, K.A., 209–210, 251, *284*
Shibano, M., 192, *278*
Shibata, T., 192, *277*
Shigeta, S., 164–165, *185*
Shinoyama, H., 194, *279*
Shin, Y.A., 70, 81, *115*
Shi, Q., 147–148, *181*
Shirahase, I., 248, *294*
Shi, W., 243–246, *292, 293*
Shi, X., 70–71, 74–75, *116*
Shiyan, S.D., 207, *283*
Shpiro, N.A., 237–238, 253, *291*
Shrivastava, H.Y., 109–110, *120*
Shulman, M.L., 207, *283*
Shuto, S., 164–165, *185*
Sierks, M.R., 200–201, 209–210, *281, 284*
Signorella, S., 71–75, 78–82, 85–90, 92–102, 104–107, 109–113, *116–120*
Sigurdskjold, B.W., 198–199, *281*
Silman, I., 229–232, 239, 252, 253, *289, 290*
Silva, R.G., 243, *293*
Sim, K.Y., 218, *286*
Sim, L., 208–209, 241–242, *283, 292*
Simmonds, M.S.J., 193, *279*
Simon, R., 159, *183*
Sinaÿ, P., 125, *179*
Sincharoenkul, L.V., 21–23, 34, *65–67*
Singh, H., 234, *290*
Singh, S.K., 164, *184*

AUTHOR INDEX

Singh, V., 243–244, *293*
Sinnot, M.L., 207, *283*
Sinnott, M.L., 196, 213, 221, *280*, *285*
Sinnwell, V., 191, *277*
Siriwardena, A., 208–209, 224, 225, 235–236, *283*, *288*, *291*
Sizun, G., 164, 165, 169–170, *185*
Skelton, B.W., 211, 222, *285*, *287*
Skrydstrup, T., 209–210, *284*
Smalley, R.E., 193, *278*
Smith, B.J., 227–228, 250, *289*
Smith, B.V., 5–6, *67*, *68*
Smith, J.W., 246–247, *294*
Smith, N.L., 234–236, *290*, *291*
Smith, P.J., 207, *283*
Smith, P.W., 192–193, 248–249, *278*, *294*
Snyder, J.P., 213–214, *285*
Snyder, N.L., 144, 147–148, 161–162, *181*, *184*
Sohoel, H., 211, 224–227, *285*, *288*
Sollogoub, M., 125, 126, *179*, 239, 272, *292*
Sommadossi, J.P., 164–167, 169, *185*
Son, J.C., 248–249, *294*
Sonti, R., 141–143, *181*
Sorensen, M.D., 164, *184*
Spangler, D.P., 213–214, *285*
Spiwok, V., 154, *182*
Spohr, U., 200–201, *281*
Springer, D., 188–189, *276*
Spruck, W., 127–128, *179*
Spurway, T.D., 247–248, *294*
Srinivasan, K., 71, *116*
Stanek, M., 164, *185*
Steffens, R., 164, *184*
Steiner, A.J., 174, *186*, 208–209, 250, 254–256, 258, 284, 295–*298*
Steinhoff, H.-J., 72, *117*
Stein, S.S., 194–195, *279*
Stenutz, R., 159–160, *184*
Stephens, O.M., 122, *179*
Stevens, J.D., 129–132, *180*
Steyaert, J., 244–245, *293*
Stick, R.V., 211, 224–229, 232–235, 237–241, 250, *284*, *285*, *288*–*291*

Stocker, B.L., 193, *278*
Stockley, T., 206, *283*
Stoddart, J.F., 154–156, *183*
Stoffer, B.B., 200, *281*
Stoltefuss, J., 248, *294*
Storer, R., 164–165, 169–170, *185*, 258, *298*
Strauss, H.L., 154, *182*, *183*
Streith, J., 227–228, *289*
Streltsov, V.A., 250, *289*
Strijland, A., 250, 252, *284*, *295*
Ströhl, D., 219–221, *287*
Strynadka, N.C.J., 227, 240–241, *289*
Stubbs, K.A., 237–238, 254–255, *291*, *292*, *297*
Sturino, C.F., 125, 143–144, *179*
Stütz, A.E., 174, *186*, *278*, *284*, 286–*289*, 295–*297*
Suami, T., 191, *277*
Suarez-Ortega, M.D., *68*
Sückfull, F., 127, *179*
Sugden, K.D., 100–101, *120*
Sugiyama, M., 109–110, *120*
Suhara, Y., 203, *282*
Sui, E.L.T., 5–6, *67*
Suits, M.D.L., 235–236, *291*
Sullivan, P.A., 230, *289*
Sumi, T., 109–110, *120*
Sundaralingam, M., 156, *183*
Surana, B., 146, 147, *181*
Susa, N., 109–110, *120*
Sussman, J.L., 229–232, 239, 252, 253, *289*, *290*
Sutcliffe, L.H., 72, *117*
Suzuki, Y., 203, 210, 256, *282*, *284*, *297*
Svensson, B., 200, *281*
Swain, T., 206, *282*
Swanson, F.J., 72, *117*
Swartz, H.M., 70–71, 74–75, *116*
Sweeley, C.C., 202, *281*
Swiatek, J., 78, 86–87, *118*
Sylvestre, I., 71, 72, 77, *116*
Szajna-Fuller, E., 71, *116*
Szakács, Z., 159–160, *183*
Szarek, M.A., 219–220, *287*

Szarek, W.A., 154, 159, *183*, 218–220, *286*, *287*
Szumiel, I., 70, 109–110, *115*

T

Tailford, L.E., 234–235, *290*
Tajmir-Riahi, H.A., 70, *115*
Takase, S., 192, *277*
Takatsuki, A., 201–202, *281*
Takeuchi, T., 192, 206, *277*, *283*
Tanaka, H., 164, *184*
Tanaka, M., 164, *185*
Tanee, Z.F., 204, *282*
Tang, L., 206, *283*
Tappel, A.L., 194, *279*
Tarelli, E., 33, 35, *63*, *64*
Tarling, C.A., 174, *186*, 224–227, 234–235, 239–241, 250, 254–256, *288–290*, *295–297*
Tarnus, C., 258, *297*
Tatsuta, K., 207, 227–228, *283*, *289*
Tauss, A., 174, *186*, 216, *286*
Taylor, E.A., 244–245, *293*
Taylor, E.J., 235, 237–238, *291*
Taylor, G.N., *65*
Taylor Ringia, E.A., 222, *287*
Taylor, S.A., *67*, *68*
Teesdale-Spittle, P., 139–140, 152, *180*, *182*
Telser, J., 77, *118*
Tempel, W., 224, 225, *288*
Terano, H., 192, *277*
Terinek, M., 214, *285*
Thatcher, G.R.J., 159, *183*
Thelwall, L.A.W., *65*
Thiem, J., 188–189, 231–233, *276*
Thompson, A.E., *65*
Thompson, A.J., 194–195, 235–236, *279*, *291*
Thonhofer, M., 254, 256, *296*, *297*
Thornton, P., 72, *117*
Tilbrook, D.M.G., 211, 224–227, 239, *285*, *288*
Tillequin, F., 204, *282*
Timmer, M.S.M., 193, *278*
Ting, L.-M., 222, 243–244, *287*, *293*

Tinsley, J., 258, *298*
Tiraidis, C., 240, *292*
Tjebbes, J., 20–21, *62*
Toker, L., 230, 239, *289*
Tong, M.K., 207, *283*
Toshima, K., 207, *283*
Toubiana, M.J., *64*
Toubiana, R., *64*
Toufeili, I.A., 21–23, *66*
Toumieux, S., 258, *298*
Tran, T.Q., 129–132, *180*
Trapero, A., 253, *296*
Traving, C., 123, *179*
Triggs-Raine, B.L., 236, *291*
Tripathi, D.M., 70, *115*
Tripathi, S., 165, *184*
Tropak, M.B., 206, 210, 254–256, *283*, *284*, *296*, *297*
Tropea, J.E., 190–192, *277*, *278*
Truscheit, E., 203–204, *282*
Trzeciak, A., 70, 109–110, *115*
Tsuchiya, A., 240, *292*
Tsujii, H., 202, *282*
Tsuruoka, T., 189, *276*
Turkenburg, J.P., 231–232, 253, *290*
Turkson, A., 149–150, *182*
Turner, P., 72, *117*
Tvaroška, I., 154, *182*
Tyler, M.J., 149–150, *182*
Tyler, P.C., 210–211, 218, 220–222, 243–246, *284*, *286–288*, *292*, *293*
Tyms, A.S., 248–249, *294*

U

Überbacher, B.J., 258, *298*
Uchida, C., 209–210, *284*
Ueda, S., 203, *282*
Ueno, S., 109–110, *120*
Uhlmann, E., 168, *186*
Ully, S., 254–255, *296*
Umeda, I., 203, *282*
Umezawa, H., 192, *277*
Umezawa, K., 206, 207, *283*
Uto, Y., 235, 253, *291*

V

Vallee, F., 233, *290*
Vallyathan, V., 71, 74–75, *116*
van Aalten, D.M.F., 237–238, 253, *291*
van Ameijde, J., 258, *297*
Van Beeumen, J., 216, *286*
van Boom, J.H., 143, *181*
van Bueren, A.L., 239, 253, *292*
van Delft, F.L., 149–150, *182*
Vandemeulebroucke, A., 244–245, *293*
van den Berg, R.J.B.H.N., 209–210, 252, 254–255, 258, *284, 295–297*
Van den Broek, L.A.G.M., 250, *295*
van den Elsen, J.M.H., 233–234, *290*
Van den Ende, W., 240–241, *292*
van den Nieuwendijk, A.M.C.H., 258, *297*
van der Burg, A.M., 250, *295*
van der Marel, G.A., 143, *181*, 252, 258, *284, 295, 297*
Van Doorslaer, S., 69–115, *117, 118, 120*
Van Holsbeke, E., 244–245, *293*
Van Laere, A., 240–241, *292*
van Maanen, J.M.S., 65
Van Opstal, O., 247, *294*
van Rixel, V.H.S., 254, *296*
Van Scherpenzeel, M., 254–255, *296*
van Soest, R.W.M., 204, *282*
van't Hart, S.M., 254, *296*
Varela, O., 141, *181*
Varghese, J.N., 227–228, 250, *289*
Varki, A., 123, *179*
Varrot, A., 214, 224–227, 239, *285, 288, 289*
Vasella, A.T., 206, 209–210, 214, 227–229, 232–235, 237–241, 247–248, 250, *282, 284, 285, 289–291, 294*
Vázquez, J.T., 159–160, *183, 184*
Veibel, S., 194–195, *279*
Verdoucq, L., 229, 247–248, *289, 294*
Verhaest, M., 240–241, *292*
Verhagen, C.P., 252, *295*
Verheyden, J.P.H., 6, 49, 50, *65*
Vernon, C.A., 194–195, *279*
Versées, W., 244–245, *293*
Vianello, P., 250, *295*
Viazovkina, E., 168, *186*
Viehmann, H., 224–227, *288*
Vincent, F., 224–227, *288*
Vinck, E., 72, 77, *117*
Virden, R., 247–248, *294*
Vocadlo, D.J., 125, 126, *179*, 196, 236–239, 272, *280, 291, 292*
Vocatlo, D.J., 237–238, *291*
Vogel, P., 125, *179*, 218, *286*
Vonhoff, S., 214, *285*
Vorbruggen, H., 167–168, *185*
Vyas, N.K., 216, *285*

W

Waffo, A.F.K., 204, *282*
Wakatsuki, S., 240, *292*
Waldmann, H., 177, *186*
Walezak, T., 70–71, 74–75, *116*
Walker, D.E., 198, *280*
Walton, J., 235, *291*
Wandji, J., 204, *282*
Wang, A.H.-J., 239, 253, *292*
Wang, B.-C., 224, 225, *288*
Wang, C.C., 243–245, 258, *293, 298*
Wang, G., 72, *117*
Wang, J.-F., 204, *282*
Wang, M.-X., 258, *297*
Wang, R.-W., 219–222, *287*
Wang, W.-B., 220, 222, *287*
Wansi, J.D., 204, *282*
Warren, R.A.J., 247–248, *294*
Wartell, R.M., 171, *186*
Watanabe, K.A., 159, *183*
Watanabe, S., 209–210, *284*
Watkin, D.J., 190–191, *277*
Watzlawick, H., 223–224, 230, 232, *288, 290*
Webman, R., 122, *179*
Weckerle, W., 188–189, *276*
Wei, A., 147–148, *181, 182*
Weigand, J., 188–189, 231–233, *276*
Weil, J.A., 72–75, *117*
Weisblat, D.I., 133, *180*
Welch, B.D., 122, *179*
Welter, A., 190, *277*

Wendeborn, S., 164, *185*
Wengel, J., 164, 171, *184*, *186*
Wennekes, T., 209–210, 252, *284*, *295*
Wenthold, P.G., 147–148, *181*
Wertz, J.E., 72–75, *117*
Wetterhahn Jennette, K., 70–71, *115*
Wetterhan, K.E., 70, 72, 100–101, *115*, *120*
Weyhermüller, T., 71, 72, 77, *116*
Weymouth-Wilson, A.C., 258, *298*
Whalen, L.J., 250, *295*
Whistler, R.L., 151, *182*, 189–190, *277*
White, A.H., 194–195, 211, *280*, *285*
Whiting, P.H., *67*
Whitworth, G.E., 236, *291*
Wicki, J., 210, 224–227, 256, *284*, *288*, *297*
Widmalm, G., 159–160, *184*
Wieghardt, K., 71, 72, 77, *116*
Wijzenbroek, M., 254, *296*
Wilds, C.J., 164, 171, *185*, *186*
Williams, A., 196, *280*
Williams, D.H., 258, *298*
Williams, J.M., *61*, *62*, 191, *277*
Williams, N.R., *66*, 150–151, *182*
Williams, R.T., 246–247, *294*
Williams, S.J., 215–216, 224–227, 235–236, *285*, *288*, *291*
Wilson, F.X., 248–249, 258, *294*, *298*
Winchester, B.G., 51, *67*, 193, 217, 218, 220–222, *278*, *286*, *287*
Windhab, N., 164, *185*
Wing, C., 244–245, *293*
Wingender, W., 203–204, *282*
Win-Mason, A.L., 193, *278*
Winters, A.L., 258, *297*
Wipke, W.T., 188, *276*
Withers, S.G., 174, *186*, 194–196, 198–199, 207–210, 213, 215–218, 220, 221, 224–227, 234–236, 239–241, 250, 253–256, *279–281*, *283–285*, *287–291*, *295–297*
Witty, D.R., 248–249, *294*
Wojewodzka, M., 70, 109–110, *115*
Wolfenden, R., 209, *284*
Wölfler, H., 250, *295*

Wolfrom, M.L., 133, 149, *180*, *182*
Wong, C.-H., 125, *179*, 216, 250, 253, *286*, *295*, *296*
Wong, J.C.Y., 125, 143–144, *179*
Wongjirad, T.M., 122, *179*
Wong, R.Y., 190–191, *277*
Woods, R.J., 135–136, *180*, 234, *290*
Woodward, R.B., 188, 224–227, 231–233, *276*
Wormald, M.R., 250, 258, *295*, *297*
Wright, D.E., 149–150, *182*
Wrodnigg, T.M., 174, *179*, *186*, *284*, *286*, *287*, *289*, *295–298*
Wu, C.-Y., 253, *296*
Wu, H.-J., 239, 253, *292*
Wu, J.-B., 70, *115*
Wu, P., 244–245, *293*
Wustman, B.A., 230, 239, *289*
Wu, X., 164, *185*, 219–220, *287*
Wu, Y.-T., 253, *296*

X

Xiang, Y., 235–236, *291*
Xia, S., 147, *181*
Xie, H., 247–248, *294*
Xinhua, Q., 125, *179*
Xu, H.-W., 204, *282*
Xu, J., 220, *287*
Xu, Y., 218, *287*

Y

Yagi, M., 190, 248, *277*, *294*
Yamada, N., 164, *184*
Yamagishi, K., 209–210, *284*
Yamamoto, K., 240, *292*
Yamane, K., 224–227, *288*
Yamashita, Y., 220, 222, *287*
Yamazaki, C., 209–210, *284*
Yang, M.S., 204, *282*
Yariv, J., 207, *283*
Ya Tsibakhashvili, N., 71, 74–75, *116*
Yde, M., 247, *294*
Yip, P., 233, *290*
Yip, V.L.Y., 194–195, *280*

Yokose, K., 203, *282*
Yoshida, T., 233, *290*
Yoshikuni, Y., 248, *294*
Yoshimura, J., 150–151, *182*
Yoshimura, T., 73, 77, 93–94, 104–105, *118*
Young, G., 209, *284*
Yu, C.-Y., 220, 222, 258, *287*, *297*
Yuen, C.T., *65*, *66*
Yu, H.-P., 253, *296*
Yu, L., 252, *295*
Yule, K.C., 189–190, *277*
Yumoto, H., 189, *276*
Yu, Z., 250, *295*
Yuzwa, S.A., 125, 126, *179*, 239, 272, *292*

Z

Zabel, M., 146, *181*
Zamboni, V., 247–248, *294*
Zamojski, A., 123, *179*
Zandenberg, W.F., 236, *291*
Zareen, S., 204, *282*
Zavilla, J., 203–204, *282*

Zechel, D.L., 195, 196, 224–229, 232–233, 235, 239, 250, *280*, *288*, *289*, *291*
Zhang, L., 71, 72, 101–102, *116*, *117*
Zhang, R., 258, *298*
Zhang, W., 258, *298*
Zhang, X., 220, 222, *287*
Zhang, Y., 125, *179*, 239, *292*
Zhao, D., 236, 239, *291*
Zheng, X., 237–238, *291*
Zhong, W., 234, *290*
Zhong, Z., 216, 219, *286*
Zhu, J., 218, *286*
Zhu, X., 209–210, 251, *284*
Zhu, Y., 235–236, *291*
Ziegler, M.O.P., 231–232, *290*
Zimmer, K.-P., 250, *295*
Zographos, S.E., 240, *292*
Zubkova, O.V., 222, *287*
Zutshi, R., 122, *179*
Zweegers, T., 254, *296*
Zweier, J.L., 72, *117*

SUBJECT INDEX

A

2-Acetamido-1,2,5-trideoxy-1,5-imino-D-glucitol, 192–193
1-C- (5-Adamantanylmethoxy)pentyl-1-deoxynojirimycin, 252
Alditols-oxo-Cr(V) complexes
  classification of, 80
  difference, EPR parameters, 81
  EPR spectra, 79–80
  five-membered vs. six-membered chelates, 78
  g and ^{Cr}A tensors, 78
  hyperfine substructure, 79–81
  isomers, 84
Aldohexoses-oxo-Cr(V) complexes
  anomeric forms, 89
  D-galactose and D-fructose ligands, 86–87
  EPR signals, 88
  furanose sugars, diol groups of, 92–93
  ligand-exchange reactions, 89
  NADH, 87
  redox reaction pathways, 87–88
  reduction mechanism, 93–94
  room-temperature CW-EPR spectra, 90–92
  six-coordinated complex formation, 88
N-Alkyl-1-deoxy-D-galactonojirimycin, 210
2-O-Alkylated iminoxylitol, 252
Allosteric glycosidase inhibitors. See Noncompetitive glycosidase inhibitors
Ambroxol, 206
Amine-based inhibitors, 208–211
6-Amino-6-deoxycastanospermine, 238
5-Amino-5-deoxy-D-glucose, 189, 212

2,3-anhydro-4,6-O-benzylidene-α-D-mannopyranosyl 4,6-O-benzylidene-α-D-glucopyranoside, 36
3,4-Anhydro-5-O-benzyl 1,2 O-isopropylidene-α-D-galactoseptanose, 132
(-)-Anisomycin, 49
Antisense oligonucleotides (AONs), 168
Antisense therapy, 171
Azepanes, 125

B

Beciparcil, 176–177
5,7-O-Benzylidene-3-deoxy-3-nitro-α-D-glycero-D-altro-septanoside, 150
5-O-Benzyl-1,2-O-isopropylidene-α-D-galactoseptanose, 132
5-O-Benzyl-1,2-O-isopropylidene-α-D-glucoseptanose, 132
5-O-Benzyl-1,2-O-isopropylidene-α-D-glucoseptanose, 132
6-O-Benzyl-2,3-O-isopropylidene-D-ribose, 143–144
5-O-Benzyl-1,2-O-isopropylidene-4-O-methyl-α-D-galactoseptanose, 132
5-O-Benzyl-1,2-O-isopropylidene-3-O-methyl-α-D-guloseptanose, 132
5-O-Benzyl-1,2-O-isopropylidene-4-O-tosyl-α-D-galactoseptanose, 132
Brevetoxin, 126
Broussonetine E, 192
N-Butyl-1-deoxynojirimycin vs. ceramide, 250
5-O-tert-Butyldiphenylsilyl-2,3-O-isopropylidene-D-ribonolactone, 146

C

Calystegine B2, 192
Carbohydrate-related glycosidase inhibitors
   competitive inhibitors, 208–211
   substrate-related covalent inhibitors, 206–208
Castanospermine, 218
   superposition of, 229, 231
Chloroethyl pyranoside, spiroacetal derivatives of, 54
4-Chloro-galacto-sucrose, 21–23
Chromium(V) complexes
   applications, 71
   carcinogenicity, 70–71
   chromium(VI) oxidation reaction, 70–71
   spectroscopic techniques, 72
Ciguatoxin, 126
Competitive glycosidase inhibitors, 208–211
Conduritol B epoxide, 230
Conformational analysis, septanose monosaccharides
   anomeric effect, 159
   ^{13}C NMR shifts, 161–162
   conformational map of, 157
   low-energy conformations, 156–160
   methyl α-D-glycero-D-ido-septanoside, 154–156
   methyl α-septanoside, crystal structure, 159–160
   observations, 157–158
   5-O-methyl septanosides, 161
   pseudorotational wheel, 157
   seven-membered rings, 154–156
   six-membered ring systems, 154
   T conformers, interconversion of, 156
Covalent glycosidase inhibitors, 203–204
[CrO(cis-O1,O2-cyclohexanediolate)2]⁻, 81
[CrO(trans-O1,O2-cyclohexanediolate)2]⁻, 81
C-6-substituted septanosides, conformational analysis, 158
C-6-unsubstituted septanosides, conformational map of, 157
Cyclophellitol, 207

D

DADMe-immucillin-H, 243–244
Database CAZy (Carbohydrate-Active EnZymes), 196
6-Deoxycastanospermine, 218
5-Deoxy-2,3:6,7-di-O-diethylidene-D-allo-heptitol., 165–167
3-Deoxy-3-fluoro-calystegin B2, 221
1-Deoxyfluorocastanospermine, 218
6-Deoxyfluorocastanospermine, 218
6-Deoxy-6-fluorocastanospermine, 223
1-Deoxy-L-fuconojirimycin, 239
1-Deoxynojirimycin. See 1,5-Dideoxy-1,5-imino-D-glucitol
Desosamine hydrochloride, Guthrie phenylhydrazine cyclization, 18–20
D-Galactose diethyl dithioacetal, 127–128
1,6-Dideoxy-6,6-difluoronojirimycin, 219–220
1,7-Dideoxy-7-fluorocastanospermine, 218
1,4-Dideoxy-4-fluoronojirimycin, 218
1,4-Dideoxy-1,4-imino-D-arabinitol, 191–193, 240
1,5-Dideoxy-1,5-imino-D-glucitol, 189–190, 212
   superposition of, 229, 231
1,4-Dideoxy-1,4-imino-D-mannitol, 192–193
1,5-Dideoxy-1,5-imino-D-mannitol, 191
2,5-Dideoxy-2,5-imino-D-mannitol, 190, 212
2,5-Dideoxy-2,5-imino-D-mannitol, 192
1,5-Dideoxy-1,5-imino-L-fucitol, 192–193
4,4-Difluoro isofagomine, 221
3,21-Di-O-acetylcichoridiol, 204
2,7-Di-O-benzyl-4-deoxy-4-nitro-α-D-glycero-D-talo-septanoside, 150
Dissociative ion-pair SN1-type mechanism, 12

SUBJECT INDEX

E
Electron paramagnetic resonance (EPR), 72–73. *See also* Oxo-Cr(V)-sugar complexes, structural characterization
Endoplasmatic Reticulum α-1,2-Mannosidase I, 233
4-*epi*-isofagomine, 210
  pKa value, 213–214
Ethyl β-D-galactofuranoside, 133

F
Fagomine. *See* 1,2,5-Trideoxy-1,5-imino-D-*arabino*-hexitol
Fischer nitromethane cyclization, 15–18
4-Fluoro-galacto-sucrose, 45

G
α-Galactosidases, 202
Glucohydroximolactam, 229
Glucotetrazole, 229
Glycon structure
  α-galactosidases, 202
  β-galactosidases, 199
  α-glucosidases, 200–202
  β-glucosidases, 197–199
  α-mannosidases, 202–203
Glycosidase inhibitors
  carbohydrate-related inhibitors
    competitive inhibitors, 208–211
    substrate-related covalent inhibitors, 206–208
  high-molecular-weight inhibitors, 203
  nonsugar inhibitors, 204–206
  small molecules, 203–204
Glycosidases, structure determination of, 216
Glycoside hydrolases
  general features of, 193–194
  inverting and retaining enzymes, 195
Golgi α-mannosidase II, 233–235
g tensor, 72–73
Guthrie phenylhydrazine cyclization, 18–20

H
Heptoses, 123
2,3,6,1′,4,′6′-Hexa-*O*-pivaloylsucrose, 24
High-molecular-weight glycosidase inhibitors, 203
Hydroxy acids-oxo-Cr(V) complexes
  aldonic acids, 102
  D-gluconic acid, 103
  D-glucuronic acid, 105–107
  l-hydroxy acids, 100–101
  uronic acids, 104–105
L-*glycero*-D-*manno*-Heptose, 127–128

I
Imino sugars
  aza-sugars, 211
  breakthrough of, 191
  catalysis visualization
    N-acetylhexosaminidases, 236–239
    fucosidases, 239–240
    D-galactosidase, 236
    D-glucosidase inhibitors, 224–233
    D-mannosidase inhibitors, 233–236
    nucleoside hydrolases, 242–246
    pyrrolidine, 240–242
  deoxyfluoro derivatives, 217–224
  discovery of, 189–192
  enzyme—inhibitor complexes, 259–274
  functional groups, significance of, 216–217
  increased hydrophobicity, 248–254
  lipophilic inhibitors, 254–256
  non-glycon binding sites, 246–256
  PDB-entries, table of, 259–274
  pKa value, 213
  stereochemical considerations, catalytic protonation, 214–216
Immucillin-GP, 243
Immucillin-H, 243
Inverting glycohydrolases, 195
  mechanism of, 195
Irreversible glycosidase inhibitors. *See* Covalent glycosidase inhibitors
Isofagomine, 209–210

K
Kifunensine, 192–193

M
Mannoimidazole, 234–236
α-Mannosidases, 202–203
Methyl 2-acetamido-3,4,5,7-tetra-*O*-acetyl-2-deoxy-α-D-*glycero*-D-*ido*-septanoside, 134–136
Methyl 2-acetamido-3,4,5,7-tetra-*O*-acetyl-2-deoxy-α-D-*glycero*-D-*ido*-septanoside, 134–135
Methyl α-D-*glycero*-D-*ido*-septanoside, 163
Methyl α-D-*glycero*-D-*talo*-septanoside, 141
Methyl β-lactoside, benzoylation of, 30–31
Methyl 3-bromo-3-deoxy-β-maltoside, 22
Methyl 2,5-di-*O*-benzoyl-3,4-*O*-isopropylidene-β-L-*ido*-septanoside, 130–132
Methyl glycosides-oxo-Cr(V) complexes
  potential binding modes, 95
  room-temperature CW-EPR signal, 94–95
  synthesis models of, 97
Methyl 2,3-*N*-acetylepimino-4,6-*O*-benzylidene-2,3-dideoxy-α-D-allopyranoside, 20–21
Methyl 2,3,4,5-tetra-*O*-acetyl-β-D-galactoseptanoside, 134–135
Methyl 2,3,4,5-tetra-*O*-benzyl-β-D-*glycero*-D-*gulo*-septanoside, 148–149
Methyl 3,4,5,7-tetra-*O*-benzyl-2-benzylamino-2-deoxy-α-D-*glycero*-D-*ido*-septanoside, 134–135
Methyl 3,4,5,7-tetra-*O*-benzyl-2-benzylamino- 2-deoxy-α-D-*glycero*-D-*ido*-septanoside, 134–135
Methyl 2,3,4,5-tetra-*O*-methyl-β-D-glucoseptanoside, 128–129
5′-Methylthioadenosine nucleosidase (MTAN) inhibitors, 210–211
Methyl 3,4,6-trideoxy-3,4-epimino-α-L-galactopyranoside, 20–21
Monosaccharide septanoses, 127–133
Mycaminose, 17

N
4-Nitrophenyl 6-deoxy-β-D-galactopyranoside, 198
4-Nitrophenyl 6-deoxy-β-D-glucopyranoside, 197
Nojirimycin, 125
Noncompetitive glycosidase inhibitors, 203–204
Nonsugar inhibitors, 204–206

O
α-1-*C*-Octyl-1-deoxynojirimycin, 252
Oligonucleotides (ONs), 122–123
Oxacycloheptatriene, 138–139
Oxepines
  aldoseptanoses synthesis, intermediates in, 141
  cyclopropanation and ring expansion, 138–141
  glycosylations, 147–149
  isomers, 145
  RCM strategy, 144–145
  ring-closing metathesis, 143–147
  types of, 145
Oxidative degradation, sugar diethylsulfonyl dithioacetals, 12
Oxo-Cr(V) complexes
  ENDOR spectrum, 75–76
  HYSCORE spectrum, 75–76
  temperature CW-EPR spectra, 74
    advantage and drawback, 73
    alditols, 77–86
    aldohexoses, 86–94
    aldopentoses, 86–94
    hydroxy acids, 100–108
    isotropic hyperfine coupling, 74–75
    methyl glycosides, 94–97
    parameters, 72–73
Oxo-Cr(V)—sugar complexes, structural characterization
  2-acetamido-2-deoxy-D-glucose, 98
  alditols
    classification of, 80
    difference, EPR parameters, 81

EPR spectra, 79–80
five-membered vs. six-membered
  chelates, 78
g and ^{Cr}A tensors, 78
hyperfine substructure, 79–81
isomers, 84
aldohexoses
  anomeric forms, 89
  EPR signals, 88
  furanose sugars, diol groups of, 92–93
  D-galactose and D-fructose ligands,
    86–87
  ligand-exchange reactions, 89
  NADH, 87
  redox reaction pathways, 87–88
  reduction mechanism, 93–94
  room-temperature CW-EPR spectra,
    90–92
  six-coordinated complex formation, 88
aldopentoses, 86–94
2-deoxy-D-*arabino*-hexose, 98
2-deoxy-D-*erythro*-pentose, 98
ENDOR spectrum, 75–76
hydroxy acids
  aldonic acids, 102
  D-gluconic acid, 103
  D-glucuronic acid, 105–107
  1-hydroxy acids, 100–101
  uronic acids, 104–105
methyl glycosides
  potential binding modes, 95
  room-temperature CW-EPR signal,
    94–95
  synthesis models of, 97
oligosaccharides, 108–114
3-*O*-methyl-D-glucopyranose, 99–100
polysaccharides, 108–114

P

Penarolide, 204
2,3,4,5,6-Penta-*O*-acetyl-1-chloro-1,1-
  dideoxy-1-ethylthio-D-galactose
  aldehydrol, 133
(-)-Pentenomycin I, 48, 50

Phosphoramidites, synthesis of, 167
Protein–septanose interactions, lectin-binding
  studies, 172–174
Purine nucleoside phosphorylase (PNP)
  inhibitors, 210–211
Pyrimethamine, 206

R

Raffinose, chlorination of, 41, 42
RCM strategy. *See* Ring-closing metathesis
  (RCM) strategy
Retaining glycohydrolases
  mechanism of, 196
  pathway, 196
Reversible glycosidase inhibitors. *See*
  Competitive glycosidase inhibitors
Richardson's rules, 57–58
  pictorial summary of, 59
Ring-closing metathesis (RCM) strategy,
  144–145
RNAseH, 171

S

Septanose carbohydrates
  fused polycyclic marine toxin ethers, 126
  investigations, 163–177
    oligonucleotides, 168–171
    protein—septanose interactions, 172–176
  polyhydroxylated azepanes, 125
  research areas, 125
  strategies, septanosides
    cyclization via C—O bond formation,
      127–138
Septanose monosaccharides, conformational
  analysis
  anomeric effect, 159
  ^{13}C NMR shifts, 161–162
  conformational map of, 157
  low-energy conformations, 156–160
  methyl α-D-*glycero*-D-*ido*-septanoside,
    154–156
  methyl α-septanoside, crystal structure,
    159–160
  observations, 157–158

Septanose monosaccharides, conformational analysis (cont.)
  5-O-methyl septanosides, 161
  pseudorotational wheel, 157
  seven-membered rings, 154–156
  six-membered ring systems, 154
  T conformers, interconversion of, 156
Septanoses synthesis, ring expansion of cyclic ketones, 150–151
Septanose sugars
  expanded monomer approach, 123
  transannular reactions of
    electrophilic intermediates, 153
    precursors, 153
    products of, 153
Septanosides, 123–125
Siastatin B, 192
Small molecules glycosidase inhibitors, 203–204
(S)-Spirobi-1,4-dioxane, 55
Substrate-related covalent glycosidase inhibitors, 206–208
Sucralose. See 4,1′6′-Trichloro-4,1′6′-trideoxy-galacto-sucrose (TGS)
Sucrose, hepta-pivaloylation of, 21–23
Sugar diethylsulfonyl dithioacetals, oxidative degradation of
  Fischer nitromethane cyclization, 15–18
  plausible mechanism, 13
  Richardson and Hough's mechanistic proposal, 14
  stereostructure of, mycaminose, 17

Swainsonine, 191, 212
(-)-Swainsonine, 48, 52

T

2,3,4,5- Tetra-O-acetyl-6-O-tosyl-D-galactose diethyl dithioacetal, 151–152
2,3,4,5-Tetra-O-methyl-D-glucose, 128–129
1,3,4,6-Tetra-O-pivaloyl-β-D-fructofuranosyl 2,3,6-tri-O-pivaloyl-α-D-glucopyranoside, 21–23
5-Thio-D-5-thio-D-xylopyranose, 189–190
5-Thio-L-idopyranose, 189–190
6-Thioseptanosides, 151
Thymidine, 169
Trehalose, 3,30-dideoxy-analogues of, 35
4,1′6′-Trichloro-4,1′6′-trideoxy-galacto-sucrose (TGS), 10–11
1,2,5-Trideoxy-2-fluoro-1,5-imino-D-glucitol, 218
1,2,5-Trideoxy-1-fluoro-2,5-imino-D-mannitol, 223
1,4,6-Trideoxy-6-fluoro-1,4-imino-D-mannitol, 220
1,4,6- Trideoxy-6-fluoro-1,4-imino-D-mannitol, 222
1,2,5-Trideoxy-1,5-imino-D-*arabino*-hexitol, 125, 189
1,4,6-Trideoxy-1,4-imino-D-mannitol, 212
Triethylamine-mediated cycloisomerization, 146
2,3,5-Tri-O-benzyl-D-arabinofuranose, 143–144

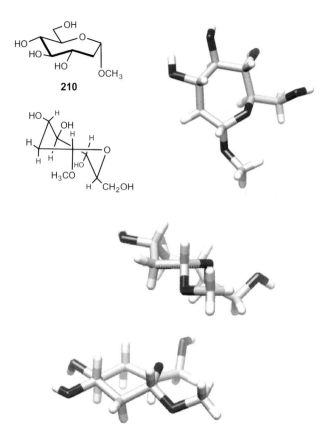

PLATE 1 Different perspectives of the methyl 2-deoxy-α-D-*gluco*-septanoside (**210**) structure from X-ray data.

PLATE 2 Computed low-energy conformations of methyl septanosides. Left: Methyl α-D-*glycero*-D-*ido*-septanoside (conf 36); Right: Methyl β-D-*glycero*-D-*gulo*-septanoside (conf 1).

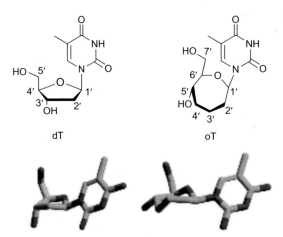

PLATE 3 Thymidine (dT) and an oxepanyl analogue (oT). Structures on the bottom (taken from ref. 149) show minimized conformers for both dT and oT, emphasizing the similarity in the disposition of the base and the C-4/C-6 hydroxymethyl group.

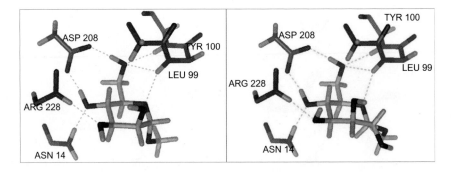

PLATE 4 Detail of the monosaccharide binding pocket with (left) pyranoside ligand **237** and (right) septanoside ligand **208** as determined from the QM/MM computations (taken from ref. 124).

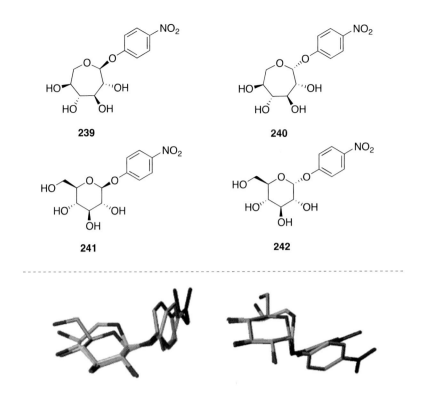

PLATE 5 Top: *p*-Nitrophenyl L-idoseptanosides **239** and **240** used to mimic D-glucopyranosides **241** and **242** as substrates for glycosidases. Bottom: Overlays of minimized structures. Left is the overlay of **239** with **240** and right is the overlay of **240** with **242** (taken from ref. 164).

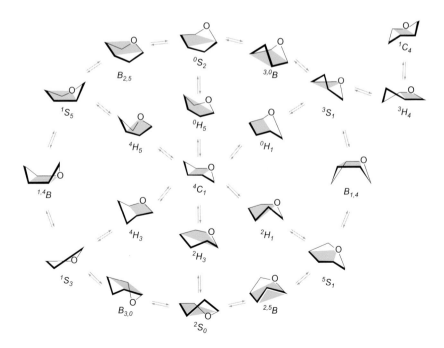

PLATE 6 View at "GPS," the "*G*lobe of *P*yranoid *S*ugar conformations."[244]

PLATE 7 Superposition of ligands 1-deoxynojirimycin **2** (yellow, PDB 2J77) and castanospermine **8** (green, PDB 2CBU) bound to *Thermotoga maritima* β-glucosidase. The ring nitrogen atoms, all hydroxyl groups in the pyranoid rings, as well as O-1 (**8**) and the primary O-6 (**2**), respectively, are closely matched.

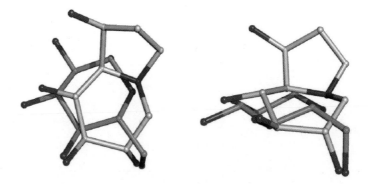

PLATE 8 Superposition from two different viewpoints of the ligands castanospermine, **8** (gray, PDB 2PWG) and 1-deoxynojirimycin **2** (green, PDB 2PWD) bound to trehalulose synthase from *Pseudomonas mesoacidophila* MX-45. The lateral inversion of **2** relative to the pyranoid ring of **8**, with O-6 of **2** matching O-6 (corresponding to O-2 in carbohydrate numbering) of **8** is clearly visible.